Mobile Security: A Comprehensive Guide to Securing Information

Mobile Security: A Comprehensive Guide to Securing Information

Editor: Alexander Castro

NY RESEARCH PRESS
New York

Published by NY Research Press
118-35 Queens Blvd., Suite 400,
Forest Hills, NY 11375, USA
www.nyresearchpress.com

Mobile Security: A Comprehensive Guide to Securing Information
Edited by Alexander Castro

International Standard Book Number: 978-1-63238-863-6 (Hardback)

Cataloging-in-Publication Data

Mobile security : a comprehensive guide to securing information / edited by Alexander Castro.
p. cm.
Includes bibliographical references and index.
ISBN 978-1-63238-863-6
1. Mobile communication systems--Security measures. 2. Wireless communication systems--Security measures.
3. Mobile computing--Security measures. 4. Data protection. I. Castro, Alexander.
TK5103.2 .M63 2022
621.384--dc23

Contents

Preface

In the modern world, perhaps the most commonly used device is the mobile phone. More and more businesses and users are using smartphones for communication, and organization and planning of their work. Such technologies are bringing about marked changes in the organization of information systems which places data at risk. This is because smartphones store and collect sensitive information that compromises the privacy and security of the individual as well as the intellectual property of the company. This makes mobile security an increasingly important aspect in mobile computing. As smartphones are a point of access to the Internet, they can also be attacked with malware. Ransomware, viruses and Trojans, and spyware are examples of malware. Many of the security threats can be managed through rootkit detectors, memory protection, file systems permissions mechanisms, antivirus and firewall, safeguards in network routing points, etc. The ever-growing need of advanced technology is the reason that has fueled the research in the field of mobile security in recent times. This book unfolds the innovative aspects of mobile security which will be crucial for the progress of this domain in the future. As this field is emerging at a rapid pace, the contents of this book will help the readers understand the modern concepts and applications of the subject.

All of the data presented henceforth, was collaborated in the wake of recent advancements in the field. The aim of this book is to present the diversified developments from across the globe in a comprehensible manner. The opinions expressed in each chapter belong solely to the contributing authors. Their interpretations of the topics are the integral part of this book, which I have carefully compiled for a better understanding of the readers.

At the end, I would like to thank all those who dedicated their time and efforts for the successful completion of this book. I also wish to convey my gratitude towards my friends and family who supported me at every step.

Editor

1

Detecting Android Malwares with High-Efficient Hybrid Analyzing Methods

Yu Liu,[1] Kai Guo,[1] Xiangdong Huang,[1] Zhou Zhou,[2] and Yichi Zhang[1]

[1] *School of Electronic Information Engineering, Tianjin University, Tianjin 30072, China*
[2] *Teachers College of Columbia University, New York, NY 10027, USA*

Correspondence should be addressed to Xiangdong Huang; xdhuang@tju.edu.cn

Academic Editor: Xiapu Luo

In order to tackle the security issues caused by malwares of Android OS, we proposed a high-efficient hybrid-detecting scheme for Android malwares. Our scheme employed different analyzing methods (static and dynamic methods) to construct a flexible detecting scheme. In this paper, we proposed some detecting techniques such as Com+ feature based on traditional Permission and API call features to improve the performance of static detection. The collapsing issue of traditional function call graph-based malware detection was also avoided, as we adopted feature selection and clustering method to unify function call graph features of various dimensions into same dimension. In order to verify the performance of our scheme, we built an open-access malware dataset in our experiments. The experimental results showed that the suggested scheme achieved high malware-detecting accuracy, and the scheme could be used to establish Android malware-detecting cloud services, which can automatically adopt high-efficiency analyzing methods according to the properties of the Android applications.

1. Introduction

According to IDC Mobility Research [1], the total shipments of smartphones and tablet computers had reached 1.84 billion in 2016. Among these devices, 81.2% of them are running on Android OS. Due to the openness of Android OS, Android-based devices are the major target of malwares. According to the investigations of third parties, the number of discovered Android malwares increased around 750,000 in Q1/2017, and about 3.5 million new malwares may be shared by various malware providers by the end of 2017 [2]. Meanwhile, many rogue companies keep developing Android Applications (Apps) that can perform malicious actions such as stealing credit card information [3]. Therefore, the detection of Android malwares has been a critical task in mobile security for both the consumers and service providers.

Given this serious threat, the Google Play App Store performs security scans on every submitted App. Unfortunately, this safety mechanism is not as effective as expected, and the market has not blocked the malware delivering. Besides, many third-party App markets without any supervision also provide conveniences for attackers to spread their malwares. Consequently, the Android malwares are constantly evolving.

As the Android Apps are always shared and downloaded with cloud service providers, cloud-based Android malware detection will be used by cloud management to improve the service qualities. A typical malware detection cloud service is shown in Figure 1. The cloud can work as file-sharing services which use software tools and machine-learning techniques to detect the Android malwares. The uploading and delivering process of Apps can be realized with common frameworks of file-sharing cloud, and the main issue of the malware-detecting cloud is how to complete accurate malware detection with high efficiency.

Constructing comprehensive and accurate Android malware detecting solutions is a hard and long-term work. In this paper, we suggest a hybrid malware-detecting scheme to improve the traditional malware-detecting methods. Some new techniques are proposed in our scheme, for example, we adopted clustering methods to unify the dimension of calling graph features into 100, so that the dimension collapsing of calling graph-based detecting analysis can be avoided, we proposed a new analyzing feature which is named as Com+ for

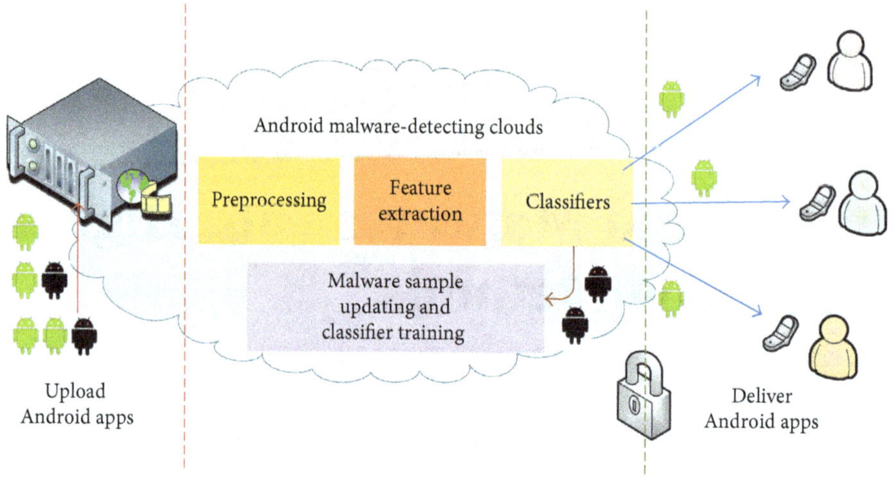

FIGURE 1: Common framework of Android malware-detecting cloud.

Apps without obfuscation strategy in static analysis, and we also designed soft tools ("*Detector*" detailed in Table 1) to improve the efficiency of the detection. With integrating and choosing proper detecting methods, our malware-detecting scheme is more comprehensive and efficient than other traditional detecting methods.

2. Relative Works

Plenty of researches have been carried out on the detection of malwares. Some of the research paid attention to the analysis of the evolution of malware ecosystem. Concerning the working mechanism and behaviors of malwares, Felt et al. [4] made a comprehensive survey on malwares of the three main platforms (Android, iOS, and Symbian) and indicated the potential bugs in the temporal counting mechanism of those systems. Di Cerbo et al. [5] suggested to classify mobile malwares as four types (viruses, botnets, worms, and Trojan horses) and further summarized six main malware behaviors such as stealing user information, intercepting/sending SMS, etc. For the security issues, factors of Android security were summarized and analyzed by Faruki et al. [6], and Sadeghi et al. [7] suggested a solution to evaluate the security of Android Apps.

For the techniques of Android malware detection, many methods have been proposed. Depending on whether the Apps need be executed in detection, the malware-detecting methods can be classified into two types: static analysis and dynamic analysis. In static analysis, researchers usually decompile the App and extract some particular features without executing them. For example, since the App Permission can reflect the App's access to the software and hardware resources of the mobile devices, Wang et al. [8] used Permission requests as the features and ranked the risk degree of samples from the malware library based on the requesting rates of Permission. Besides, API calls (officially defined packages and classes of Android OS) could be chosen as another feature for static analysis. Using API calls as features, Nishimoto et al. [9] recorded all the invocations in the Apps by inserting Log.v method into API calls. Using API calls as the features, Chan and Song [10] selected 19 typical API calls and achieved good malware-detecting results. Aafer et al. [11] proposed API feature with tuning machine learning model to classify Apps as normal (benign) or abnormal (malicious), and the performance of this solution could be improved by constant updating of malware samples. Except using Permission and API calls as features for malware detection, some researchers also used function-call relationships in static malware analysis. Wei et al. [12] used function-relation graphs to measure the structural similarity of Apps and used *DroidExec* to recognize malwares. Gascon et al. [13] generated the embedded function call graph and proved the effectiveness of this method in malware detection. In our previous work [14], we also suggested a method for static analysis based on the API calls.

TABLE 1: Analyzing software tools.

Name	Function	Platform
Apk tool	Most popular decompiling tool for Android Apps	Windows
Androguard	A powerful and comprehensive integrated analyzing and operating tool based on python	Linux
Monkey	Generates pseudorandom streams of user events such as clicks and touches	All
Strace	A system call tracer which can print out all the Linux kernel system calls of Android	All
TaintDroid	Modified Android OS which can monitor private information by attaching tags	Linux
DroidBox	Dynamic analysis tool for Android Apps working on emulator	Linux
Detector	A self-designed soft tool to find newly executed App and trace and log the operations of Apps	Linux

Dynamic analysis involves executing the Apps in a controlled and isolated environment so that the operations of the Apps can be traced. Many schemes based on different dynamic analysis tools have been proposed. Burguera et al. [15] designed the Crowdroid in 2011, which used the *Strace* tool to collect App's system call logs and send them to the cloud server for further analysis. Ham et al. [16] analyzed App's system call patterns and established a pattern library of benign and malicious Apps, so that malwares could be recognized by calculating the matching patterns. Zou et al. [17] used shared library injection and hooking techniques to intercept corresponding libc function calls. Besides, taint tracking was usually used as another way of dynamic analysis. In 2010, Enck et al. [18] designed and built the *TaintDroid*, which was a customized Android OS that could attach tags on the user privacy information, and these tags could be recognized by the system when the privacy information was leaked.

The abovementioned detecting techniques can be deemed as traditional malware-detecting methods which are mainly dependent on analyzing of the App data. Recently, subjective information such as user comments and ranking was used by cloud processing to improve the performance of the malware detection. Rahman et al. [19] presented a system named as FairPlay, which could detect malware based on the analysis of review activities and search rank fraud; Narayanan et al. [20] suggested a malware-detecting scheme using the security-sensitive information of Apps. Compared with traditional detecting methods, these solutions could achieve good performance in malware detection. But, they always demand time-consuming data collection and complex computation. Thus, improvement of traditional detecting methods is still desired by practical applications.

As the Android platform develops rapidly, the types and patterns of malwares are constantly evolving to avoid detections. Plenty of the newly generated malwares can invalidate existing malware-detecting methods. Thus, malware-detecting techniques should be constantly improving with the evolution of malwares, and an up-to-date malware dataset should also be maintained to evaluate the performance of the detecting techniques.

A new Android malware-detecting scheme is proposed by this paper. Compared with some existing traditional methods which only apply unified processing for each App (e.g., [8] only can be used for Apps without obfuscation strategy and [12, 13] only use complex call graph data for analysis), our scheme could provide more comprehensive analysis for any Android Apps. Besides, as our scheme provide some new detecting techniques and different processing methods are chosen according to the properties of each App, the analyzing efficiency is also enhanced. The rest of this paper is organized as follows: In Section 3, we introduce our malware-detecting scheme and corresponding analyzing tools. Section 4 presents the methods of feature extraction and malware classification. Section 5 presents the malware dataset and corresponding experimental results, Section 6 discusses the limitation and future works of the proposed scheme, and Section 7 concludes the paper.

3. The Proposed Android Malware-Detection Scheme

As discussed in the previous section, malware detection can be classified as static or dynamic analysis. As executing the Apps is not required by static analysis, this approach is simpler than dynamic analysis. However, if the App cannot be decompiled, static analysis becomes ineffective because the features such as Permission cannot be extracted from the Apk file (execution file of Android Apps). In this case, dynamic analysis should be used by executing the App in a sandbox. Dynamic analysis can record log files for some operations of the App running in the sandbox, and the data of log files can be used as features for malware detection. Though dynamic analysis outperforms static analysis in terms of applicability, it also has drawbacks such as computational complexity and low efficiency. Thus, the malware detection should choose proper analyzing methods based on the properties of Apps to enhance the detecting efficiency.

In this paper, we propose a hybrid-detecting scheme for Android malwares that employs several static/dynamic analysis methods to complete malware detection. Our hybrid scheme can choose the appropriate method according to the properties of the App so that a good balance can be achieved between the detecting accuracy and complexity. Figure 2 shows the overview process of our malware-detecting scheme which employs many software tools, and Table 1 summarizes those tools with descriptions. As shown in Figure 2, our scheme extracts the features of the malware classifiers via four paths with different complexities, and the classifiers such as k-NN are used to judge whether the App is malicious. As the detecting accuracy is highly related to the effectiveness of the features, our scheme employs four types of methods (denoted by color and index) to derive the features according to the attributes of the Apps.

Table 2 lists the application scopes and descriptions of the four paths. The four paths are constructed by two static (Path A and Path B) and two dynamic (Path C and Path D) detecting methods. For App which does not adopt obfuscation strategy (such as malwares of [21] dataset which are delivered before 2012), our Com+ feature (new feature proposed by our scheme, detailed in following sections) can efficiently reflect the behaviors of the App, and Path A in Figure 2 should be selected for feature extraction as static analysis. In Path A, the analyzing complexity is very low as the Com+ features can be extracted directly from the Apk files (source file of Apps). With the updating of Android OS, many Apps adopt obfuscation strategies to avoid clearly declaring Com+ feature. For Apps with obfuscation strategy, Path B which adopts function call graph as features should be used. As the complexity of analyzing call graph may be very high, we also proposed clustering methods (detailed in following sections) to decrease the complexity and avoid the analyzing collapsing caused by the unlimited feature dimensions.

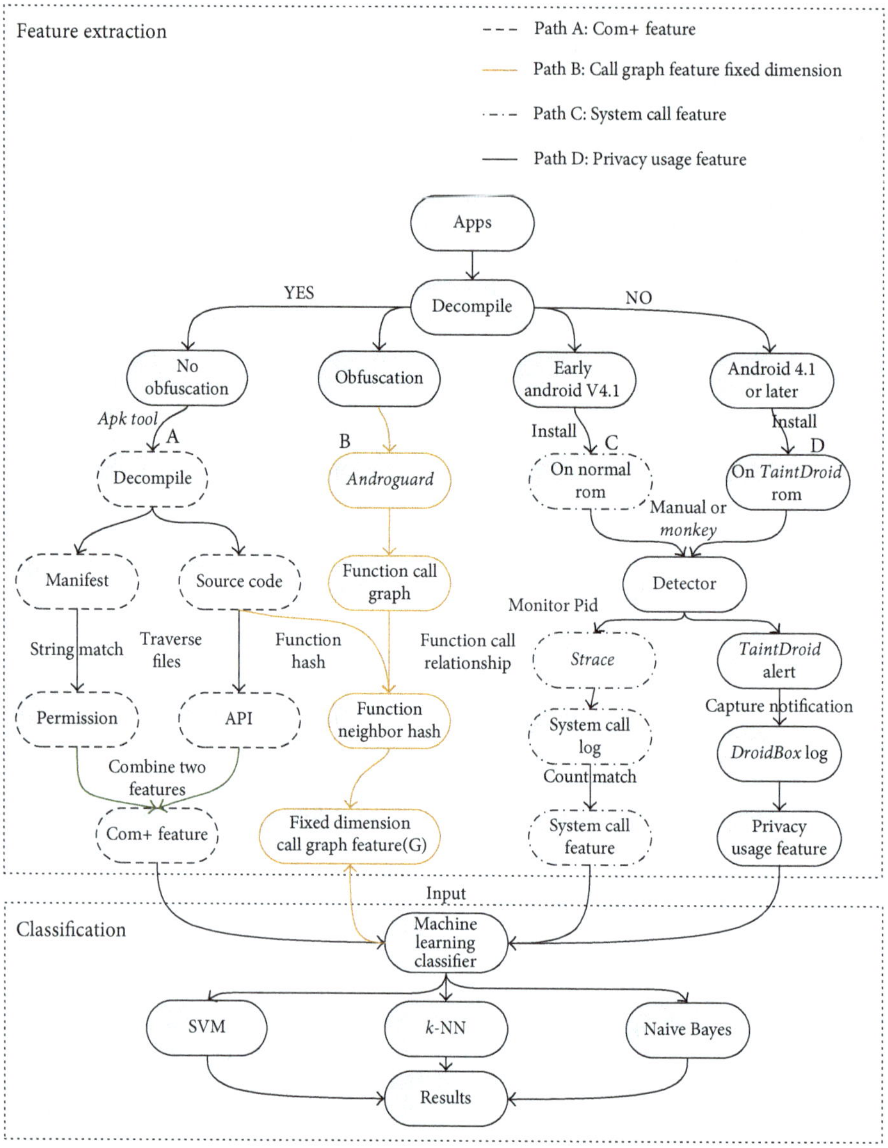

FIGURE 2: The proposed hybrid-detecting scheme.

Using Path A and Path B, malwares which could be decompiled can be analyzed with low complexity. However, Apps may use encryption techniques which prevent the App from being decompiled. If the malwares are encrypted, dynamic analysis (Path C and Path D) should be applied in malware detection. In order to monitor the behaviors of Apps, a soft tool named as *DroidBox* which can be used for Apps delivered with Android 4.1 or later is adopted by our scheme to extract running features of Apps. For Apps delivered before Android 4.1, Linux soft tool *Strace* should be adopted to extract features, because any Android versions are running on Linux kernel. Compared with static analysis, though dynamic analysis can complete malware detection for any Apps, it also demands more complex processing. Thus, the malware-detecting scheme should complete the analysis with flexible methods to obtain high analyzing efficiency.

Using the four paths of Figure 2, our malware-detecting scheme can complete the malware detection of any Apps. Thus, our scheme is more comprehensive than methods such as in [8–10, 12, 13], which can be used under some special conditions. Besides, our scheme is more efficient than that in [11, 15–20, 22], as our scheme can choose proper processing path according to the properties of each App. Generally speaking, our scheme can obtain good balance between the detecting accuracy and computation complexity, and it is a good solution for cloud-based Android malware-detecting services.

4. Features Extraction and Classification Methods of the Proposed Scheme

As shown in Figure 2, our scheme adopts four paths which are the Com+ feature (Path A), the function call graph feature (Path B), the system call feature (Path C), and the privacy usage feature (Path D) to extract features of the Apps. In the classification processes, *k*-NN, SVM, and Naive

Table 2: Four paths of feature extraction in our malware-detecting scheme.

Path	Description	Complexity	Conditions
Com+ feature	A classical presentation of Apk file	Low	Decompilable, nonobfuscation
Function call graph	A structural-level feature containing the invoking relationships	Super high	Decompilable
System call feature	A Linux kernel level dynamic feature suited for most Android versions	High	Apps earlier than Android 4.1
Privacy usage	An Android-specified dynamic feature indicating privacy leakage based on TaintDroid	High	Apps of Android 4.1 or later

Input: $S=\{s_i\}$, the set of all experimental n Apps.
Output: $C=\{c_{i,j}\}$, the feature set of experimental Apps, the ith row vector present the feature of the ith App.
begin procedure
for i = 1 to n **do**
  decompile the s_i
  generate the Permission feature p(151-bit) from Manifest;
  generate the API feature a(3262-bit) from source codes;
end for
for j = 1 to 151 **do**
  $c_{i,j}=a_j$;
end for
for j = 152 to 3413 **do**
  $c_{i,j}=p_j$;
end for
end procedure

Algorithm 1: Com+ feature extraction.

Bayes are used as classifiers. The details of the processing are presented below.

4.1. Feature Extraction

4.1.1. The Com+ Feature. Android OS has its own security approach based on Permission mechanism. For example, Android 4.1 adopts totally 151 kinds of Permissions which allow Apps to access software and hardware resources of the devices [22]. Normal Apps can retrieve all the messages using READ_SMS Permission while malwares may misuse this Permission to perform malicious actions. Therefore, some Permissions can be used as features for malware detection.

Each Apk file has a Manifest.xml file, which declares all the Permissions that the App needs to use. So we can extract Permissions from that file and generate corresponding Permission features. Each App can be represented by a binary vector, namely, P, where P_i = 1 if the ith Permission is used and P_i = 0 if the ith Permission is not declared in the Manifest.xml file. Thus, any App can be represented by a 151-dimension vector which is named as Permission feature.

However, using Permissions as features may encounter errors as some Permissions are not really employed during the execution of the App (i.e., fake Permission declaration). Thus, our scheme also employs API calls as a feature of static analysis to decrease the impact of fake Permission declaration.

Android OS provides the framework API that can be invoked by Apps so that Apps can interact with the kernel of Android OS. The official framework API consists of a set of packages and classes such as Android/app/Notification/Action and Android/view/Window Manage, which conveys substantial semantics about the App's behaviors [11]. The framework API provides 3262 kinds of API calls in total. If one App adopts an API call in the code, it means this App will execute the action of the corresponding API.

Since API calls reliably reflect the behaviors of the App, we can combine API calls together with Permissions to form a more representative and comprehensive feature, which is called the Com+ feature in this paper. The procedure of the Com+ feature extraction for one App is summarized in Algorithm 1. Using this procedure of Path A, a 3413-bit (151 Permissions features + 3262 API call features) Com+ feature can be formed for each App in our malware-detecting scheme with very low complexity.

4.1.2. Function Call Graph Feature with Fixed Dimension. With the development of malicious technology, obfuscation strategies have been widely used by malwares which can obscure the Permission and API call features. The detecting methods used in virus detection of personal computers show that structural-level properties, such as function call graphs, can offer a robust representation of virus actions [23]. Thus, we believe that function call graphs may also be

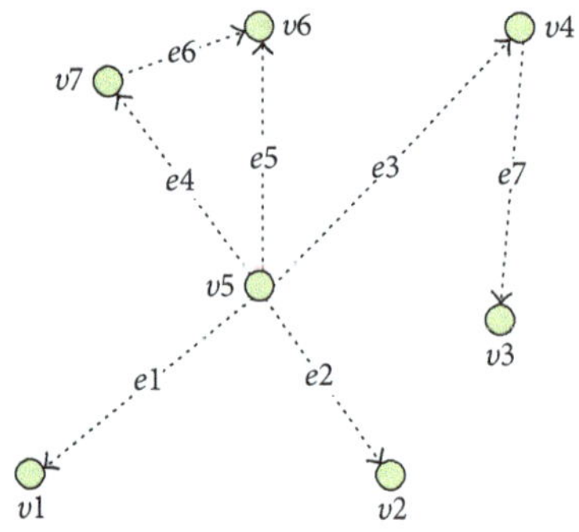

FIGURE 3: An example of the function call graph.

effective in Android malware detection even when obfuscation strategies are used by malwares. Using some software tools, the function call graph feature with limited dimension can be extracted from the Apps in our malware detecting scheme (Path B of Figure 2).

(a) *Function call graph information extraction*: We can get a ".gexf" file from the Apk using the "*Androguard*" tool of Table 1. This directed graph contains nodes of every function and edge direction from callers to callees. Then, the function call graph can be represented as a 4-tuple $G = \{V, E, L, l\}$, where V is a finite set of nodes, and each node is associated with one of the App functions. $E \subseteq V \times V$ denotes the set of calling directed edges, where an edge from a node v_1 to a node v_2 indicates a call from the function represented by v_1 to the function represented by v_2. As shown in Figure 3, the structural-level properties of the function call graphs can be expressed by labeling V and E.

Further, labeling function $l : V \rightarrow L$ can be used to label each node with the properties of the function. According to Dalvik [10] specification of Android OS, we chose 15 distinct categories of instructions based on their functionalities as shown in Table 3, and the l can be defined as

$$l(v) = \{b_1(v), b_2(v), b_3(v), \ldots, b_m(v)\}, m = 15. \quad (1)$$

If node v contains the ith instruction in Table 3, b_i is set to 1; otherwise b_i is set to 0. Then, each function node of Apps can be labeled by a 15-bit vector.

However, only considering single function is not enough for malware detection. So, we further explore the invoking relations among all the functions. Reference [24] suggested a kernel operating on an enumerable set of subgraphs in a labeled graph. Each given node v and the set of neighbor nodes V_v contain its own subgraph. The generation of the subgraph is defined by the operation as

$$F = r(l(v)), \left(\bigoplus_{z \forall V_v} l(z) \right), \quad (2)$$

where $\oplus$ represents a bit-wise XOR on the binary labels and r denotes a 15-bit left-shifting operation. Using (2), we can generate a new 30-bit feature F which contains the subgraph

TABLE 3: List of instruction categories and corresponding bit.

Category	Bit
nop	1
move	2
return	3
monitor	4
test	5
new	6
throw	7
jump	8
branch	9
arrayop	10
instanceop	11
staticop	12
invoke	13
unop	14
binop	15

features, and the F could represent both the content and invoking information of a function node.

(b) *Unify call graph features into fixed dimension*: For each App, we can extract n nodes and corresponding F features. As the number of subgraphs (n) is not certain, different Apps contain unequal amount of F features. If we directly use the $n \times F$ data as features in the machine learning process, the analysis may be collapsed by the high dimension of $n \times F$. As many tranditional softwares are transferring from the PC to Android devices, Android Apps are providing more and more functions. For example, if one App contains 10,000 functions, the dimension of $n \times F$ will reach 300,000. As huge computation will be demanded in the machine learning process which uses the data of $n \times F$ as features, the malware-detecting service may be collapsed. Hence, we need to unify the dimension of these features to meet the requirement of the classification algorithms.

In our scheme, we adopt Laplacian score feature selecting method of [25] to select 100 typical F features as the centers of k-means clustering. Then, k-means which can group objects into k classes according to the measured intrinsic characteristics is used to complete the feature unification [26]. Using the k-means clustering method, our scheme can convert the $n \times F$ features into a fixed 100-dimension feature for classification, and the analysis collapsing caused by the high dimension of function call graph can be avoided. In this paper, we named the new call graph feature with fixed dimension as G features in this paper. Figure 4 presents the processing of $n \times F$ features $\rightarrow G$ features. For an App which contained n nodes, $n \times F$ features could be extracted from this App using labeling and hashing methods presented before. Then, these $n \times F$ features are clustered by the 100 centering F features which are generated by feature selection, and the number of F which are clustered as F_i is recorded as g_i. Further, feature G could be constructed by g_i as follows:

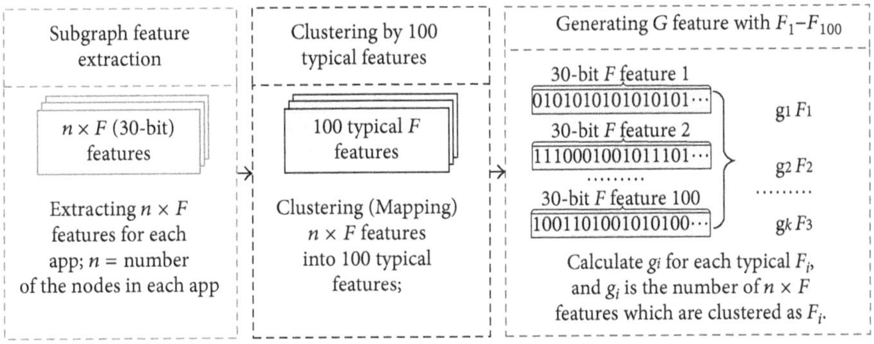

FIGURE 4: Unification of $n \times F$ features into G feature.

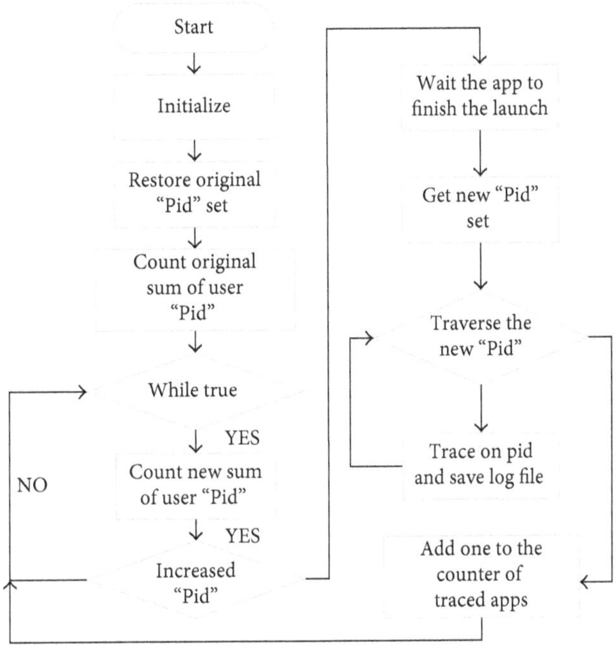

FIGURE 5: Flow of the *Strace*-based analysis.

$$G = \{g_1, g_2, \ldots, g_k\}, s.t. \sum_{i=1}^{k} g_i = n, k = 100. \quad (3)$$

4.1.3. Feature Extraction with Detector in Dynamic Analysis. If Apps use encrypting techniques, these Apps could not be decompiled, and dynamic analysis should be applied in the malware detection. In our malware-detecting scheme, we designed a special software tool *Detector* to monitor the running Apps' status and the newly started testing Apps. Besides, our *Detector* also collects and formats the log data provided by tracing soft tools, so that the dynamic features can be sent to the machine learning process of our scheme (descriptions about the *Detector* are listed in Table 1). In our scheme, Apps which prepared for malware detection are run with interval of 30 seconds, and the *Detector* can create feature files (log files) from the tracing data of *Strace* or *TaintDroid*.

The processing flow diagram of the *Detector* is shown in Figure 5. The initializing work includes setting root path and some counters. The original "Pid" [27] setting of all running user Apps is restored. The amount of "Pid" is monitored and counted in every iteration of the Apps. If the amount of "Pid" is increasing, *Detector* will deem that the testing App is turned on. Then, recording actions will be executed to trace the testing Apps and generate log files. If Apps start several "Pid," our Detector will execute multiple instances to save all the log data for each "Pid."

4.1.4. Extracting Privacy Usage Feature with DroidBox and TaintDroid. DroidBox [28] is a dynamic analysis tool which can monitor the App's privacy usage such as network connection, reading/writing operations, information leaks, SMS messages, and phone calls. To our knowledge, the *DroidBox* obtained good performance for Android 4.1 or later. As the malwares always complete some actions related with privacy usage, for Apps which are delivered with

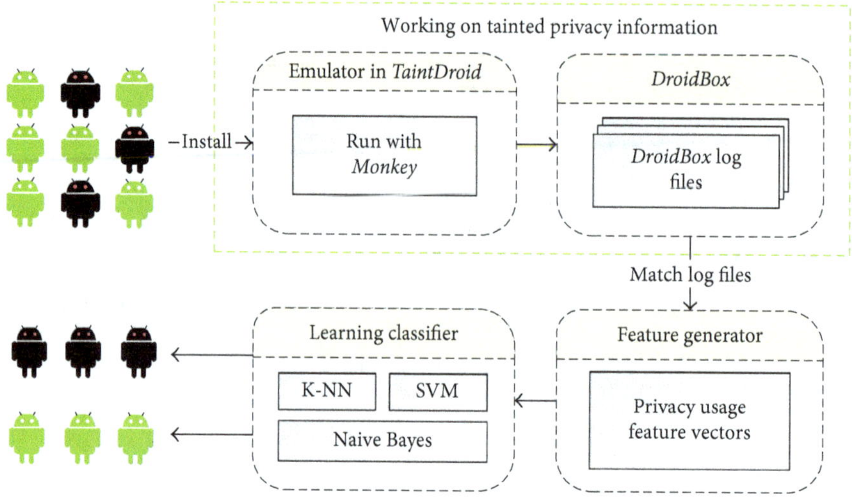

FIGURE 6: The framework of *DroidBox* analysis.

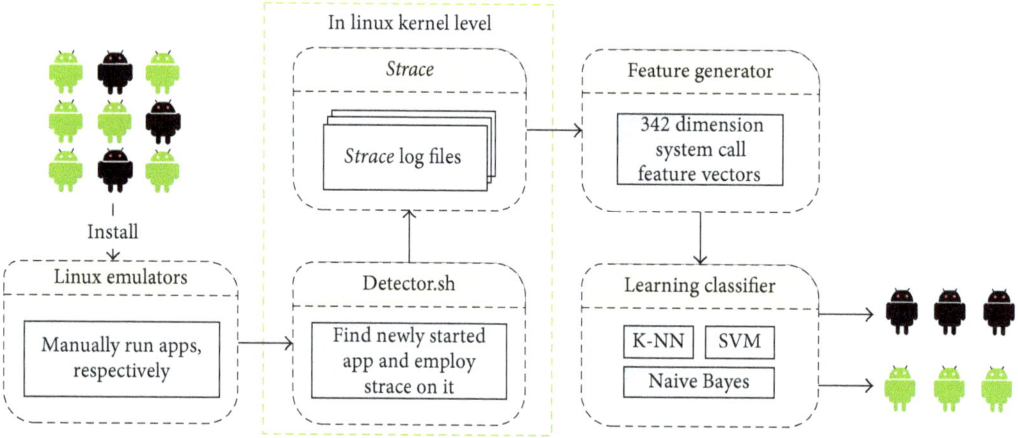

FIGURE 7: The framework of the *Strace* analysis.

Android 4.1 or later, *DroidBox* and *TaintDroid* in Table 1 should be used to trace the behaviors of Apps. Figure 6 shows how the *DroidBox* (Path D in Figure 2) path extracts the privacy usage features. It is achieved by shell scripts and python programs working together on the Linux platform. The Linux emulator is running a *TaintDroid* rom which will clean all user data when it is restarted. The *DroidBox* tool will automatically install/start the Apps on the emulator, and the *Monkey* tool [29] in Table 1 is used to complete initial operations. During the execution of Apps, our scheme can catch all the Apps' privacy usage status and generate the log files with *Detector*.

4.1.5. Extracting System Call Feature with Strace. For Apps delivered before Android 4.1, the *DroidBox* and *TaintDroid* cannot be applied for the App behavior tracing. As Android OS is built on the Linux kernel, our scheme employed *Strace* to trace the behavior of the Apps. The process is illustrated in Figure 7. After the Apps are uploaded, we install and execute all the Apps on the Linux emulating environment. Then, *Strace* (Path C in Figure 2), which is a system call tracer and debugging tool, is used to trace the Linux kernel system calls made by processes or programs. After the system call information is extracted by *Strace*, the feature log files are generated with our *Detector*.

Using the Path A–Path D, our malware-detecting scheme can extract features from any Android Apps. According to the processing of the four Paths, the complexities of feature extraction are different. Normally, Path A demands the lowest computation, Path B and Path D require similar complexity, and Path C always demands the highest computation. As our scheme can automatically deliver the Apps to the four Paths according to the properties of Apps, we can achieve good performance on the analyzing complexity. Compared with some existing malware-detecting methods, our scheme is more comprehensive and efficient, and Table 4 lists the comparisons.

4.2. Classification Methods. After the feature extraction, our scheme analyzes different features by three classification

TABLE 4: Comparisons of our scheme with some existing malware-detecting methods.

Methods	Required using conditions	Efficiency
References [8–10]	Apps can be decompiled, nonobfuscation	Low
References [12, 23]	Apps can be decompiled, obfuscation strategy	High
References [10, 18]	Apps cannot be decompiled, Android 4.1 or later	High
References [16, 17]	Apps cannot be decompiled, before Android 4.1	High
Ours	None	Median

methods: k-NN, SVM, and Naive Bayes. We believe that different classifiers may obtain various results, and the results will be impacted by parameter tuning of different classifiers. In this paper, we only adopt three classifiers which are common in machine learning process. The k-Nearest Neighbors algorithm (k-NN) is a nonparametric method used for classification [30]. In k-NN classification, the output is a class membership. An object is classified by a majority vote of its neighbors, so that it will be assigned to the most common class among its k nearest neighbors (k is a positive odd integer). If $k = 1$, the object is simply assigned to the nearest neighbor class. As the malware detection only need to divide the malwares from benigns, we set k as 1 to construct 2-type k-NN classifier.

SVM can divide the n-dimensional space representation of the data into different regions by using a hyperplane, which intends to maximize the margin between the two classes. The margin is computed using distances between the closest instances of both classes, which are called support vectors [31]. In our malware-detecting scheme, SVM with RBF kernel is also used in our malware classification process.

Naive Bayes classifier is a kind of simple probabilistic classifier based on applying Bayes' theorem with strong (naive) independence assumptions between the features. Concerning the variety and independence of different Apps, Naive Bayes can also be used to classify the malicious and benign Apps.

The above three classified methods are adopted in our experiments without parameter tuning. As the performances of the three classifiers are similar, we believe the features extracted by our detecting scheme are effective.

5. Experimental Results and Analysis

5.1. Testing Malware Datasets. In order to evaluate the performance of the proposed malware detecting scheme, two datasets were applied in our experiments. The first dataset was an open-access dataset which was built by Jiang [21] in 2012. This dataset collected 1260 Apps from August 2010 to October 2011, and these samples were classified into 46 Android malware families. To our knowledge, the dataset of [21] was the most popular and open-access malware library so far, and lots of research used this library for evaluation. Thus, we randomly selected 1000 Apps from malware and benign libraries to form a dataset before 2012, which was referred to as the older dataset in our experiments.

TABLE 5: Information and comparison of experimental datasets.

Type	Older dataset		Newer dataset	
	Malicious	Benign	Malicious	Benign
Source	Reference [21]	Wandoujia	VirusShare	Wandoujia
Time	2011	2016	2016	2016
Amount	1000	1000	1000	1000

As the Android platform is updating rapidly, the security strategies of Android is also changing. So, old malicious actions may not be active any more, and new malwares may perform different malicious functions. Considering this, we established an up-to-date dataset which provided open-access in [25], and the details malicious actions of the malwares were also shared by excel format. The malicious Apps of our new dataset were collected from VirusShare [32], one of the major virus App-sharing websites. Because some Apps of VirusShare was ineffective, we spent several months to pick the appropriate malicious samples from top-ranked malwares. Eventually, about 1000 malicious Apps delivered between September 2012 and December 2016 were selected as testing dataset. Benign Apps of our new dataset were gathered from an App market called "Wandoujia" [33] in July 2016. We obtained more than 1000 benign Apps from the top popular ranking Apps, so that these samples could cover most types of Android Apps. We believed that top-ranked popular Apps would not perform any illegal activities, otherwise the management of Apps market and the App users would not rank the Apps as top popular applications. The diversity of the benign Apps could ensure the comprehensiveness of the behaviors and features of benign sample set. Using these up-to-date malicious and benign Apps, we formed a new testing dataset which was referred as newer dataset of Table 5.

5.2. Evaluation Matrices. The performance is evaluated by three metrics: accuracy (ACC), true positive rate (TPR), and false positive rate (FPR). These metrics are defined as

$$\begin{aligned} \mathrm{TPR} &= \frac{\mathrm{TP}}{\mathrm{TP}+\mathrm{FN}}, \\ \mathrm{FPR} &= \frac{\mathrm{FP}}{\mathrm{FP}+\mathrm{TN}}, \\ \mathrm{ACC} &= \frac{\mathrm{TP}+\mathrm{TN}}{\mathrm{TP}+\mathrm{FP}+\mathrm{FN}+\mathrm{TN}}. \end{aligned} \tag{4}$$

TP (true positive) represents the number of malwares that are correctly detected. FN (false negative) denotes the number of Android malwares that are not detected (predicted as benign App). TN (true negative) is the number of benign Apps that are correctly classified, and FP (false positive) is the number of benign Apps that are incorrectly detected as Android malwares. The TPR can reflect the percentage of malwares in the whole samples which are detected as benign Apps, the FPR can reflect the percentage

Table 6: Results of static analysis with different feature sets on older dataset.

Algorithm	Permission feature			API feature			Permission and API		
	ACC	TPR	FPR	ACC	TPR	FPR	ACC	TPR	FPR
k-NN	95.76	97.31	5.21	98.42	97.30	0.43	98.61	97.21	0.14
SVM	96.65	95.31	2.65	99.10	98.51	0.34	99.21	98.38	0.14
Bayes	93.37	94.62	7.31	94.13	99.57	11.01	94.41	99.42	10.7

of benign Apps in the whole samples which are detected as malware, and ACC can evaluate the overall detecting performance. Details about these evaluating metrics could be found at [34].

For each classification method, we randomly selected the training and testing dataset at the ratio of 7 : 3. The malicious and benign features are labeled as 1 and 0, respectively. Then, we performed the classifying algorithm 100 times and calculated the average value of each measurement metrics as the final results, so that the reliability of the experiments is guaranteed.

5.3. *Experiment Results and Discussion*

5.3.1. Older Dataset. For the older dataset, we used both the Com+ feature for static analysis and the system call feature for dynamic analysis (Path A and Path C in Figure 2) to execute the detection [14].

Table 6 summarizes the classification results of the different static features, where the Com+ feature achieves a good result. The accuracy ranges from 93% to 99%. The best performance in ACC (99.28%) is achieved by the SVM classifier using the Com+ feature. Meanwhile, this algorithm has the second highest TPR (98.68%) and the lowest FPR (0.12%). Therefore, the SVM algorithm is a suitable method for the Com+ feature of static detection.

For Apps which cannot be decompiled, we used *Strace* to dynamically extract their system call features. As shown in the experimental results in Table 7, all three classifiers achieve 85% or higher in accuracy. And the Naive Bayes has the highest ACC rate (90.00%) and TPR rate (90.85%), as well as the lowest FPR (10.85%). So, we consider Naive Bayes as the good choice for the system call feature of dynamic analysis.

5.3.2. Newer Dataset. Since Android is updated continually and the latest Apps mostly adopt obfuscation on source code, it is difficult to extract the Com+ feature from Apps in the newer dataset. Meanwhile, the performance of the system call feature had also decreased. So, we employed new methods (Path B and Path D in Figure 2) to deal with Apps in the newer dataset.

The function call graph feature (Path B) is used as static analysis for the newer dataset, and Table 8 shows the comparison between different classifiers using the function call graph feature. The accuracy of all these algorithms is higher than 86%, which proves that the function call graph is an excellent feature for static analysis for the newer dataset. The *k*-NN and Naive Bayes classifications

Table 7: Results of *Strace* analysis on older dataset.

Algorithms	ACC	TPR	FPR
k-NN	87.82	89.51	13.32
SVM	85.41	83.88	11.36
Naive Bayes	90.31	90.41	10.65

produced similar results. SVM obtained the highest TPR but the worst FPR, which means it detected 96.92% of the malwares but misjudged 24.54% of the benign Apps in the meantime.

As mentioned in Section 3, *Strace* can be used on most types of Linux kernel-based system (like Android), which means it can work on different Android versions. But its weakness is also obvious such as lacking of pertinence. In contrast, *DroidBox* (Path D), which is a software tool designed especially for Android version 4.1 or higher, can provide more specific analysis on privacy information leaking according to the Android mechanism.

The experimental results of *DroidBox* analysis on new dataset are shown in Table 9. Among the three different classifiers, the best accuracy (74%) of all these algorithms is achieved by the *k*-NN classifier, and the *k*-NN classifier also obtains the highest TPR and the lowest FPR. Therefore, it is the best classifier for the privacy leaking of dynamic analysis. However, as *DroidBox* employs an emulator which is not actual Android devices, the *Monkey* tool can only simulate parts of the human operations. The performance of *DroidBox* may be further improved in our future work.

5.3.3. Discussion. As shown by the experiments described above, our scheme provides four paths to complete malware detection, namely, the Com+ feature, function call graph with limited dimension, system call feature, and privacy usage feature. Given any Android App, our scheme can select the most suitable path according to its properties in order to achieve comprehensive and efficient analysis.

Many existing Android malware-detecting methods (such as [8–10, 19, 20]) were evaluated by their testing dataset which are not shared. Thus, it was very hard to complete fair comparison between different malware-detecting methods. To our knowledge, dataset of [21] are open-access and popular malware dataset, and we used this dataset to evaluate our scheme. For the new dataset, as other existing methods have not provided open-access detecting software, it is very hard for us to complete the experiments with other up-to-date approaches.

Table 8: Results of function call graph analysis on newer dataset.

Algorithms	ACC	TPR	FPR
k-NN	86.56	86.48	13.51
SVM	86.11	96.32	24.65
Naive Bayes	86.51	84.54	11.34

Table 9: Results of *DroidBox* analysis on newer dataset.

Algorithms	ACC	TPR	FPR
k-NN	74.23	77.77	28.41
SVM	66.33	73.87	41.74
Naive Bayes	71.54	73.31	30.12

We also checked the malware samples which were detected by our scheme (detecting errors). We found that most detecting errors occurred in the game Apps with malicious action of private information leaking. We believed that the reason of the wrong detecting may be caused by the complex operation of game Apps, as these Apps always require operations such as registering private user information and payment for game tools. Some illegal activities could be covered by the processing of purchasing tools of game (e.g.,, credit card information leaking). Thus, the illegal activities were hard to be detected.

As shown in Tables 6–8, the proposed Android malware-detecting scheme obtained good detecting accuracy. Besides, the average malware-detecting complexities are decreased by our flexible feature extracting Paths. Thus, our scheme achieved good balance between the malware detecting accuracy and complexity, and it is highly suitable for cloud service providers which need to improve the security of their cloud services.

6. Limitation and Future Work

The detecting scheme proposed in this paper could be applied to construct a basic malware-detecting platform. However, the scheme is limited in some aspects, and some issues should be further improved to make the detecting scheme more comprehensive.

(i) Like many traditional malware detecting methods, the proposed scheme only can complete objective detection with software. However, some subjective methods which involve reviewing comments of App users have been used in malware detection, such as [19, 20]. With the developments of deep learning, effective user feedbacks may be extracted from huge number of comments, and we believe these feedbacks are very valuable for malware detection. Thus, our future malware detecting scheme should integrate the user's feedbacks as some features in the detecting process to achieve better detecting accuracy.

(ii) The proposed detecting scheme still demands some manual processing; thus, the scheme is hard to be applied in practical applications which demand real-time processing of huge number of App samples. As many Apps are delivered every day, more automatic processing should be developed to upgrade these manual processing, so that the proposed scheme could be adopted more easier in practical applications.

(iii) Our scheme still has not adopted self-updating mechanisms to constantly renew the testing dataset. The testing dataset used in our experiments are constructed by ourselves, and the samples only cover parts of the Android Apps. Meanwhile, the Android OS is continuously updating, and the malwares are also evolving. Thus, some auto-updating methods should be employed in our scheme to complete auto-updating of testing dataset and auto-training of classifiers.

(iv) The current detecting scheme only used some common Linux or Android soft tools to analyze the App samples. If malwares adopt some techniques to cheat these common soft tools, the detecting scheme may be invalided. Thus, some new tracing and analyzing soft tools should be integrated into our scheme.

7. Conclusion

Malware detection is very important for mobile applications. This paper proposed a high-efficiency detecting scheme for Android malwares using static and dynamic analyzing methods, such as Permission, API calls, function call graph with limited dimension, and privacy usages. Several software tools are designed and integrated to extract malware analyzing features including Com+ feature, system call, and privacy leaking information. We also proposed some new techniques such as Com+ feature and *Detector*" to improve the performance of malware detection. Using these techniques, the proposed scheme can effectively detect malwares, regardless whether the encrypting or obscuring techniques are used by malwares. Besides, our scheme can achieve good balance between the complexity and detecting accuracy as several methods are applied to complete feature extracting based on the properties of the Apps. Plenty of experiments verified the high performance of the scheme with our new open-access up-to-date malware dataset.

Acknowledgments

The authors would like to thank the anonymous reviewers for their comments and suggestions. This work was supported in part by the National Natural Science Foundation of China (Grant nos. 61373102 and 61771338) and by Yunnan Academician Funding.

References

[1] http://www.idc.com/prodserv/4Pillars/mobility#Press.

[2] https://www.magzter.com/news/488/1242/052017/0jt0j.

[3] "Kaspersky: forget lone hackers, mobile malware is serious business," February 2014.

[4] A. P. Felt, M. Finifter, E. Chin, S. Hanna, and D. Wagner, "A survey of mobile malware in the wild," in *Proceedings of the 18th ACM Conference on Computer and Communications Security*, pp. 3–14, Swissôtel Chicago, Chicago, IL, USA, October 2011.

[5] F. Di Cerbo, A. Girardello, F. Michahelles, and S. Voronkova, "Detection of malicious applications on Android OS," in *Proceedings of the 4th International Workshop on Computational Forensics (IWCF)*, pp. 138–149, Springer, Berlin, Heidelberg, Germany, November 2010.

[6] P. Faruki, A. Bharmal, V. Laxmi et al., "Android security: a survey of issues, malware penetration, and defenses," *IEEE Communications Surveys and Tutorials*, vol. 17, no. 2, pp. 998–1022, 2015.

[7] A. Sadeghi, H. Bagheri, J. Garcia, and S. Malek, "A taxonomy and qualitative comparison of program analysis techniques for security assessment of Android software," *IEEE Transactions on Software Engineering*, vol. 43, no. 6, pp. 492–530, 2017.

[8] W. Wang, X. Wang, D. Feng, J. Liu, Z. Han, and X. Zhang, "Exploring permission-induced risk in Android applications for malicious application detection," *IEEE Transactions on Information Forensics and Security*, vol. 9, no. 11, pp. 1869–1882, 2014.

[9] P. P. Chan and W. K. Song, "Static detection of Android malware by using permissions and API calls," in *Proceedings of Security and Privacy in Communication Networks. SecureComm 2013*, vol. 1, pp. 82–87, Sydney, Australia, September 2013.

[10] P. P. Chan and W. K. Song, "Static detection of Android malware by using permissions and API calls," in *Proceedings of International Conference on Machine Learning and Cybernetics*, vol. 1, pp. 82–87, Lanzhou, China, July 2014.

[11] Y. Aafer, W. Du, and H. Yin, "DroidAPIMiner: mining API-level features for robust malware detection in Android," in *Proceedings of Security and Privacy in Communication Networks. SecureComm 2013*, pp. 86–103, Sydney, Australia, September 2013.

[12] T. E. Wei, H. R. Tyan, A. B. Jeng, H. M. Lee, H. Y. M. Liao, and J. C. Wang, "DroidExec: root exploit malware recognition against wide variability via folding redundant function-relation graph," in *Proceedings of 17th International Conference on Advanced Communication Technology (ICACT)*, pp. 161–169, Seoul, South Korea, July 2015.

[13] H. Gascon, F. Yamaguchi, D. Arp, and K. Rieck, "Structural detection of Android malware using embedded call graphs," in *Proceedings of the 20th ACM Conference on Computer and Communications Security*, pp. 45–54, Berlin, Germany, November 2013.

[14] Y. Liu, Y. Zhang, H. Li, and X. Chen, "A hybrid malware detecting scheme for mobile Android applications," in *Proceedings of the 2016 IEEE International Conference on Consumer Electronics (ICCE)*, Las Vegas, NV, USA, pp. 155–156, January 2016.

[15] I. Burguera, U. Zurutuza, and S. Nadjm-Tehrani, "Crowdroid: behavior-based malware detection system for Android," in *Proceedings of the 18th ACM Conference on Computer and Communications Security*, pp. 15–26, Swissôtel Chicago, Chicago, IL, USA, October 2011.

[16] Y. J. Ham, D. Moon, H. W. Lee, J. D. Lim, and J. N. Kim, "Android mobile application system call event pattern analysis for determination of malicious attack," *International Journal of Security and Its Applications*, vol. 8, no. 1, pp. 231–246, 2014.

[17] S. Zou, J. Zhang, and X. Lin, "An effective behavior-based Android malware detection system," *Security and Communication Networks*, vol. 8, no. 12, pp. 2079–2089, 2015.

[18] W. Enck, P. Gilbert, S. Han et al., "TaintDroid: an information-flow tracking system for realtime privacy monitoring on smartphones," *ACM Transactions on Computer Systems*, vol. 32, no. 2, 29 pages, 2014.

[19] M. Rahman, M. Rahman, B. Carbunar, and D. H. Chau, "Search rank fraud and malware detection in Google Play," *IEEE Transactions on Knowledge and Data Engineering*, vol. 29, no. 6, pp. 1329–1342, 2017.

[20] A. Narayanan, M. Chandramohan, L. Chen, and Y. Liu, "Context-aware, adaptive, and scalable Android malware detection through online learning," *IEEE Transactions on Emerging Topics in Computational Intelligence*, vol. 1, no. 3, pp. 157–175, 2017.

[21] X. Jiang, *An Evaluation of the Application ("App") Verification Service in Android 4.2*, December 2012, http://www.malgenomeproject.org/.

[22] N. Peiravian and X. Zhu, "Machine learning for Android malware detection using permission and API calls," in *Proceedings of the IEEE 25th International Conference on Tools with Artificial Intelligence*, pp. 300–305, Herndon, VA, USA, November 2013.

[23] X. Hu, T. C. Chiueh, and K. G. Shin, "Large-scale malware indexing using function-call graphs," in *Proceedings of ACM Conference on Computer and Communications Security*, pp. 611–620, Chicago, IL, USA, November 2009.

[24] D. Haussler, "Convolution kernels on discrete structures," Tech. Rep., vol. 646, Department of Computer Science, University of California at Santa Cruz, Santa Cruz, CA, USA, 1999.

[25] X. He, D. Cai, and P. Niyogi, "Laplacian score for feature selection," in *Proceedings of 18th International Conference on Neural Information Processing Systems*, Vancouver, BC, Canada, December 2005.

[26] A. K. Jain, "Data clustering: 50 years beyond K-means," *Pattern Recognition Letters*, vol. 31, no. 8, pp. 651–666, 2010.

[27] https://en.wikipedia.org/wiki/Process_identifier.

[28] https://github.com/pjlantz/droidbox.

[29] http://developer.Android.com/intl/zh-cn/tools/help/monkey.html.

[30] N. S. Altman, "An introduction to kernel and nearest-neighbor nonparametric regression," *The American Statistician*, vol. 46, no. 3, pp. 175–185, 1992.

[31] V. N. Vapnik, *The Nature of Statistical Learning Theory*, Springer Science & Business Media, New York, NY, USA, 2013.

[32] https://virusshare.com.

[33] https://www.wandoujia.com.

[34] https://en.wikipedia.org/wiki/Sensitivity_and_specificity.

2

Integrated Packet Classification to Support Multiple Security Policies for Robust and Low Delay V2X Services

Jaehyeong Wee and Wooguil Pak

Department of Computer Engineering, Keimyung University, Daegu 704-701, Republic of Korea

Correspondence should be addressed to Wooguil Pak; wooguilpak@kmu.ac.kr

Guest Editor: Safdar H. Bouk

One of the key applications in the 5G system is Vehicle-to-Everything (V2X). Ultra-low delay communication is essential for the safety of users and pedestrians in V2X. However, as sophisticated and various cyberattacks are increasing, it becomes hard to satisfy low delay constraints. To protect networks from such attacks, even single network security equipment provides multiple security functions, resulting in the inevitable additive delay in packet processing. In this paper, we suggest a new packet classification paradigm to resolve this issue. The proposed algorithm integrates multiple policy rule-sets into a single rule-set and classifies incoming packets using the integrated rule-set. Thus, it has a unique feature providing high classification performance regardless of the number of security policies. Through extensive performance evaluations, we confirm that the performance improvement is also increased with the total rule-set number increasing without the significant overhead of memory cost. We expect that it will mitigate the delay issue of existing network equipment for upcoming services such as V2X.

1. Introduction

Vehicle-to-Everything (V2X) service is one of the most promising applications in the 5G system. It frequently exchanges information between drivers, pedestrians, vehicles, and transportation infrasystems [1–6], and the information should be delivered with low delay and high reliability for the safety of involved persons.

Modern cyberattacks have become more sophisticated and diverse, and as a result, security functions installed in modern security equipment also become more complex and various. For protecting networks from various cyberattacks, single multifunction network equipment has been introduced [7]. For example, unified threat management (UTM) supports multiple rule-sets using multiple policy tables as shown in Figure 1. Such integrated network equipment has advantages in security but disadvantages in strict delay requirements of V2X. Each security policy is implemented by complicate packet classification that searches a matching rule with the highest priority by comparing each field of every rule with the incoming packet header. As the integrated equipment should independently perform packet classification for each policy rule-set, the classification cost increases as the number of rule-sets increases [8–14]. Multiple classifications are a bottleneck of network performance, especially in terms of the delay [15–20]. Therefore, the high performance and scalable packet classification is essential for supporting V2X.

In this paper, we propose a new packet classification algorithm that has a distinct feature against other competitors. Although most existing classification algorithms suffer from deteriorated performance as the total rule-set number increases, the proposed algorithm can achieve high classification performance regardless of the total rule-set number. It can effectively support reliable and low delay V2X services. Figure 2 shows the overall architecture of the software-defined networking (SDN) for V2X. Packet classification is a basic function of OpenFlow controller and SDN switch.

To increase the performance of packet classification, high-end SDN switches adopt the expensive hardware-based solution. However, the OpenFlow controller usually adopts software-based packet classification since hardware solution cannot achieve high flexibility to support various security requirements from customers. This algorithm targets OpenFlow controllers and software-based SDN switches to reduce the burden of packet classification. It can play a very

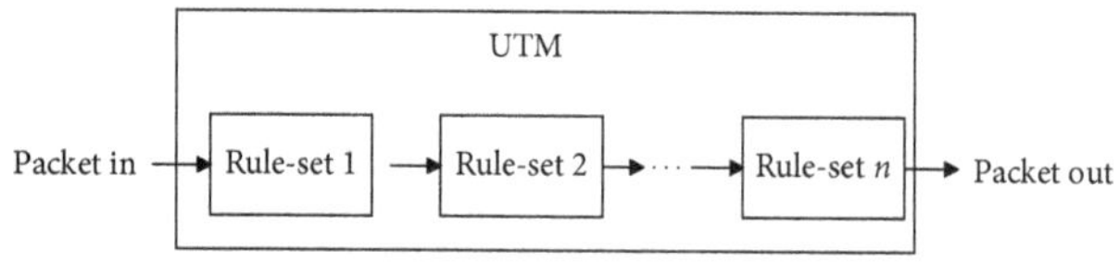

FIGURE 1: Multiple packet classification using multiple rule-sets in UTM equipment.

important role in them to provide high performance and high security, simultaneously.

The remainder of this paper is organized as follows: Section 2 briefly presents related work, and the motivation of this research is explained in Section 3. In Section 4, the proposed algorithm is described in detail. The performance evaluation results are compared with those of competitors in Section 5. Finally, Section 6 concludes.

2. Related Work

Although many factors can be used to evaluate the performance of packet classification algorithms, packet classification speed and memory requirements are most important factors. However, most algorithms cannot support high classification speed with low memory requirement.

Packet classification is classified into hardware- and software-based approaches [21–27]. Hardware-based packet classification can achieve very high classification speed that is impossible for the software-based one. Most modern network equipment adopts hardware packet accelerators to provide 100 Gbps performance with multiple rule-sets. However, the hardware should be redesigned to satisfy the various requirements of users such as adding a new field in the rule structure. Moreover, hardware-based solutions usually adopt expensive memory called ternary-content addressable memory (T-CAM) for classification. Since supported rule-set size is determined by the size of T-CAM, it costs very high to support large rule-sets.

The strongest advantage in the software-based approach is flexibility. If the structure of field should be changed, it can be easily supported by modifying software. Another merit of the software-based approach is cost. If larger rule-set is needed, the user can increase the rule-set capacity of the network equipment by just adding much cheaper dynamic random-access memory (DRAM) compared to T-CAM.

Well-known algorithms belonging to the software-based approach are exhaustive search, cross-producting-based classification, tuple space search, and decision tree-based algorithms [21]. Now, we will briefly describe each software-based algorithm.

Exhaustive search linearly compares each rule with keys from highest to lowest priority to find the matching rule. Due to the searching procedure, the packet classification performance is degraded as the rule-set size increases. However, it requires smallest memory among all packet classifications and supports very fast update. Most of all, it can be easily implementable. As a result, it is suitable for a system with a small rule-set.

Cross-producting-based classification independently performs searching for each field, and it merges intermediate results [28–33]. This procedure is repeated until the final matching rule is found. It is one of fastest classification algorithms but it requires a huge amount of memory and time to build a classification table. Since it cannot support incremental update, it needs to rebuild entire table whenever a rule-set is updated. Although it has such critical weaknesses, it can support the classification performance almost similar to that of the hardware-based approach. Therefore, a lot of research is still going on to improve the weaknesses.

Tuple space search probes each sub-rule-set called tuple space to find the matching rule [34–37]. A tuple is defined by combination of each prefix length for five tuples, and the set of tuples are called tuple space. Since each rule of a rule-set belongs to only one of the tuples, tuple space has a good scalability in terms of a rule-set size. Although it achieves a moderate classification performance, it supports fast update, i.e., inserting or deleting a rule. Therefore, it has been adopted in Open vSwitch [38]. However, the classification performance is decreased proportional to the number of tuples, thus requiring further research to improve the performance.

Decision tree-based algorithm recursively chooses a child node according to the predefined policy on decision tree built based on a rule-set until it reaches a leaf node [39–47]. If it reaches, it searches the matching rule with the highest priority among all rules stored in the leaf node. The overall classification performance is known to have log complexity in terms of the rule-set size.

Decision tree-based algorithm provides a comparable classification performance with that of cross-producting-based classification algorithm but requires much smaller memory size. Thus, the decision tree-based algorithm is one of the most actively researched algorithms at present. When a decision tree-based algorithm partitions a rule-set into multiple sub-rule-sets, partitioning criteria is controlled by two factors: *space factor*, the maximum ratio of the sum of all rules belong to all sub-rule-sets to the original rule-set size, and *binth*, the maximum allowed rule size in the leaf node. Hence, the classification performance and the table size can be adjusted according to the requirements of applications.

Large *space factor* increases partitioning number but decreases the height of a decision tree, resulting in fast classification performance. However, the total number of duplicate rules is increased, and therefore, generating a large decision tree. On the other hand, large *binth* reduces partitioning number, so the tree size is decreased but the searching cost in the leaf node increases, and thus, providing low classification performance.

Well-known algorithms belonging to the decision tree-based approach are HiCuts and HyperCuts [39, 40]. Although they provide high classification performance, they still suffer from a large decision tree due to significant rule duplications. Recently, EffiCuts was introduced to decreasing rule duplications [41]. EffiCuts is based on HyperCuts but groups rules by fields with wildcard and generates a separate tree for each group. This approach significantly reduces rule duplications, so the total tree size is greatly decreased. However, separate tree deteriorates the classification performance. As a mitigation, trees with similar wildcard characteristics can be merged to increase classification performance while the overall tree

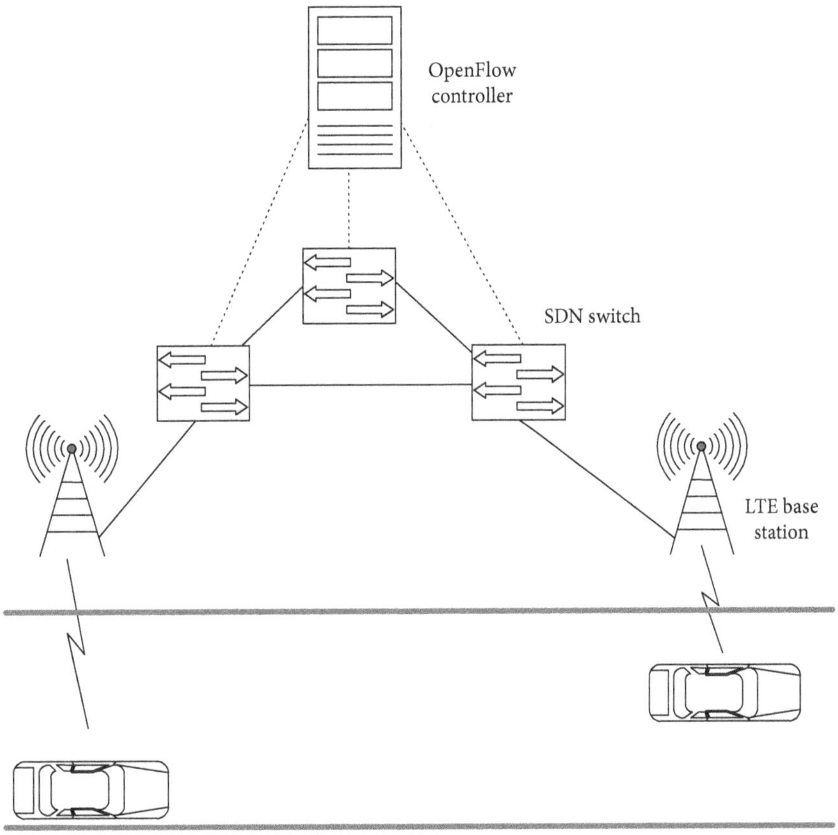

FIGURE 2: Overall architecture of V2X networks.

size is almost the same. EffiCuts is known to support fast updating [48, 49].

We will describe the operation of EffiCuts in detail. At first, EffiCuts splits the total rule-set into some predefined categories according to how many wildcard field each rule contains, where wildcard field is a field on which the rule has a large matching range, typically at least 50% of the total range of the field. For 5-tuple rule-set, we have four cases as follows:

(i) Category 1: four wildcard field rules

(ii) Category 2: three wildcard field rules

(iii) Category 3: two wildcard field rules

(iv) Category 4: one or zero wildcard field rules

For example, assuming that matching ranges of a rule for source IP, destination IP, source port, destination port, and protocol are ANY, ANY, 0 to 32768, 80, 0 to 128, it has four wildcard fields except for destination port, and therefore belonging to Category 1. Since each category contains similar rules only, EffiCuts builds a decision tree for sub-rule-set belonging to the same category and reduces replicated rules during building a decision tree. Although EffiCuts generates multiple decision trees, the total tree size is very small compared to the original HyperCuts. However, the number of decision tree affects the total classification performance. To reduce the number of trees, EffiCuts merges similar categories. This tree merging process increases the total tree size but it can still avoid excessive replication of rules. By doing so, EffiCuts achieves high classification performance and low memory requirement, simultaneously.

Table 1 summarizes each feature of packet classification algorithms.

3. Motivation

As shown in Table 1, the software-based approach consumes much memory to achieve high classification performance. However, high complexity of memory requirement results in low scalability in terms of rule-set size. Although decision tree-based algorithms have a high complexity of memory requirement, i.e., $O(N^D)$, latest decision tree algorithms show very low memory requirements, where N and D denotes the total dimension number and the rule-set size, respectively.

To verify the memory requirement, we performed the following experiment. We synthesized multiple firewall rule-sets whose size is from 20K to 100K using ClassBench [50]. Then, we built the total decision tree and calculated the ratio of the tree size to the rule-set size for each rule-set, where space factor and binth were configured to the best values. Figure 3 shows the experimental results obtained from EffiCuts. EffiCuts shows almost the same ratio regardless of the rule-set size, which means EffiCuts achieves almost $O(N)$ for memory requirement. Thus, it can decrease the decision tree size by 100 times for 100,000 rules compared to HiCuts or HyperCuts [41].

Figure 4 shows the ratio of the average memory access number and the rule-set size on the same configuration.

TABLE 1: Comparison features of software-based packet classification algorithms according to algorithm type.

Algorithm	Classification performance	Memory requirement	Incremental update
Exhaustive	$O(N)$	$O(N)$	Possible
Cross-producting	$O(DW)$	$O(N^D)$	Impossible
Tuple space	$O(N)$	$O(N)$	Possible
Decision tree	$O(D)$	$O(N^D)$	Partially possible

Note: rule-set size, total bit length of all keys, and the size of dimension of rule are denoted by N, W, and D, respectively

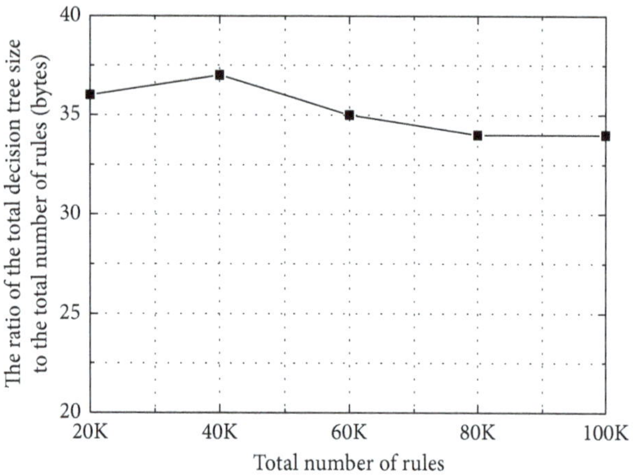

FIGURE 3: The ratio of the total decision tree size for EffiCuts to the total number of rules as the number of rules increases.

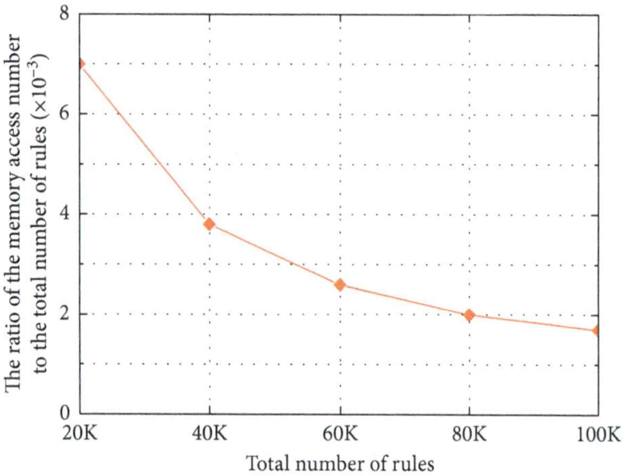

FIGURE 4: The average ratio of memory access number to the total number of rules as the number of rules increases.

We synthesized the packet data using each rule-set and searched the decision tree to find every packet in the data. We counted the total number of memory accesses during searching process and obtained the ratio of the total number and the total packet number. As the rule-set size increases, the number of memory access for EffiCuts also increases in Figure 4. However, the ratio of the access number to the rule-set size decreases as shown in Figure 4. From Figures 3 and 4, we can finally find two characteristics as follows:

Characteristic 1: $M(R_1) + M(R_2) > M(R_1 + R_2)$, where $M(R)$ is the average memory access number for rule-set R.

Characteristic 2: $S(R_1) + S(R_2) \cong S(R_1 + R_2)$, where $S(R)$ is the size of decision tree for rule-set R.

For example, we can see that $M(T_{20K}) + M(T_{40K}) > M(T_{20K} + T_{40K}) \cong M(T_{60K})$ from Figure 4 and $S(T_{20K}) + S(T_{40K}) \cong S(T_{20K} + T_{40K}) \sim S(T_{60K})$ from Figure 3, respectively, where T_n denotes the testing rule-set with a size of n used in the experiment and where K means 1,000.

Until now, existing research studies for packet classification focus on classification with a single rule-set. However, network systems with multiple rule-sets become popular, and fast classification algorithm oriented on a single rule-set has limitation to achieve high performance for multiple rule-sets. Thereby, it is required to consider multiple rule-sets for designing high performance classification algorithms. Hence, Characteristics 1 and 2 suggest a new guideline for developing packet classification algorithms. According to Characteristic 1, if a system has multiple rule-sets, it is advantageous to integrate them into one rule-set to construct a decision tree for improving classification speed. Characteristic 2 also implies that the size of the decision tree for integrated rule-sets is not larger than the sum of sizes for each decision tree for all rule-sets.

We finally conclude that packet classification algorithm based on integrated rule-sets has many advantages and suggest a new classification algorithm utilizing the features of integrated rule-sets.

4. Proposed Algorithm

The proposed algorithm performs packet classification using an integrated rule-set that combines all rule-sets in a system. At first, we briefly show the features of the proposed algorithm, and then, describe the algorithm in detail. For simple explanation, we assume that the rule consists of five tuples but it can be easily extended to more field cases.

4.1. Features of the Proposed Algorithm

4.1.1. Minimized Classification Cost. The proposed algorithm can complete total packet classification for all rule-sets with one search. Therefore, it can minimize the increased overhead due to the repetitive classification. Since it can maintain the high packet classification performance regardless of the number of rule-sets, it is very important feature of the proposed algorithm.

4.1.2. Early Packet Drop. Integrated rule-set has not only advantage to decrease classifying overhead but also to remove unnecessary classifying. For example, Figure 5 shows existing and proposed packet classifications. Assume that an incoming packet is allowed by rule-sets 0 to $k - 1$, but it is rejected by rule-set k. In this case, packet classifications for rule-sets 0 to $k - 1$ are eventually unnecessary since the packet cannot be forwarded due to rule-set k. However, packet classification for each rule-set is performed in sequence, so it cannot avoid

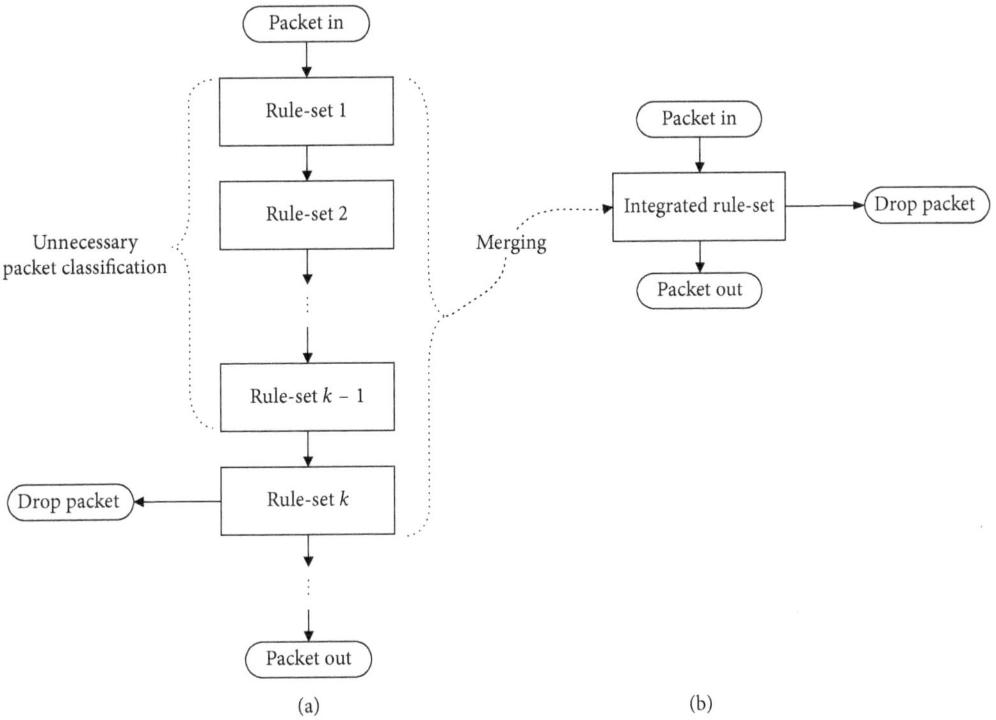

FIGURE 5: Packet classifications of existing and proposed algorithm. (a) Previous packet. (b) Proposed packet.

the unnecessary classifications for rule-sets 0 to $k - 1$ for existing classification. For the proposed classification algorithm, all rule-sets are integrated into one larger rule-set, making almost the same effect as searching multiple rule-sets, simultaneously. Thereby, the problem of existing classification is mitigated in the proposed one.

4.2. Building Decision Tree. The proposed algorithm builds a decision tree using EffiCuts after merging each rule-set into a large rule-set. However, it needs unique procedure called "fast rule skipping" and "early drop marking" in each leaf node for improving the searching performance.

4.2.1. Fast Rule Skipping. The proposed algorithm requires an additional table called "rule-set starting index table" to store all indexes of the first rule in each rule-set. If we reach a leaf node during traversing the tree, we should find matching rules for each rule-set. Original EffiCuts linearly searches matching rules, so it will take a long time. To increase searching performance, we need to skip unvisited rules in rule-set k and go to the next rule-set $k+1$ when we find matching rule in the rule-set k. It is called "rule skipping." For example, if we find a matching rule r_2 for ACL rule-set in the leaf node as shown in Figure 6, we do not need to check rules r_3 and r_5 anymore. In this case, we can find the index number for firewall rule-set, i.e., 4, and skip r_3 and r_5. Thus, we can directly start searching the matching rule for firewall rule-set.

4.2.2. Early Packet Drop Marking. Assume that we build a node of a decision tree. Each node corresponds to disjoint hypercube searching space. Let us define some notations for describing "early packet drop marking":

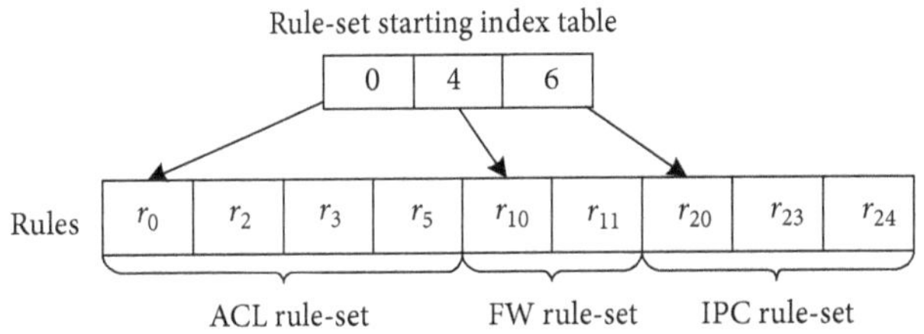

FIGURE 6: An example of a leaf node that contains 9 rules, i.e., 4, 2, and 3 rules for access control list (ACL), firewall (FW), and IP chain (IPC) rule-sets, respectively. Rules in each rule-set are arranged in order of increasing priority.

(i) Δ_v: the searching space for node v

(ii) k: the total rule-set number

(iii) n_p^v: the total number of rules for rule-set p belong to node v

(iv) $r_p^v[i]$: ith rule of rule-set p belonging to node v when the rules are sorted in the order of decreasing priority

(v) $S(\cdot)$: a set of all matching keys with given rule

(vi) $S_p^v[i] \triangleq S(r_p^v[i]) \cap \Delta_v$

We define $D_p^v(j)$ as a set of all keys matching with action "drop" from first to ith rules of the rule-set p belonging to node v when the rules are sorted in the order of decreasing priority. Then, it is recursively defined as

$$D_p^v(j) = \begin{cases} D_p^v(j-1) \cup S_p^v[j] & \text{if action of } r_p^v[i] \text{ is "drop"}, \\ D_p^v(j-1) - S_p^v[j] & \text{otherwise}, \end{cases} \quad (1)$$

where $D_p^v(0) = \varnothing$.

Assume that an incoming packet is, respectively, matched with $r_p^v[i]$ and $r_q^v[j]$ for rule-sets i and j, where the actions of $r_p^v[i]$ and $r_q^v[j]$ are "allow" and "drop." In this case, the packet should be dropped by the rule-set j. If all packets matching with $r_p^v[i]$ are always matched with rules in other rule-set with action "drop," it will be very helpful to know that the packet will be dropped for increasing classification performance. This idea can be generalized as follows.

If $S_p^v[i] \subset D_q^v(n_q^v)$, any packet matched with $r_p^v[i]$ in node v is dropped, where $p \neq q$. Thus, while building a node v, the proposed algorithm finds any rule $r_p^v[i]$ s.t. $S_p^v[i] \subset D_q^v(n_q^v)$, where $p < q$, and mark $r_p^v[i]$ with "early packet drop." If a packet matches with a rule that has a mark "early packet drop" during searching, the searching procedure is finished and the packet is dropped. This "early packet drop marking" significantly increases the performance.

4.3. Proposed Packet Classification Performance Analysis. The proposed algorithm merges multiple rule-sets into an integrated one and constructs a decision tree. Now, we will show numerical analysis results for our algorithm. Assume that rules are homogeneous, and the decision tree is perfectly balanced B-tree for easy analysis. Let us define some notations as follows:

(i) k: the total rule-set number.

(ii) B: binth, the maximum allowed rule size in the leaf node.

(iii) c: the child number of each node. For easy analysis, we assume that c is fixed.

(iv) s: space factor. The maximum ratio of the sum of all rules belongs to all sub-rule-sets to the original rule-set size.

(v) N: the total rule number for each rule-set size. We also assume that N is fixed.

4.3.1. Total Packet Classification Cost Analysis. Assume that EffiCuts has N rules in a root node. If it has c child nodes and the space factor is s, the first level child node has at most sN/c rules. In a similar way, we can calculate the rule number in the leaf node as follows:

$$\frac{s^h N}{c^h} \leq B, \tag{2}$$

and it should be equal to or less than B, where the height of the decision tree is h_{EffiCuts}. From (2), we can find the height as follows:

$$h_{\text{EffiCuts}} \cong \log_{(c/s)}\left(\frac{N}{B}\right). \tag{3}$$

For the proposed algorithm, we can similarly obtain the height as

$$h_{\text{proposed}} \cong \log_{(c/s)}\frac{kN}{B}. \tag{4}$$

Thus, the total packet classification cost for EffiCuts is approximated as follows:

$$kh_{\text{EffiCuts}} \cong k \cdot \log_{(c/s)}\frac{N}{B}. \tag{5}$$

Now, we calculate the difference between two costs:

$$\begin{aligned} kh_{\text{EffiCuts}} - h_{\text{proposed}} &= k \cdot \log_{(c/s)}\frac{N}{B} - \log_{(c/s)}\frac{kN}{B}, \\ &= (k-1)\cdot\log_{(c/s)}\frac{N}{B} \\ &\quad - \log_{(c/s)}k > 0 \because k > 1 \text{ and } \frac{N}{B} \gg k. \end{aligned} \tag{6}$$

Then, we can conclude that the proposed algorithm can always provide higher classification performance than EffiCuts.

4.3.2. Total Decision Tree Size Analysis. Since we assume that the decision tree is a perfectly balanced B-tree, EffiCuts requires at most $2^{h_{\text{EffiCuts}}} - 1$ nodes for one rule-set, so the total number of nodes is $k(2^{h_{\text{EffiCuts}}} - 1)$. Similarly, the proposed algorithm requires $2^{h_{\text{proposed}}} - 1$. If we calculate the difference between two node sizes,

$$\begin{aligned} &k\left(2^{h_{\text{EffiCuts}}} - 1\right) - \left(2^{h_{\text{proposed}}} - 1\right) \\ &= k\left(2^{\log_{(c/s)}(N/B)} - 1\right) - \left(2^{\log_{(c/s)}(kN/B)} - 1\right) \\ &= k\left(2^{\log_{(c/s)}(N/B)} - 1\right) - \left(2^{\log_{(c/s)}(kN/B)} - 1\right) \\ &= k\left(2^{\log_{(c/s)}(N/B)} - 1\right) - \left(2^{\log_{(c/s)}k + \log_{(c/s)}(N/B)} - 1\right) \\ &= 2^{\log_{(c/s)}(N/B)}\left(k - 2^{\log_{(c/s)}k}\right) - k + 1 \\ &= 2^{\log_{(c/s)}(N/B)}\left(k - k^{\log_{(c/s)}2}\right) - k + 1. \end{aligned} \tag{7}$$

Since $2^{\log_{(c/s)}(N/B)}(k - k^{\log_{(c/s)}2}) - k + 1 < 0$, if $1 < (c/s) < 2$, the proposed algorithm creates larger tree than EffiCuts, where $1 < (c/s) < 2$. However, we found that $c \gg s$ for most nodes in a decision tree. It means that the proposed algorithm builds a tree which size is not significantly large compared to EffiCuts.

5. Performance Evaluation

We compared the performance of the proposed algorithm with EffiCuts. Since EffiCuts is almost an unique decision tree-based packet classification algorithm to support fast classification and large rule-set size simultaneously, we choose it as a competitor. We measured average and worst case classification memory access numbers, and decision tree size using the optimal bucket size and space factor for each evaluation. Since the average classification memory access number defines the overall performance of the network equipment, it is the most important metric. The worst case classification memory access number represents the maximum queuing delay required to guarantee in-order packet forwarding. Last, the total decision tree size is also a critical factor to represent the scalability in terms of rule-set size. Considering modern network traffic increases

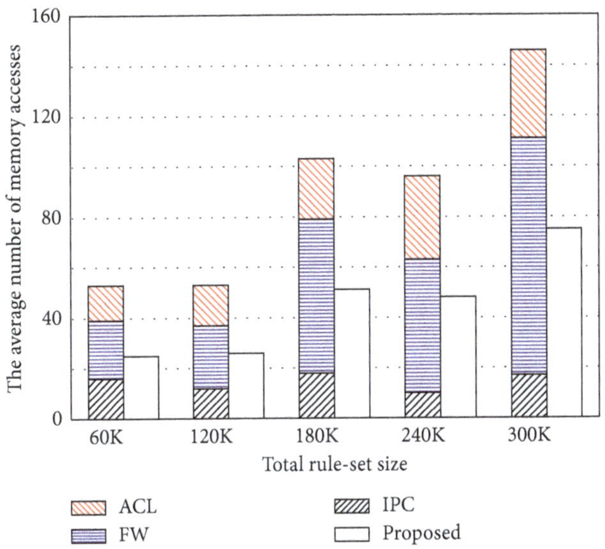

FIGURE 7: The comparison results of EffiCuts and the proposed algorithm for the average number of memory accesses as total rule-set size increases, where three rule-sets are used.

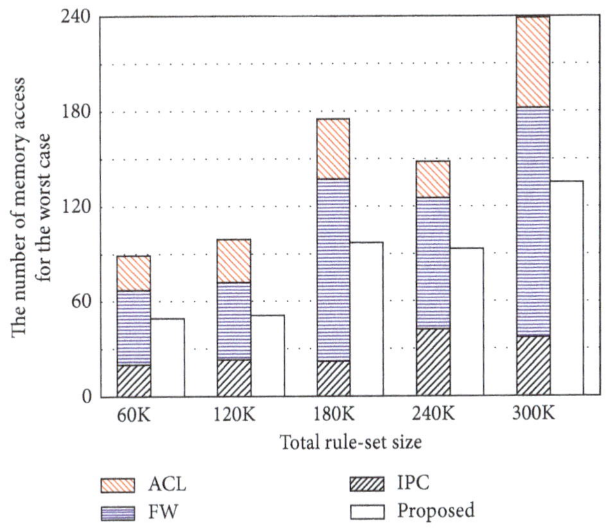

FIGURE 8: The comparison results of EffiCuts and the proposed algorithm for the worst case number of memory accesses as total rule-set size increases, where three rule-sets are used.

exponentially and rule-set becomes larger and more complicated to support various services, we choose these three metrics for performance evaluation.

For evaluating performance of the proposed algorithm, multiple rule-sets are needed. Thus, three rule-set types such as FW, ACL, and IPC were generated using Classbench [50]. Each rule consists of five tuples, and the rule-set size was set to 20K to 100K increasing by 20K where K means 1,000. Thus, the integrated rule-set size was to 60K to 300K. For each evaluation, binth and space factor were set to the optimal values, i.e., 30 and 2, respectively.

Figure 7 shows the average classification performance in terms of the average number of memory accesses according to the size of the total integrated rule-set. The proposed algorithm achieves about 2.5 times lower memory access number regardless of the rule-set size compared to EffiCuts. It is almost similar to the memory access number of each rule-set. It confirms that integrated rule-set has many benefits to increase the classification performance.

Figure 8 shows results for the worst case packet classification performance. The proposed algorithm decreases the memory access number by 2.2 times regardless of the total rule-set size compared to competitor. Although the improvement is slightly smaller than that for average memory access number, it also confirms that the proposed algorithm is very effective to increase the packet classification performance for the worst case.

The worst case performance actually affects the packet processing delay since most network equipment should guarantee that the packets are processed in sequence, keeping that the orders of incoming and outgoing packets are the same. As the worst classification performance is improved, it can efficiently provide in-order packet forwarding while minimizing packet queuing delay.

Figure 9 shows the comparison results between proposed and EffiCuts for decision tree size. As mentioned earlier as

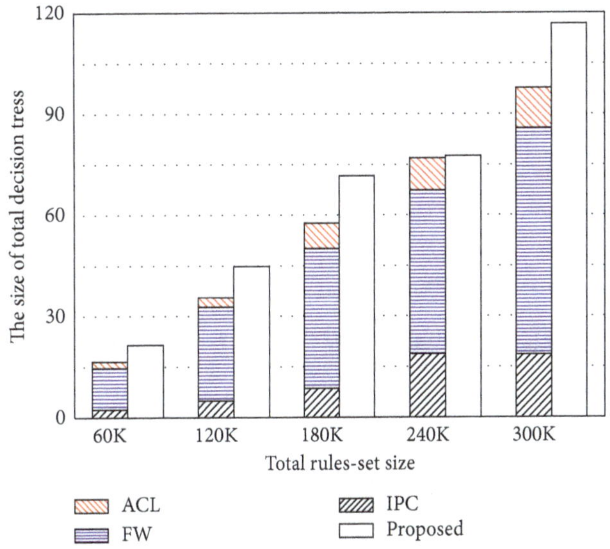

FIGURE 9: The comparison results of EffiCuts and the proposed algorithm for the decision tree size as total rule-set size increases, where three rule-sets are used.

Characteristic 2, the proposed algorithm generates a decision tree whose size about 20% is larger than that of EffiCuts for 300K rules. Therefore, the proposed algorithm does not suffer from significantly increased tree size caused by rule-set integration.

Although we used three rule-sets for most performance evaluations, it is also important to investigate the performance as the rule-set number increases for evaluating scalability in terms of the rule-set number. Figure 10 shows the ratio of the results of EffiCuts to those of the proposed algorithm for the memory access and the decision tree sizes as the rule-set size increases from 1 to 10.

As shown in Figure 10, the decision tree size of the proposed algorithm is almost the same to that of EffiCuts

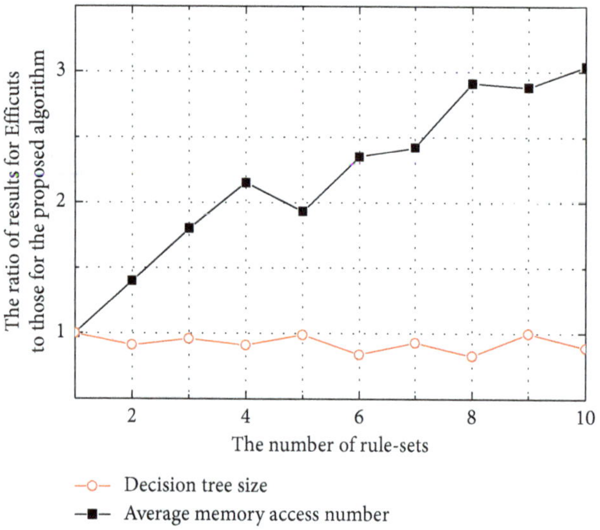

Figure 10: The ratio of results for EffiCuts to those for the proposed algorithm for decision tree and average memory access number as the number of rule-set increases, where each rule-set size is fixed to 20K.

regardless of rule-set size but the memory access size is decreased fast compared to EffiCuts. For 10 rule-sets, the proposed algorithm achieves 3 times higher classification performance while the decision tree size is just increased by 10%. Thus, we can see that our proposed algorithm can provide high classification performance without any cost of decision tree size.

6. Conclusions

In this paper, we proposed a new packet classification algorithm to achieve high packet classification performance without significant increasing of memory requirement. It can be adopted in modern high performance network equipment that use various classification rule-sets such as routing, switching, QoS, and other rule-sets. Existing network equipment with multiple rule-sets independently perform classification for each rule-set, thus resulting in deteriorated performance as the rule-set number increases. Our algorithm combines each rule-set and achieves high performance that cannot be provided by existing algorithms. We expect that it will help to enable robust and low delay V2X services in modern networks.

References

[1] O. O. Fagbohun, "Comparative studies on 3G, 4G and 5G wireless technology," *IOSR Journal of Electronics and Communication Engineering*, vol. 9, no. 2, pp. 133–139, 2014.

[2] H. Tullberg, P. Popovski, Z. Li et al., "METIS system concept: the shape of 5G to come," *IEEE Communications Magazine*, vol. 54, no. 12, pp. 132–139, 2016.

[3] P. K. Agyapong, M. Iwamura, D. Staehle, W. Kiess, and A. Benjebbour, "Design considerations for a 5G network architecture," *IEEE Communications Magazine*, vol. 52, no. 11, pp. 65–75, 2014.

[4] T. Janevski, "A system for PLMN-WLAN internetworking," *Journal of Communications and Networks*, vol. 7, no. 2, pp. 192–206, 2005.

[5] M. Hata, "Fourth generation mobile communication systems beyond IMT-2000," in *Proceedings of Fifth Asia-Pacific Conference on Communications and Fourth Optoelectronics and Communications Conference (APCC/OECC '99)*, pp. 765–767, Beijing, China, October 2002.

[6] A. Gohil, H. Modi, and S. K. Patel, "5G technology of mobile communication: a survey," in *Proceedings of International Conference on Intelligent Systems and Signal Processing ISSP-2013*, pp. 288–292, Anand, Gujarat, March 2013.

[7] *The Benefits of Multiple Flow Tables and TTPs Version Number 1.0 February 2*, 2015, https://www.opennetworking.org/images/stories/downloads/sdn-resources/technical-reports/TR_Multiple_Flow_Tables_and_TTPs.pdf.

[8] Z. Chen, Y. Wu, J. Ge, and E. Yuepeng, "A new lookup model for multiple flow tables of OpenFlow with implementation and optimization considerations," in *Proceedings of IEEE International Conference on Computer and Information Technology (CIT)*, pp. 528–532, Xi'an, China, September 2014.

[9] N. Gude, T. Koponen, J. Pettit et al., "Nox: towards an operating system for networks," *ACM SIGCOMM Computer Communication Review*, vol. 38, no. 3, pp. 105–110, 2008.

[10] Y. Kanizo, D. Hay, and I. Keslassy, "Palette: distributing tables in software-defined networks," in *Proceedings of IEEE INFOCOM*, pp. 545–549, Turin, Italy, April 2013.

[11] N. Sarrar, S. Uhlig, A. Feldmann, R. Sherwood, and X. Huang, "Leveraging zipf's law for traffic offloading," *ACM SIGCOMM Computer Communication Review*, vol. 42, no. 1, pp. 16–22, 2012.

[12] Z. Wu, Y. Jiang, and S. Yang, "An efficiency pipeline processing approach for OpenFlow switch," in *Proceedings of 41st IEEE Conference on Local Computer Networks (LCN)*, pp. 204–207, Dubai, UAE, November 2016.

[13] *Flow Table Explosion with Openflow 1.0 (And Why We Need Openflow 1.3)*, http://blog.ipspace.net/2013/10/flow-table-explosion-with-openflow-10.html.

[14] X.-N. Nguyen, D. Saucez, C. Barakat, and T. Turletti, "Rules placement problem in OpenFlow networks: a survey," *IEEE Communications Surveys & Tutorials*, vol. 18, no. 2, pp. 1273–1286, 2016.

[15] M. Zolanvari, "SDN for 5G," http://www.cs.wustl.edu/~jain/cse570-15/ftp/sdnfor5g.pdf.

[16] *OpenFlow Switch Specification Version 1.5.0*, 2014, https://www.opennetworking.org/images/stories/downloads/sdn-resources/onf-specifications/openflow/openflow-switch-v1.5.0.noipr.pdf.

[17] N. Landage and T. Dhope, "Survey of SDN based packet classification techniques," *International Journal of Science and Research (IJSR)*, vol. 5, no. 7, pp. 1113–1115, 2016.

[18] N. McKeown, T. Anderson, H. Balakrishnan et al., "OpenFlow: enabling innovation in campus networks," *ACM SIGCOMM Computer Communication Review*, vol. 38, no. 2, pp. 69–74, 2008.

[19] S. Agarwal, M. Kodialam, and T. Lakshman, "Traffic engineering in software defined networks," in *Proceedings of IEEE INFOCOM*, pp. 2211–2219, Turin, Italy, April 2013.

[20] T. Inoue, T. Mano, K. Mizutani, S.-I. Minato, and O. Akashi, "Rethinking packet classification for global network view of software-defined networking," in *Proceedings of IEEE 22nd International Conference on Network Protocols (ICNP)*, pp. 296–307, Raleigh, NC, USA, October 2014.

[21] A. Yahya, D. Al-Nejadi, and N. Shaikh-Husin, "Survey on multi field packet classification techniques," *Research Journal of Recent Sciences*, vol. 4, no. 2, pp. 98–106, 2015.

[22] P. Gupta and N. McKeown, "Algorithms for packet classification," *IEEE Network*, vol. 15, no. 2, pp. 24–32, 2002.

[23] K. Pagiamtzis and A. Sheikholeslami, "Content-addressable memory (CAM) circuits and architectures: a tutorial and survey," *IEEE Journal of Solid-State Circuits*, vol. 41, no. 3, pp. 712–727, 2006.

[24] K. G. Pérez, X. Yang, S. Scott-Hayward, and S. Sezer, "A configurable packet classification architecture for Software-Defined Networking," in *Proceedings of 27th IEEE International System-on-Chip Conference (SOCC)*, pp. 353–358, Las Vegas, Nevada, USA, September 2014.

[25] K. G. Perez, X. Yang, S. Scott-Hayward, and S. Sezer, "Optimized packet classification for software-defined networking," in *Proceedings of IEEE International Conference on Communications (ICC)*, pp. 859–864, Sydney, Australia, June 2014.

[26] H. Lim, "Survey and proposal on packet classification algorithms," in *Proceedings of International Conference on High Performance Switching and Routing (HPSR)*, Richardson, Texas, USA, June 2010.

[27] D. E. Taylor, "Survey and taxonomy of packet classification techniques," *ACM Computing Surveys*, vol. 37, no. 3, pp. 238–275, 2005.

[28] P. Gupta and N. McKeown, "Packet classification on multiple fields," in *Proceedings of the Conference on Applications, Technologies, Architectures, and Protocols for Computer Communication*, pp. 147–160, Cambridge, MA, USA, August 1999.

[29] B. Xu, D. Jiang, and J. Li, "HSM: a fast packet classification algorithm," in *Proceedings of 19th International Conference on Advanced Information Networking and Applications (AINA, 2005)*, pp. 1–6, Taipei, Taiwan, March 2005.

[30] D. E. Taylor and J. S. Turner, "Scalable packet classification using distributed crossproducting of field labels," in *Proceedings of IEEE 24th Annual Joint Conference of the IEEE Computer and Communications Societies (INFOCOM, 2005)*, pp. 269–280, New York, NY, USA, March 2005.

[31] G. S. Jedhe, A. Ramamoorthy, and K. Varghese, "A scalable high throughput firewall in FPGA view document," in *Proceedings of 16th International Symposium on Field-Programmable Custom Computing Machines*, pp. 43–52, Palo Alto, California, USA, April 2008.

[32] Y. Pan, B. Chen, and T. Xu, "A timesaving recursive flow packet classification algorithm," in *Proceedings of International Conference on Networks Security, Wireless Communications and Trusted Computing (NSWCTC '09)*, pp. 442–445, Wuhan, China, April 2009.

[33] P.-C. Wang, C.-L. Lee, C.-T. Chan, and H.-Y. Chang, "Hardware-based packet classification made fast and efficient," in *Proceedings of 11th International Conference on Parallel and Distributed Systems*, Fuduoka, Japan, July 2005.

[34] V. Srinivasan, S. Suri, and G. Varghese, "Packet classification using tuple space search," *ACM SIGCOMM Computer Communication Review*, vol. 29, no. 4, pp. 135–146, 1999.

[35] M. Alutoin and P. Raatikainen, "Diagonal tuple space search," in *Proceedings of Global Telecommunications Conference (GLOBECOM '04)*, pp. 719–724, IEEE, Dallas, Texas, USA, November 1999.

[36] M. Alutoin and P. Raatikainen, "Diagonal tuple space search in two dimensions," in *Proceedings of 3rd International Networking Conference Networking (IFIP-TP6)*, pp. 308–319, Athens, Greece, May 2004.

[37] P.-C. Ting, Y.-S. Hsu, and T.-H. Lee, "Fast tuple space based packet classification algorithm for two-field conflict free filters," in *Proceedings of IEEE International Conference on Communications (ICC '03)*, pp. 325–331, Anchorage, AK, USA, May 2003.

[38] B. Pfaff, J. Pettit, T. Koponen et al., "The design and implementation of open vSwitch," in *Proceedings of 12th USENIX Symposium on Networked Systems Design and Implementation (NSDI '15)*, pp. 117–130, Oakland, CA, USA, May 2015.

[39] P. Gupta and N. McKeawn, "Classifying packets with hierarchical intelligent cuttings," *IEEE Micro*, vol. 20, no. 1, pp. 34–41, 2000.

[40] S. Singh, F. Baboescu, G. Varghese, and J. Wang, "Packet classification using multidimensional cutting," in *Proceedings of ACM Conference Applications, Technologies, Architectures, and Protocols for Computer Communications (SIGCOMM '03)*, pp. 213–224, Karlsruhe, Germany, Auguest 2003.

[41] B. Vamanan, G. Voskuilen, and T. N. Vijaykumar, "Efficuts: optimizing packet classification for memory and throughput," *ACM SIGCOMM Computer Communication Review*, vol. 40, no. 4, pp. 207–218, 2010.

[42] H. Lu and S. Sahni, "O(logW) multidimensional packet classification," *IEEE/ACM Transactions on Networking*, vol. 15, no. 2, pp. 462–472, 2007.

[43] G. Antichi, C. Callegari, A. W. Moore, S. Giordano, and E. Anastasi, "JA-trie: entropy-based packet classification," in *Proceedings of IEEE 15th International Conference on High Performance Switching and Routing, HPSR 2014*, vol. 15, pp. 32–37, Vancouver, BC, Canada, July 2014.

[44] Y.-K. Chang and Y.-H. Wang, "CubeCuts: a novel cutting scheme for packet classification," in *Proceedings of 26th International Conference on Advanced Information Networking and Applications Workshops (WAINA)*, pp. 274–279, Fukuoka, Japan, March 2012.

[45] H. Song and J. S. Turner, "ABC: adaptive binary cuttings for multidimensional packet classification," *IEEE/ACM Transactions on Networking*, vol. 21, no. 1, pp. 98–109, 2013.

[46] M. Kathuria and S. Gambhir, "Genetic binary decision tree based packet handling schema for WBAN system," in *Proceedings of Recent Advances in Engineering and Computational Sciences (RAECS)*, Chadigarh, India, March 2014.

[47] W. Pak and Y.-J. Choi, "High performance and high scalable packet classification algorithm for network security systems," *IEEE Transactions on Dependable and Secure Computing*, vol. 14, no. 1, pp. 37–49, 2017.

[48] T. S. Urmila, "Decision tree based network packet classification algorithms," *International Journal of Advanced Networking and Applications (IJANA)*, Special issue, pp. 28–34, 2015.

[49] B. Vamanan and T. N. Vijaykumar, "TreeCAM: decoupling updates and lookups in packet classification," in *Proceedings of the Seventh Conference on emerging Networking Experiments and Technologies*, Tokyo, Japan, December 2011.

[50] D. E. Taylor and J. S. Turner, "ClassBench: a packet classification benchmark," *IEEE/ACM Transactions on Networking (TON)*, vol. 15, no. 3, pp. 499–511, 2007.

3

Location Privacy Protection Research based on Querying Anonymous Region Construction for Smart Campus

Ruxia Sun,[1] **Jinwen Xi**,[1] **Chunyong Yin**,[1] **Jin Wang**,[2] **and Gwang-jun Kim**[3]

[1]*School of Computer and Software, Jiangsu Engineering Center of Network Monitoring, Nanjing University of Information Science & Technology, Nanjing, China*
[2]*School of Computer & Communication Engineering, Changsha University of Science & Technology, Changsha, China*
[3]*Department of Computer Engineering, Chonnam National University, Gwangju, Republic of Korea*

Correspondence should be addressed to Gwang-jun Kim; kgj@jnu.ac.kr

Academic Editor: Jaegeol Yim

Along with the rapid development of smart campus, the deployment of novel learning applications for smart campus requires full consideration of information security issues. Location privacy protection is one of the most important issues, which considers the privacy protection and guarantees the quality of service. The existing schemes did not consider the area of the querying regions for location-based service provider (LSP) during the construction of the anonymous regions, which led to the low quality of service. To deal with this problem, the user's query range was introduced to present a novel anonymous region construction scheme. In the proposal, the anonymous server first generated the original anonymous subregions according to the user's privacy requirements, and then merged these subregions to construct the anonymity regions submitted to LSP based on the size of corresponding querying regions. The security and experiment analysis show that the proposed scheme not only protects the user's privacy effectively but also decreases the area of LSP querying regions and the region-constructing time, improving the quality of service for smart campus.

1. Introduction

With the development and construction of novel learning applications for smart campus, smart devices and services, smart meters, smart terminals, and the like are widely applied to offer real-time learning feedback to students through continuously monitoring and analyzing the status and activities of students with various devices and platforms. As a large number of smart meters and intelligence appliances are accessed, incorporating various technologies and enabling world-changing learning, the network border further extends to the user side. Security risks on the user side for smart campus will become more and more prominent. Data privacy issues, especially location privacy protection and the quality of service, must be considered [1].

Users' location privacy threats refer to the risks that an attacker can obtain unauthorized access to raw location data by locating a transmitting device and identifying the subject (person) using it. Examples of such risks include spamming users with unwanted advertisements, drawing sensitive inferences from victims' visits to various locations (e.g., students and teachers' offices), and learning sensitive information about them (identity, religious and political affiliations, etc.). Hence, location privacy protection for smart campus is becoming a critical issue [2].

However, location information is consistently sent to service providers without protection when users query LBSs, allowing location providers to collect location information from all users. The collected location information may expose users to customized advertisement or even be sold to third parties. A worse scenario is location information may be leaked to adversaries with criminal intents. Therefore, many researchers focus on creating location protection algorithms to protect the location privacy of users [3].

The European Union's Information Protection Supervision Organization recently said that high-tech equipment such as smart meters that monitor household energy consumption will pose a huge threat to personal privacy. Smart meters may track personal information, and the vast amounts of information collected can have serious consequences for consumers.

With the popularization of mobile devices and location technology, location-based services (LBSs) are widely used in real life, which refers to the user accesses to its designated location information query and entertainment services through the mobile device [4]. However, the location-based service provider (LSP) may also collect and abuse user's service information while providing the convenient LBS for the user, to illegally obtain the user's confidential information. The location privacy protection in LBS has attracted the extensive attention of researchers [5–8].

In view of the popularization of mobile positioning devices, if these novel learning applications are used in smart campus, the combination of location information and services at different moments in personal privacy may reveal sensitive information such as the user's behavior habits and work nature. For example, if a mobile user is collected near a hospital, the user may be presumed to have any disease or health condition. If the user's starting location and ending location are analyzed since the last few days, the user's home address or work unit, the nature of work, and so on can be speculated. Therefore, the location data of mobile object brings convenience to people and brings about the threat of revealing privacy, which may contain other sensitive information such as home address, personal preference, personality habit, health status, working property, personal income, etc. If this information falls into the hands of normal institutions, it is a tool for information protection, if it falls into the hands of illegal institutions, it will be the weapon of innocent destruction. What we can do is to seek transparency in the use of personal information and to protect user's location privacy not to be exploited by unscrupulous businesses and illegal agencies.

As the most commonly used LBS privacy protection method, the basic idea of k-anonymity [9] is that when a user sends a LBS query, he will send his real location and querying content to a trusted anonymous server, then the anonymous server will remove the user's identification information and generate anonymity regions containing other $k-1$ users for the real location, finally sending them (all the generated anonymity regions and the real location) along with the querying content to the LSP. Compared with other LBS privacy protection methods (such as pseudo location [10], fuzzification [11], differential privacy [12], and cryptography-based methods [13, 14]), k-anonymity has the following advantages: (1) users can get accurate querying results; (2) the user's cost of computing and communication is small; and (3) this method can confuse the relevance between users and LBS queries. So, k-anonymity is widely used in LBS privacy protection [15, 16].

In smart campus, we can query the nearby points of interest. There are many sensors to be addressed for enabling such novel learning applications and services, which aims to enhance the service quality of novel learning applications. When k-anonymity method is used to protect the privacy of users in the above query, if the anonymous region generated by the anonymous server is too large, the querying cost of the location-based service provider (LSP) will increase and the service quality deteriorates [17–19]. To solve this problem, the existing methods [20–26] obtain n disjoint anonymous subregions by removing the part that does not contain the users in the regions, to reduce the area of anonymous regions and improve the service quality as shown in Figure 1. However, the quality of service in LBS queries based on k-anonymity is not only related to the size of the anonymous regions but also to the user's query range. If we use the existing methods to construct the anonymous regions, the quality of service cannot be effectively improved. As shown in Figure 2, when using the existing division methods to divide the initial anonymous region, LSP will repeatedly search the points of interest in some regions, reducing the quality of service, and r is the query radius. This paper also proves this point through experiments.

This paper proposes the LBS privacy protection scheme based on querying anonymous region construction. In which, firstly, the anonymous server generates k initial anonymous subregions according to the privacy protection requirements of users and merges the anonymous regions according to the corresponding querying regions so that the anonymous regions finally submitted to the LSP can reduce the querying cost of LSP and improve the service quality without reducing the user's privacy protection level. This is the first k-anonymity privacy protection scheme based on constructing the user querying anonymous regions. The main contributions of this paper are as follows:

(1) Based on the theoretical analysis, it is concluded that the existing anonymous region demarcation method cannot reduce the LSP querying cost and improve the service quality, and we prove it through experiments.

(2) We think that the area of querying regions is the judgment criterion to merge the anonymous subregions and propose an anonymous region construction scheme based on the user's query range. Security analysis shows that the proposed scheme can effectively protect the users' location privacy.

(3) A large number of experiments show that this scheme can effectively reduce the querying cost of LSP and improve the service quality, without causing a large computational cost for anonymous servers.

2. Related Work

At present, the international research on the privacy protection for smart campus is still in the initial stage. The discussion on the privacy protection focuses more on the risk analysis of exposing personal privacy for the wireless applications and devices. Research in the United States is the leader, which publishes relevant documents on this issue.

Privacy laws in the United States do not explicitly address the smart campus and its related data, which is same as the regulations of the existing national Internet of Energy

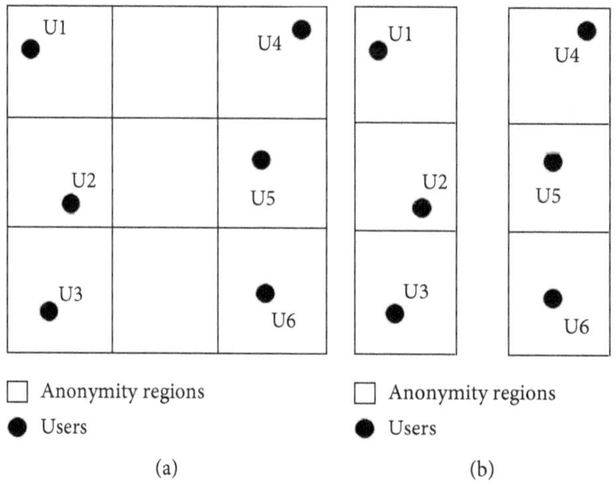

FIGURE 1: The existing anonymous regions division method. (a) Initial anonymity regions. (b) Anonymity regions after division.

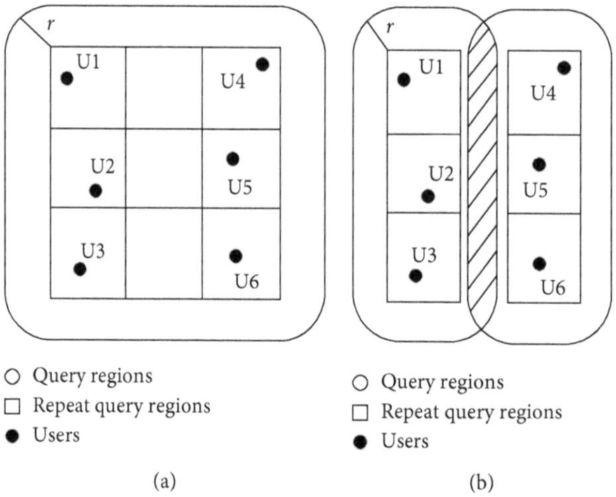

FIGURE 2: The querying regions of nearby points of interest. (a) Initial query regions. (b) Query regions after division.

and power transmission [27]. The existing laws and regulations need to be revised to appeal to the smart campus. At the same time, new data items in the Internet of Energy, as well as new ways of using existing data, require more research and public opinion to adapt to current laws or shape new laws.

US Internet of Energy Information Security is very concerned about privacy issues. NIST (National Institute of Standards and Technology) released "Guidelines for Smart Grid Cyber Security: Vol. 2, Privacy and Smart Grid" [28] in August 2010 and preliminarily analyzed the privacy issues of smart campus. For location privacy protection, many researchers have done a great deal of work.

Grutese applied k-anonymity to the field of location privacy protection for the first time [4]. The anonymous server divides the area with a quadtree structure and stores the users in the corresponding region in the node. When the user requests a service, a quadtree is retrieved upward from a leaf node corresponding to the user's location. If the current leaf node does not satisfy the k-anonymity, the parent node will be retrieved until no less than other $k-1$ users to obtain the anonymous region. However, if the number of users in a leaf node does not satisfy the privacy requirements, it needs to retrieve its parent node, which will result in a four-fold increase in the anonymity area, thereby degrading the user's service quality. To deal with the above problem, Mokbel et al. [29] improved the anonymous region construction method in [4]. In their scheme, if the current leaf node does not satisfy k-anonymity, firstly its sibling nodes will be retrieved, if it still does not satisfy the user's privacy requirements, then it retrieves the parent node. To further solve the problem of low quality of service due to the large anonymous region based on the quadtree structure, Li et al. [30] proposed to reduce the area of anonymous regions and improve the service quality by suppressing part of users' requests and deleting the most distant trails.

Subsequently, the researchers proposed Clique Cloak [31] and Hilbert Cloak [32], respectively, to construct the anonymous regions by searching for the nearest users who meet the privacy requirements. In [31], users can customize the privacy protection requirements, but the scheme uses an undirected graph to construct an anonymous region and the cost of the solution is too large. It may happen that the anonymous region has not been successfully constructed but beyond the anonymous period. In [32], the anonymous server maps all users in two-dimensional space to a one-dimensional array according to the Hilbert curve and divides the users into several sets according to the value of k. When a user makes a service request, the anonymous region is constructed by using the user set to which the user belongs.

Recently, Jin et al. [2] thought the anonymization server could lead to a new class of attacks called location injection attacks which can successfully violate users' in-distinguishability (guaranteed by k-anonymity) among a set of users. Therefore, they propose and characterize location injection attacks by presenting a set of attack models and quantify the costs associated with them. Then, they propose and evaluate k-Trustee, which is resilient to location injection attacks and guarantees a low bound on the user's in-distinguishability. The experimental results show that the proposed cloaking algorithm guaranteeing k-Trustee is effective against various location injection attacks, but the quality of location services is poor and cannot be guaranteed.

In all the above solutions, if the k users used to construct the anonymous regions are far away from each other, the anonymous regions may still be too large, and the service quality may be low. Tan et al. [33] were the first ones to apply the idea of regionalization to the construction of anonymous regions, which divided the users in the anonymous regions into separate groups through the Hilbert space filling curve. When a user makes a server request, the anonymous server will use the locations of other users in the group to which it belongs to construct the anonymous region. Subsequently, Li and Zhu [34] also used the method of regionalization to research on reducing the area of anonymous regions and improving the quality of service. The anonymous server firstly constructs an anonymous region containing k users and then removes the anonymous regions that do not contain the users according to relationship among the users' locations to form multiple anonymous subregions that do not intersect each other.

However, the existing anonymous region construction schemes [35–37] ignore the impact of the users' query range on LBS querying service quality. When using these methods to construct the anonymous regions, the LSP will repeatedly search the points of interest in the partial regions, reducing the service quality.

3. Improved k-Value Location Privacy Protection Method

3.1. System Structure. This paper uses the centralized system architecture [38], composed of three parts: user, anonymous server, and LSP. When the user requests a service, the anonymous server blurs the real location of the user into an anonymous region and sends it to the LSP, which contains not only the real user but also other at least $k-1$ users. In this case, the correct rate that LSP correlates the service query with the right user will be not more than $1/k$, which achieves k-anonymity. The system structure is shown in Figure 3.

Assume that there is a secure communication channel between the user and the anonymous server. When the user queries the point of interest nearby, the secure channel is used to send the query request $q = \langle \text{ID}, L(x,y), r, \text{POI}, p\rangle$ to the trusted anonymous server. In which, ID represents the identity of the user; $L(x,y)$ represents the location coordinates of the user; r represents the radius of the user query; POI represents the point of interest of the user query; p represents the privacy protection requirement of the user current query; k represents the anonymous region generated by the anonymous server contains at least other $k-1$ users; and $A_{\min}$ represents the minimum area of anonymous regions generated by anonymous server.

After receiving the user's request, the trusted anonymous server will determine the identity through authentication and find other $k-1$ users to generate anonymous regions that the area is not less than $A_{\min}$ according to the user's privacy protection requirements $p = (k, A_{\min})$ and then send the query request $Q = \langle \text{CR}, r, \text{POI}\rangle$ obtained from the anonymization to the semitrusted LSP. CR represents the anonymous region generated by the anonymous server for the current user making the service query.

After receiving the anonymous query request sent by the anonymous server, the LSP will search in the database and return all the query candidate results to the anonymous server. After the anonymous server receives the query result from the LSP, it selects the query result according to the location $L(x,y)$ of the user, and finally returns the accurate query result to the user.

In this system model, we treat the LSP directly as an attacker. The attacking purposes are as follows: (1) the real location of the user could be identified in the anonymous region sent from the anonymous server and (2) the real user will be speculated from the query request.

In addition, in the above model, LBS query quality of service is mainly affected by the following four factors: (1) the time required by the anonymous server to generate anonymous regions; (2) the time that the anonymous server sends the anonymous query request to LSP; (3) the time required by the LSP to retrieve the database for the anonymous query request sent by the anonymous server; and (4) the time it takes for the LSP to send the retrieval result to the anonymous server, and the time required by the anonymous server to finalize the query result. Because the time taken for the LSP to send search results to the anonymous server and the time required for the anonymous server to finalize the query results is affected by the distribution of points of interest, the time required for the anonymous server to send the anonymous query request to the LSP is affected by the transmission bandwidth. Therefore, this paper evaluates the quality of service based on k-anonymous LBS queries only by the time it takes the anonymous server to generate the anonymous regions and the LSP to retrieve the database.

3.2. Querying Region of LSP. After receiving the anonymous query request $Q = \langle \text{CR}, r, \text{POI}\rangle$ sent by the anonymous server, the LSP first calculates the querying region QAR according to the anonymous region CR and the query radius

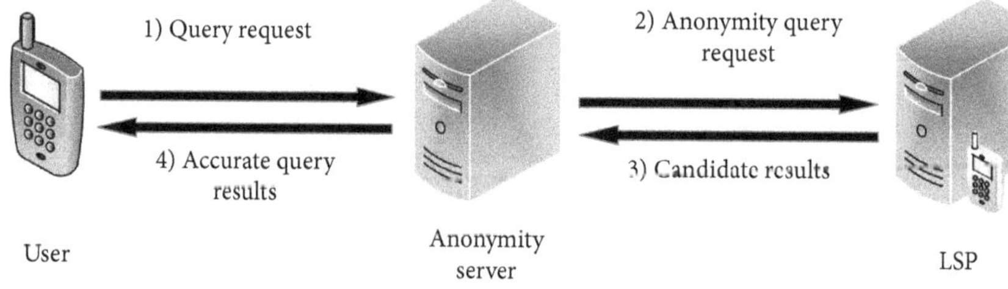

FIGURE 3: System structure.

r and then searches the interest point of the user query in the region QAR. The querying regions corresponding to the anonymous regions are as shown in Figure 4.

When the anonymous region is a circle, the area S_c of the querying region is

$$S_c = \pi(\tilde{r} + r)^2, \tag{1}$$

where $\tilde{r}$ represents the radius of the circular anonymous region and πr^2 is the area of the anonymous region.

When the anonymous region is a rectangle, the area S_r of the querying region is

$$S_r = ab + 2(a + b)r + \pi r^2, \tag{2}$$

where a, b is the side length of the rectangular anonymous region and ab is the area of the anonymous region CR.

It can be learned from the above analysis that after receiving the anonymous query request sent by the anonymous server, the time required by the LSP to retrieve the database is not only related to the size of the anonymous region CR generated by the anonymous server but also to the radius of the user's query, in other words, it is determined by the querying region. However, the existing anonymous region partitioning schemes only take the size of the anonymous regions into account, which makes these solutions not effectively improve the LBS query service quality and even further reduces the service quality.

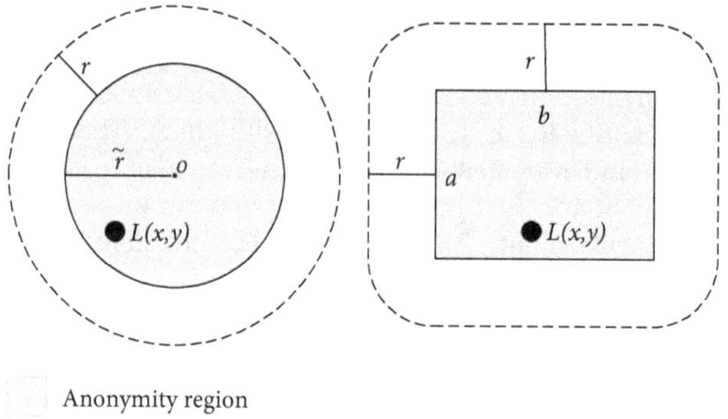

FIGURE 4: Querying region of LSP.

4. Anonymous Region Construction Scheme Based on Query Range

In this paper, the anonymous server first generates k initial anonymous subregions according to the privacy protection requirements of the users and constructs the corresponding querying regions by calculating and then merges the anonymous subregions to finally obtain the anonymous subregion set. The solution can reduce the LSP query cost and improve the service quality without lowering the user privacy protection level.

4.1. Generation of Initial Anonymous Subregions. After receiving the service request sent by the users, the anonymous server will start to search the other $k-1$ users according to the user's privacy protection requirement $p = (k, A_{\min})$ and obtain their location information $L_1(x_1, y_1), \ldots, L_{k-1}(x_{k-1}, y_{k-1})$. Then, the server will generate k disjoint initial anonymous subregions $\mathrm{AR}_0, \mathrm{AR}_1, \ldots, \mathrm{AR}_{k-1}$ to satisfy

$$\begin{cases} \mathrm{centre}(\mathrm{AR}_i) \neq L_i(x_i, y_i), \\ \mathrm{area}(\mathrm{AR}_i) \neq A_{\min}, \end{cases} \tag{3}$$

where AR_i represents the initial anonymous subregion containing the i-th location $L_i(x_i, y_i)$, $0 \le i \le k-1$; $L_0(x_0, y_0)$ represents the location of the user who sent the query request; $\mathrm{centre}(\mathrm{AR}_i)$ represents the central location of initial anonymous subregion AR_i; and $\mathrm{area}(\mathrm{AR}_i)$ represents the area of initial subanonymous region AR_i.

4.2. Merger of Anonymous Subregions. After generating the k initial anonymous subregions $\mathrm{AR}_0, \mathrm{AR}_1, \ldots, \mathrm{AR}_{k-1}$, the anonymous server will calculate the area $\mathrm{area}(\mathrm{QAR}_0)$, $\mathrm{area}(\mathrm{QAR}_1), \ldots, \mathrm{area}\ (\mathrm{QAR}_{k-1})$ of the corresponding querying regions $\mathrm{QAR}_0, \mathrm{QAR}_1, \ldots, \mathrm{QAR}_{k-1}$, respectively, according to the query radius of the user and judge whether the anonymous subregions need to be merged on the basis of the area of the querying regions. In order to effectively reduce the cost, LSP retrieves the points of interest, and after the merger of anonymous subregions, it should ensure

$$S_{\min} = \min \sum_{i=0}^{l} \mathrm{area}(\mathrm{QAR}_i'). \tag{4}$$

That is to say, the sum of the area of each querying region QAR_i' corresponding to each anonymous subregion AR_i' in the anonymous regions CS is the smallest. AR_i' represents the i-th anonymous subregion after the merger of anonymous subregions, $0 \le i \le l$, $0 \le l \le k-1$.

The merger process of anonymous subregions in the anonymous server is as follows:

(1) Filtering the anonymous subregions that require region consolidation

To ensure that the querying area of the anonymous region is minimized, the anonymous server chooses the anonymous subregions AR_i and AR_j to merge if the area of querying region is the smallest after the merger of them. The selected anonymous subregions AR_i and AR_j are needed to satisfy

$$\begin{aligned} \forall i, j \in [0, k-1], \mathrm{QAR}_{i,j} &= \operatorname{argmin}\{\operatorname{area}(\mathrm{QAR}_{i,j})\}_{i \neq j}, \\ \operatorname{centre}(\mathrm{QAR}_{i,j}) &\neq L_i(x_i, y_i), \\ \operatorname{centre}(\mathrm{QAR}_{i,j}) &\neq L_j(x_j, y_j). \end{aligned} \tag{5}$$

(2) Selecting anonymous subregions to merge

For the two anonymous subregions AR_i and AR_j, $AR_{i,j} = \text{gen}(AR_i, AR_j)$ represents the new anonymous subregion formed by merging the anonymous subregion AR_i with AR_j and the corresponding querying region is $QAR_{i,j}$. If $\text{area}(QAR_i) + \text{area}(QAR_j) \leq \text{area}(QAR_{i,j})$, it does not merge AR_i with AR_j. Else it merges AR_i with AR_j to make the new anonymous subregion $AR_{i,j}$.

(3) Repeating the process until there is no need to merge anonymous subregions. In this case, the anonymous server will obtain the anonymous set

$$\mathrm{CS} = \{AR'_0, AR'_1, \ldots, AR'_l\}. \tag{6}$$

In this scheme, when the anonymous server merges the anonymous subregions, it chooses the anonymous regions AR_i and AR_j to make the area of the querying region the smallest after merging. Therefore, after every merger of the anonymous subregions, the area of the new querying region $AR_{i,j}$ is the smallest, which ensures that the querying area of the final anonymous region set CS is the smallest.

In this paper, the locations of k users are used as the input to generate the corresponding anonymous subregions $AR_0, AR_1, \ldots, AR_{k-1}$ and merge them to obtain the final anonymous subregion set CS, which is submitted to the LSP (Algorithm 1).

5. Scheme Analysis and Experimental Results

5.1. Analysis of Security. In our scheme, we construct the disjoint initial anonymous subregions $AR_0, AR_1, \ldots, AR_{k-1}$ that the area of each initial anonymous subregion is less than $A_{\min}$ according to k real locations $L_0(x_0, y_0)$, $L_1(x_1, y_1), \ldots, L_{k-1}(x_{k-1}, y_{k-1})$ and ensure every real location is not located in the center of the initial anonymous subregion for smart campus, that is, $\text{centre}(AR_i) \neq L_i(x_i, y_i)$. If the anonymous server directly sends the k initial anonymous subregions to the LSP, which cannot correctly identify the real location $L_i(x_i, y_i)$ of the user from the anonymous region AR_i. If the anonymous server uses the proposed scheme to merge the anonymous subregions and sends the final anonymous region set CS to the LSP, the area of every anonymous subregion AR_i satisfies $\text{area}(AR_i) = A_{\min}$, so the area of every merged anonymous subregion AR'_i will satisfy no less than $A_{\min}$ and for any location $L_i(x_i, y_i) \in AR'_i$, and it satisfies that the center of AR'_i is not $L_i(x_i, y_i)$. In addition, the user's identity has been removed from the anonymous query request sent by the anonymous server. Though the LSP received the anonymous query request $Q = \langle \mathrm{CR}, r, \mathrm{POI} \rangle$, it is unable to know the user who made the service request. Therefore, this scheme proposed in this paper can effectively protect the privacy of users.

5.2. Analysis of Computational Complexity. When the anonymous server receives the service query request sent by the user and uses this solution proposed in this paper to generate an anonymous region for it, it firstly generates k initial anonymous subregions according to the privacy protection requirement of the user. The computational complexity of generating the initial anonymous subregions is $O(k)$. When the anonymous server determines whether any two anonymous subregions AR_i and AR_j need to be merged, it first needs to use the anonymous subregions AR_i and AR_j to generate a new anonymous region $AR_{i,j}$. In this case, the number of new anonymous regions that need to be generated is $C_k^2 = k(k-1)/2$, and the computational complexity is $O(k^2)$. Subsequently, the anonymous server will calculate the area of the querying region corresponding to each newly generated anonymous region $AR_{i,j}$, which requires $k(k-1)/2$ times of computation, and the computational complexity is $O(k^2)$. Finally, comparing the area of the querying region AR_i and AR_j with that of the newly generated anonymous region $AR_{i,j}$ to determine whether the both initial anonymous subregions are needed to merge, which requires $k(k-1)/2$ times of computation, and the computational complexity is $O(k^2)$. So the computational complexity of merging all the initial anonymous subregions is $O(k^2) + O(k^2) + O(k^2) = O(k^2)$. After obtaining the new anonymous region $AR_{i,j}$ merged from the initial anonymous subregions AR_i and AR_j, it also needs to judge whether the new anonymous subregions need to be merged again with other anonymous subregions. During the merging process of the anonymous subregions, the best case is that any one of k initial anonymous subregions does not need to be merged with others. In this case, the computational complexity required to implement this solution is

$$O_{\text{best}} = O(k) + O(k^2) = O(k^2). \tag{7}$$

In contrast, the worst case is that any one of k initial anonymous subregions needs to be merged with others and finally merged to an anonymous region. In this case, it needs to repeat the above anonymous region merger and judge $k-1$ time, and the computational complexity required to implement this solution is

$$O_{\text{worst}} = O(k) + O((k-1)k^2) = O(k^3). \tag{8}$$

5.3. Experimental Results and Analysis. In this paper, a network-based generator of moving objects (NGMO) [39] is used to generate the experimental data. This generator is often

Input: k pieces of locations $L_0(x_0, y_0), L_1(x_1, y_1), \ldots, L_{k-1}(x_{k-1}, y_{k-1})$; the query radius r; the privacy requirement A_{min};
Output: anonymous region set CS;
for $i = 0$ to $k - 1$ do
  $AR_i \leftarrow Gen(L_i(x_i, y_i))$;
  $centre(AR_i) \neq L_i(x_i, y_i), Area(AR_i) = A_{min}$,
  $CS \leftarrow (AR_i)$;
  $QAR_i \leftarrow Gen(AR_i, r)$, calculating $Area(QAR_i)$;
end for
for $i, j = 0$ to ij and $i \neq j$ do
  $AR_{i,j} \leftarrow Gen(AR_i, AR_j)$;
  $CR \leftarrow AR_{i,j}$;
  $QAR_{i,j} \leftarrow Gen(AR_{i,j}, r)$, calculating $Area(QAR_{i,j})$;
  if $Area(QAR_i) + Area(QAR_j) \leq Area(QAR_{i,j})$ then
  $CS \leftarrow CS \backslash \{AR_{i,j}\}$;
  end if
  if $Area(QAR_i) + Area(QAR_j) > Area(QAR_{i,j}), QAR_{i,j} = \text{argmin}\{Area(QAR_{i,j})\}_{i \neq j}$, $centre(QAR_{i,j}) \neq L_i(x_i, y_i)$ and $centre(QAR_{i,j}) \neq L_j(x_j, y_j)$ then
  $CS \leftarrow CS \backslash \{AR_i, AR_j\}$;
  end if
end for
return CS

ALGORITHM 1: Anonymous subregion generation algorithm based on querying region division.

used in the existing research of LBS privacy protection, which is based on the Oldenburg map of German city and simulates the location information of users by setting parameters such as the number of moving objects. We set the privacy requirement as k, the generated initial anonymous subregion is rectangular, and its area satisfies $A_{min} = 160000$ m^2. At the same time, to assess the LSP query cost, this paper simulates 500000 points of interest, such as restaurants, hotels, hospitals, parking lots, and so on. In addition, R-tree structure is used to access these points of interest because R-tree is the best-balanced tree for storing high-dimensional data, which can effectively improve the searching efficiency in high-dimensional space [40]. The experimental environment is Intel (R) Core(TM) i7-6700HQ CPU @ 2.60 GHz, 20.0 GB RAM. The algorithms are programmed by Python and the programs run in the Windows 10 Enterprise. The experimental data sets are shown in Table 1.

5.3.1. Problems Existing in Real Anonymity Region Constructions Schemes. To prove that the existing anonymous region construction scheme cannot effectively improve the service quality of LBS query, this paper selects Casper scheme [29] and Fragment scheme [34], respectively, to search nearby points of interest. As the most commonly used anonymous region construction method, the Casper method generates at least an anonymous region that contains all k users. Fragment is to deal with the anonymous region generated by the Casper scheme to reduce the area of the anonymous regions by removing the part that does not contain the user's location according to the location of the user in the anonymous region.

This paper compares the time required to generate the anonymous region, the area of the anonymous region, and the area of the querying region of LSP in the above two schemes to prove that the existing anonymous region division methods will further reduce the quality of service when there is an overlap querying region between both anonymous regions. In this part of the experiment, this paper sets the user's query radius $r = 500$ m. The experimental results are shown in Figures 5 and 6.

As is shown in Figure 5, compared with the Casper scheme, the Fragment scheme uses a region division method to reduce the area of anonymous regions, for example, when $k = 25$, the area of anonymous regions generated by the Casper scheme is 5.71×10^7 m^2, and the time of generating the anonymous regions is 180.275 ms. The area of anonymous regions generated by the Fragment scheme is 3.41×10^7 m^2, and the time of generating the anonymous regions is 175.331 ms. However, in the Casper and Fragment schemes, the time required for the LSP to query the points of interest is, respectively, 9.940 s (the corresponding area of querying regions is 7.234×10^7 m^2) and 10.501 s (the corresponding area of querying regions is 7.329×10^7 m^2). Therefore, when the anonymous server adopts the Casper and Fragment schemes to generate the anonymous regions, the cost time when the user obtains the accurate query results (without considering the transmission delay) is $9.940 + 0.180 = 10.120$ s and $10.501 + 0.175 = 10.676$ s, respectively. When the anonymous server adopts the existing region division schemes to construct the anonymous regions, the time for the user to obtain the query result increases instead.

5.3.2. Analysis of Our Scheme's Effectiveness. In this section, we compare our proposed scheme with Casper scheme and prove that the proposed scheme can effectively reduce the query cost of the LSP and improve the query service quality in the smart campus. The experimental results are shown in Figures 5 and 6.

TABLE 1: Experimental data sets.

Status	Id	Reward number	Type id	Time stamp	X-axis	Y-axis	Speed
New point	105	1	0	0	4229.0	16335.0	298.0
New point	106	1	0	0	19065.0	9922.0	132.0
New point	107	1	0	0	3670.0	20230.0	298.0
New point	108	1	1	0	5565.0	18047.0	298.0
New point	109	1	0	0	10567.0	17947.0	298.0
Point	109	2	0	1	10275.4	17638.7	672.0
Point	108	2	1	1	5487.7	17812.9	504.5

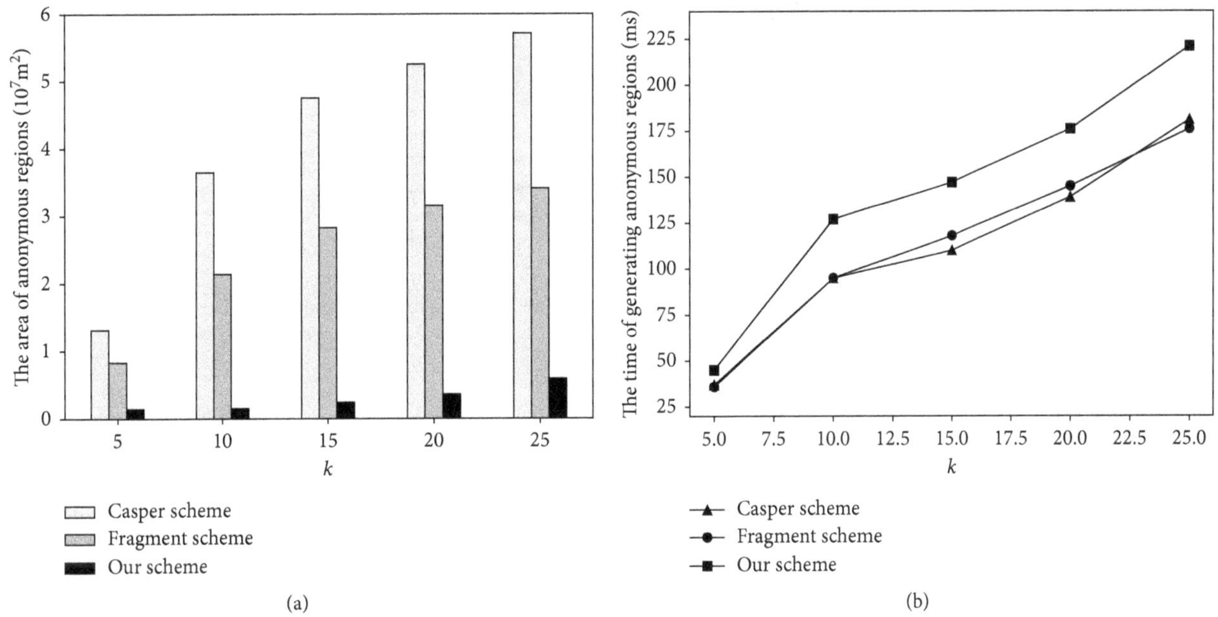

FIGURE 5: Computational cost of anonymity server. (a) The area of anonymous regions. (b) The query cost of LSP.

As is shown in Figures 5 and 6, compared with the existing scheme, the anonymous regions and querying regions generated by our proposed scheme significantly reduced and the same as the time required for LSP to query the points of interest. For example, when $k = 25$, the area of anonymous regions generated by our scheme is 5.89×10^6 m^2, which decreases by 2.80×10^7 m^2 compared to the Casper scheme, and the time to generate the anonymous regions increases from 180.275 ms to 221.034 ms, increasing by 40.759 ms. The area of querying regions generated by our scheme is 2.493×10^7 m^2, which decreases by 4.741×10^7 m^2 compared to the Casper scheme, and the time for query processing increases from 9.940s to 1.961s. Therefore, compared with the Casper scheme, our scheme can effectively improve LBS query service quality.

As can been seen from Figures 5(b) and 6(b), compared with the Casper scheme, as the value of k increases, our scheme needs more extra time to construct the anonymous regions (the time difference between our scheme and the Casper scheme), but less time for query processing. At the same time, the time that the anonymous server generates the anonymous regions is in the milliseconds level, which has little impact on the user delay. The time of query processing is in the second level, which greatly affects the user delay. The query processing time of the LSP determines the user's service quality to a great extent.

The paper also has made a brief analysis about the influence of the user-specified query radius on the scheme. The experimental results are shown in Figures 7(a) and 7(b). As the user query radius r increases, the number of anonymous subregions that need to be merged also increases, increasing the computational cost of the anonymous server. As a result, the area of querying regions and the time of LSP query processing also increase, which obviously influences the LBS query service quality.

In summary, the proposed scheme not only reduces the computational cost of the anonymous server but also significantly reduces the area of the LSP querying regions and its query time, which effectively improves the LSB query service quality for smart campus.

6. Conclusions

In this paper, we merged novel learning applications technology with mobile technology to research on the location privacy protection technology for smart campus and proposed a new location privacy protection scheme based on querying anonymous region construction which faces the

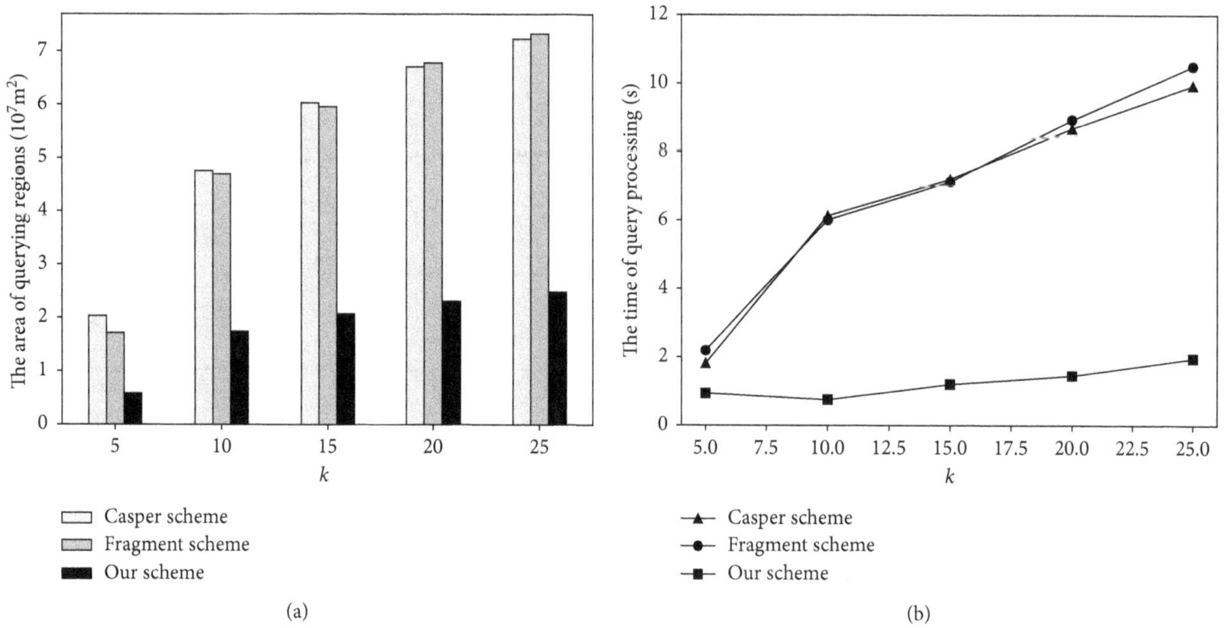

FIGURE 6: The area of querying regions and query cost of LSP. (a) The area of querying regions. (b) The query cost of LSP.

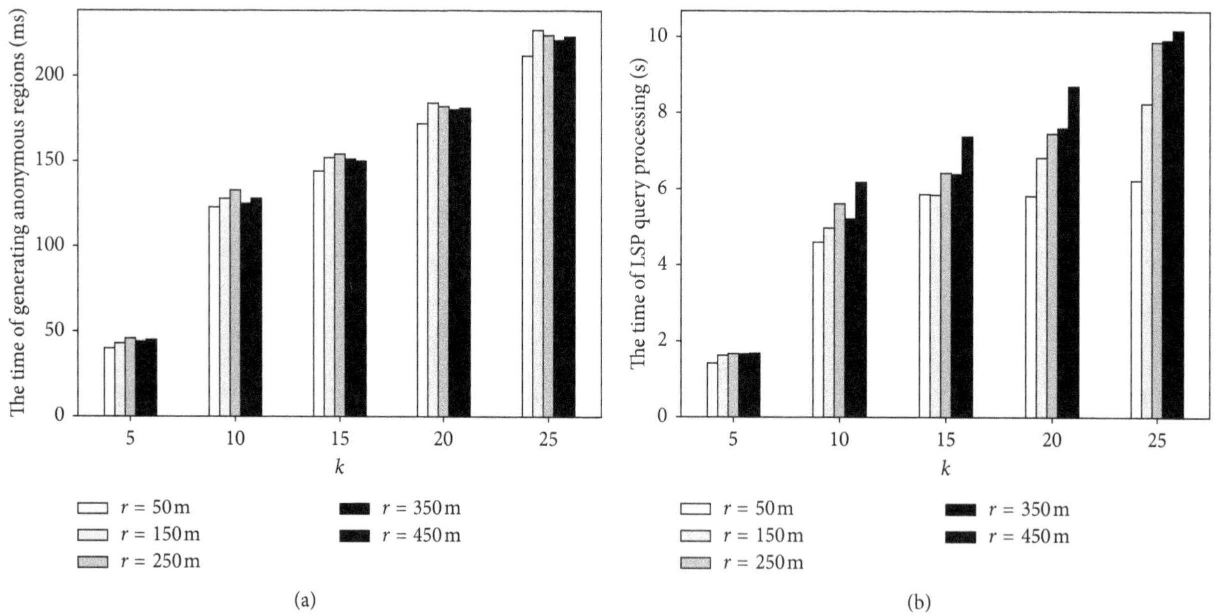

FIGURE 7: The impact of user query radius on the scheme of this paper. (a) The time of generating anonymous regions. (b) The time of LSP query processing.

challenges to achieve the higher quality of smart campus services. Through theoretical analysis and experimental results in the location privacy protection based on k-anonymity, this paper proves that the existing anonymous region construction schemes based on the idea of region division cannot effectively improve the LBS query service quality. The root cause of this problem is that the service quality of sensors in the smart campus is not only related to the anonymous area size constructed by the anonymous server but also to the range of the LSP query. To deal with the above problem, this paper introduces the user's query range into the construction process of the anonymous region. The anonymous server first generates k initial anonymous subregions based on the user's privacy protection requirement, and then the anonymous subregions will be judged if it needs to be merged based on the size of the querying region. The scheme analysis shows that our proposed scheme can effectively reduce the query cost of LSP and improve the service quality while protecting user's location privacy for smart campus.

Acknowledgments

This work was funded by the National Natural Science Foundation of China (61772282, 61772454, and 61811530332). It was also supported by the Priority Academic Program Development of Jiangsu Higher Education Institutions (PAPD), Postgraduate Research & Practice Innovation Program of Jiangsu Province (KYCX18_1032), Natural Science Foundation of Jiangsu Province (BK20150460), and Jiangsu Collaborative Innovation Center on Atmospheric Environment and Equipment Technology (CICAEET). It was also funded by the open research fund of Key Lab of Broadband Wireless Communication and Sensor Network Technology (Nanjing University of Posts and Telecommunications), Ministry of Education.

References

[1] J. S. Turner, "New directions in communications," *IEEE Journal on Selected Areas in Communications*, vol. 13, no. 1, pp. 11–23, 1995.

[2] L. Jin, C. Li, B. Palanisamy, and J. Joshi, "k-trustee: location injection attack-resilient anonymization for location privacy," *Computers & Security*, vol. 78, pp. 212–230, 2018.

[3] Y. Huang, Z. Cai, and A. G. Bourgeois, "Search locations safely and accurately: a location privacy protection algorithm with accurate service," *Journal of Network and Computer Applications*, vol. 103, pp. 146–156, 2018.

[4] S. Xiao, "Consideration of technology for constructing Chinese smart grid," *Automation of Electric Power Systems*, vol. 9, no. 33, pp. 1–4, 2009.

[5] T. Peng, Q. Liu, and G. Wang, "Enhanced location privacy preserving scheme in location-based services," *IEEE Systems Journal*, vol. 11, no. 1, pp. 219–230, 2017.

[6] X. Chen and Y. Mu, "Preserving user location privacy for location-based service," in *Proceedings of 11th International Conference on Green, Pervasive, and Cloud Computing (GPC 2016)*, pp. 290–300, Xi'an, China, May 2016.

[7] R. Schlegel, C. Y. Chow, Q. Huang, and D. S. Wong, "User-defined privacy grid system for continuous location-based services," *IEEE Transactions on Mobile Computing*, vol. 14, no. 10, pp. 2158–2172, 2015.

[8] K. G. Shin, X. Ju, Z. Chen, and X. Hu, "Privacy protection for users of location-based services," *Wireless Communications IEEE*, vol. 19, no. 1, pp. 30–39, 2012.

[9] M. Gruteser and D. Grunwald, "Anonymous usage of location-based services through spatial and temporal cloaking," in *Proceedings of the 1st International Conference on Mobile Systems, Applications and Services*, pp. 31–42, ACM, San Francisco, CA, USA, May 2003.

[10] B. Niu, Z. Zhang, X. Li, and H. Li, "Privacy-area aware dummy generation algorithms for location-based services," in *Proceedings of 2014 IEEE International Conference on Communications (ICC)*, pp. 957–962, IEEE, Sydney, NSW, Australia, June 2014.

[11] X. Shu, D. Yao, and E. Bertino, "Privacy-preserving detection of sensitive data exposure," *IEEE Transactions on Information Forensics and Security*, vol. 10, no. 5, pp. 1092–1103, 2015.

[12] C. Yin, J. Xi, R. Sun, and J. Wang, "Location privacy protection based on differential privacy strategy for big data in industrial Internet-of-Things," *IEEE Transactions on Industrial Informatics*, vol. 14, no. 8, pp. 3628–3636, 2017.

[13] Z. Fu, F. Huang, K. Ren, J. Weng, and C. Wang, "Privacy-preserving smart semantic search based on conceptual graphs over encrypted outsourced data," *IEEE Transactions on Information Forensics and Security*, vol. 12, no. 8, pp. 1874–1884, 2017.

[14] Z. Fu, K. Ren, J. Shu, X. Sun, and F. Huang, "Enabling personalized search over encrypted outsourced data with efficiency improvement," *IEEE Transactions on Parallel & Distributed Systems*, vol. 27, no. 9, pp. 2546–2559, 2016.

[15] C. Yin, S. Zhang, J. Xi, and J. Wang, "An improved anonymity model for big data security based on clustering algorithm," *Concurrency and Computation: Practice and Experience*, vol. 29, no. 7, pp. 1–13, 2017.

[16] K. Vu, R. Zheng, and J. Gao, "Efficient algorithms for K-anonymous location privacy in participatory sensing," in *Proceedings of 31st Annual IEEE International Conference on Computer Communications (INFOCOM, 2012)*, pp. 2399–2407, IEEE, Orlando, FL, USA, March 2012.

[17] T. Ma, Y. Zhang, J. Cao et al., "KDVEM: a k-degree anonymity with vertex and edge modification algorithm," *Computing*, vol. 97, no. 12, pp. 1165–1184, 2015.

[18] R. Zhang, Y. Zhang, and C. Zhang, "Secure top-k query processing via untrusted location-based service providers," in *Proceedings of 31st Annual IEEE International Conference on Computer Communications (INFOCOM, 2012)*, pp. 1170–1178, IEEE, Orlando, FL, USA, March 2012.

[19] R. Zhang, J. Sun, Y. Zhang, and C. Zhang, "Secure spatial top-k query processing via untrusted location-based service providers," *IEEE Transactions on Dependable and Secure Computing*, vol. 12, no. 1, pp. 111–124, 2015.

[20] C. Y. Chow, M. F. Mokbel, and W. G. Aref, "Casper: query processing for location services without compromising privacy," *ACM Transactions on Database Systems (TODS)*, vol. 34, no. 4, pp. 1–48, 2009.

[21] C. Yin, J. Xi, and R. Sun, "Location privacy protection based on improved-value method in augmented reality on mobile devices," *Mobile Information Systems*, vol. 2017, Article ID 7251395, 7 pages, 2017.

[22] C. Yin and J. Xi, "Maximum entropy model for mobile text classification in cloud computing using improved information gain algorithm," *Multimedia Tools and Applications*, vol. 76, no. 16, pp. 16875–16891, 2017.

[23] H. Rong, T. Ma, M. Tang, and J. Cao, "A novel subgraph K+-isomorphism method in social network based on graph similarity detection," *Soft Computing*, vol. 22, no. 8, pp. 2583–2601, 2017.

[24] B. Gu, X. Sun, and V. S. Sheng, "Structural minimax probability machine," *IEEE Transactions on Neural Networks and Learning Systems*, vol. 28, no. 7, pp. 1646–1656, 2017.

[25] B. Gu, V. S. Sheng, K. Y. Tay, W. Romano, and S. Li, "Incremental support vector learning for ordinal regression," *IEEE Transactions on Neural Networks & Learning Systems*, vol. 26, no. 7, pp. 1403–1416, 2015.

[26] R. Lu, X. Lin, X. Liang, and X. Shen, "A dynamic privacy-preserving key management scheme for location-based services in VANETs," *IEEE Transactions on Intelligent Transportation Systems*, vol. 13, no. 1, pp. 127–139, 2012.

[27] A. R. A. Bouguettaya and M. Y. Eltoweissy, "Privacy on the web: facts, challenges, and solutions," *IEEE Security & Privacy*, vol. 99, no. 6, pp. 40–49, 2003.

[28] A. Lee, "Guidelines for smart grid cyber security," NIST Interagency/Internal Report (NISTIR)-7628, National Institute of Standards and Technology, Gaithersburg, MD, USA, 2010.

[29] M. F. Mokbel, C. Y. Chow, and W. G. Aref, "The new Casper: a privacy-aware location-based database server," in *Proceedings of IEEE 23rd International Conference on Data Engineering (ICDE 2007)*, pp. 1499-1500, IEEE, Istanbul, Turkey, April 2007.
[30] X. Li, E. Wang, W. Yang, and J. Ma, "DALP: a demand-aware location privacy protection scheme in continuous location-based services," *Concurrency and Computation: Practice and Experience*, vol. 28, no. 4, pp. 1219–1236, 2016.
[31] B. Gedik and L. Liu, "Protecting location privacy with personalized k-anonymity: architecture and algorithms," *IEEE Transactions on Mobile Computing*, vol. 7, no. 1, pp. 1–18, 2008.
[32] P. Kalnis, G. Ghinita, K. Mouratidis, and P. Dimitris, "Preventing location-based identity inference in anonymous spatial queries," *IEEE Transactions on Knowledge & Data Engineering*, vol. 19, no. 12, pp. 1719–1733, 2007.
[33] K. W. Tan, Y. Lin, and K. Mouratidis, "Spatial cloaking revisited: distinguishing information leakage from anonymity," in *Proceedings of 11th International Symposium on Spatial and Temporal Databases*, pp. 117–134, Aalborg, Denmark, July 2009.
[34] T. C. Li and W. T. Zhu, "Protecting user anonymity in location-based services with fragmented cloaking region," in *Proceedings of 2012 IEEE International Conference on Computer Science and Automation Engineering (CSAE)*, pp. 227–231, IEEE, Zhangjiajie, China, May 2012.
[35] D. Steiert, D. Lin, Q. Conduff, and W. Jiang, "Poster: a location-privacy approach for continuous queries," in *Proceedings of the 22nd ACM on Symposium on Access Control Models and Technologies*, pp. 115–117, ACM, Indianapolis, IN, USA, June 2017.
[36] E. H. Yilmaz, E. Ferhatosmanoglu, and R. C. Aksoy, "Privacy-preserving aggregate queries for optimal location selection," *IEEE Transactions on Dependable and Secure Computing*, vol. 2017, no. 99, p. 1, 2017.
[37] I. Memon, "Authentication user's privacy: an integrating location privacy protection algorithm for secure moving objects in location based services," *Wireless Personal Communications*, vol. 82, no. 3, pp. 1585–1600, 2015.
[38] Y. Wang, D. Xu, and F. Li, "Providing location-aware location privacy protection for mobile location-based services," *Tsinghua Science and Technology*, vol. 21, no. 3, pp. 243–259, 2016.
[39] T. Brinkhoff, "A framework for generating network-based moving objects," *GeoInformatica*, vol. 6, no. 2, pp. 153–180, 2002.
[40] M. Hadjieleftheriou, Y. Manolopoulos, Y. Theodoridis, and V. J. Tsotras, "R-trees–a dynamic index structure for spatial searching," in *Encyclopedia of GIS*, pp. 993–1002, Springer, Boston, MA, USA, 2008.

4

TinyDroid: A Lightweight and Efficient Model for Android Malware Detection and Classification

Tieming Chen,[1] **Qingyu Mao**,[1] **Yimin Yang**,[1] **Mingqi Lv**,[1] **and Jianming Zhu**[2]

[1]*College of Computer Science and Technology, Zhejiang University of Technology, Hangzhou 310023, China*
[2]*College of Information Engineering, Central University of Finance and Economics, Beijing 100081, China*

Correspondence should be addressed to Tieming Chen; tmchen@zjut.edu.cn

Academic Editor: Prosanta Gope

With the popularity of Android applications, Android malware has an exponential growth trend. In order to detect Android malware effectively, this paper proposes a novel lightweight static detection model, *TinyDroid*, using instruction simplification and machine learning technique. First, a symbol-based simplification method is proposed to abstract the opcode sequence decompiled from Android Dalvik Executable files. Then, N-gram is employed to extract features from the simplified opcode sequence, and a classifier is trained for the malware detection and classification tasks. To improve the efficiency and scalability of the proposed detection model, a compression procedure is also used to reduce features and select exemplars for the malware sample dataset. *TinyDroid* is compared against the state-of-the-art antivirus tools in real world using Drebin dataset. The experimental results show that *TinyDroid* can get a higher accuracy rate and lower false alarm rate with satisfied efficiency.

1. Introduction

With the fast development of mobile Internet, the popularity of mobile devices, and the rapid growth of mobile applications, smartphones have become the most popular tools for people to access the Internet. The statistics from Gartner show that more than 400 million smartphones were sold globally in the final quarter of 2015, with Android operating system accounting for 80.7% [1]. As of February 2016, the number of Android apps in Google Play is approaching 2 million. Meanwhile, internal application vulnerabilities and malware are the primary threat to the security of Android applications. Since application developers lack the necessary knowledge and skills about security code audit, it is easy to create vulnerable code in every development step in the software life cycle. Once the attackers exploit these vulnerabilities, they will damage the confidentiality, integrity, and availability of applications. The creator of an Android malware usually inserts a small malicious code to a popular application and spreads the malware through some third-party app stores without security management [2, 3]. According to a new report from the mobile department of Alibaba, 18% of the Android devices have ever been infected with viruses, and 95% of the popular mobile applications have been counterfeited and kinds of malware have evolved quickly [4]. The malicious activities of malware typically include system failure, information leakage, and data corruption, so malicious mobile code will damage the user's benefits.

The existing methods of malware detection include signature-based detection methods and behaviour-based detection methods. Signature-based detection methods detect malware by comparing the binary code of software with a database that stores all the signatures of the known malware. Although the signature-based detection methods have the advantages of simplicity and efficiency with high accuracy, they cannot detect an unknown malware and need to maintain a vast signature database.

On the other hand, behaviour-based detection methods detect malware by comparing the behaviour pattern of software with that of known malware. Although this method leads to a high rate of false alarm, it can effectively detect unknown malware and compensate for the disadvantages of signature-based detection methods. Two types of analysis could be performed on behaviour-based detection, that is, static analysis [5–8] and dynamic analysis [9–14]. The dynamic analysis can analyze object-code obfuscation to some extent by monitoring the run-time situations of an application, which runs in a secure sandbox mode, but it requires a significant amount of computing resources and has low code coverage. Instead, static analysis can get features based on analysis on sequences of instructions obtained using reverse engineering. In a word, static analysis has the advantages of efficiency and high code coverage, and it is a relatively lightweight process as compared to the dynamic analysis.

In this paper, we firstly identify the problem of Android malware detection and summarize a model of threats faced by developers and researchers. Then, we propose a detection method named *TinyDroid* using instruction simplification and machine learning technique. The main contributions of this paper are twofold as follows:

(i) N-gram direct on the reduced symbolic Dalvik opcode sequences instead of the original full instructions

A simple symbol set is introduced to simplify the original Dalvik instructions sequence, where one series of instructions with similar function can be assigned as one symbol. N-gram technique is employed to handle the symbolic sequence.

(ii) Reduction on both features and number of samples for building a lightweight model

In order to achieve high efficiency of detection and classification, a further reduced scheme is proposed to largely cut down the number of N-gram items and training samples. Specifically, information gain is employed for attribute reduction and affinity propagation is used for sample selection.

2. Related Works

In this section, we review related work in two areas: usage of opcodes and N-gram to detect malware and lightweight approaches to detect Android malware.

2.1. Usage of Opcodes and N-Gram to Detect Malware. Some references study the effectiveness of opcodes for detecting or characterising malware. Rad et al. [15] use the histogram of opcodes to classify metamorphic virus family variants. This method takes into account the frequency distribution of opcodes but ignores the sequential patterns of opcodes.

References [16–18] have investigated the effectiveness of N-gram opcodes extracted from the disassembled application and represented programs as N-gram density that made a robust indicator of malware. However, these methods have a large number of features and do not consider efficiency.

Wolfe et al. [19] extracted Java bytecode from the Android application and computed the N-gram frequencies of the bytecode. Then, they applied dimensionality reduction on the N-gram frequencies using principal components analysis (PCA). This approach, where the dimensionality of the feature set is very large and difficult to handle due to the unrestricted selection of features, goes to the other extreme when compared to the permission approach.

2.2. Lightweight Approaches to Detect Android Malware. Some static and lightweight mechanisms for detecting Android malware have been introduced. For example, in 2012, Zhao et al. [20] proposed a lightweight framework named RobotDroid that uses active learning on Android applications to induce an accurate detection model with minimal labeled samples.

In 2013, Santos et al. [21] proposed an efficient model, which is based on the frequency of the appearance of opcode sequences. Also, they provided empirical validation which shows that opcode sequence is feasible to detect unknown malware.

In 2014, Arp et al. [22] designed a lightweight method of Android malware detection. They apply linear-time static analysis and learning techniques for efficiency.

In 2017, DroidSieve [23] relies on several syntactic and resource-centric features, which are robust and computationally inexpensive to detect Android malware.

These methods have not considered the issue of scalability. This problem produces excessive storage requirements, increases time complexity, and impairs the general accuracy of the models.

3. Threat Model

Before introducing our detection method, it is necessary to summarize the threats we currently encounter in Android malware detection. Four threats arise from the Android malware detection:

Repacking and Reflection. A more recent trend we have observed is the increasing prevalence of Android malware leveraging packing technology. The application can use the secondary packaging to modify the source code or add a small amount of malicious code due to insufficient reinforcement. Java's reflection mechanism is expressed as a way to dynamically get information and call objects. Malware creators use reflection mechanisms as an important way to hide malicious behaviour in software. Malicious applications, in order to evade static analysis, can transmit malicious code by invoking sensitive methods at reflection at runtime.

Malware in MultiDEX. Android programs are compiled into Dalvik Executable (DEX) files. A typical Android app has a single DEX file, and some complex applications contain multiple DEX files. Therefore, some malware would include malicious code in multiple DEX files. If only a single DEX file is analyzed, it may not be able to distinguish whether it is malicious. This simple step can serve as an evasion technique against static analysis.

Run-Based Malware. Instant Run is a new feature in Android system. It allows developers to quickly deploy patches to a debug application .zip file into the application. Some malware authors hide the malware payload portion of their app in code fragments that are hidden in the zip file used by Instant Run. This approach to detection evasion can only be used on Android Lollipop and later SDK levels. Although it cannot be used on apps in Google Play, it was still possible to spread in the third application markets.

Native Code. To avoid static detectors at the bytecode level, malicious code relies on some special native API functions and kernel system functions to infect, spread, and hide. Moreover, sometimes, malware also embeds malicious payload within the native content.

Each threat can evade some methods of malicious detection. To cope with some of the above threats, *TinyDroid* is proposed to detect Android malware using instruction simplification and machine learning technique.

4. Design of TinyDroid

4.1. Overview. To solve the aforementioned problems, this paper focuses on lightweight machine learning-based detection of Android malware using Dalvik instructions simplification, exemplar selection, and optimization.

We develop a detection system *TinyDroid*, which has a high accuracy on detection and classification of Android malware. *TinyDroid* includes two procedures. The first one is to create the detection model and classification model. The second is to predict Android malware. The main steps are shown, respectively, in Figures 1 and 2.

Training set for the detection model is divided into two subsets: malware apps and benign apps. An APK file always contains several DEX files, which can be executed in Dalvik Virtual Machine. The APK file can be disassembled into smali codes by Apktool [24], where smali is an explanation for the Dalvik bytecode and can be simplified into symbolic instructions by our method. The detection model is supported by N-gram, exemplar selection method, and machine learning algorithm. Similarly, the classification model has the same steps as in Figure 1, except that it needs to divide the training set into different family subsets.

As shown in Figure 2, the apps to be predicted are firstly preprocessed and the reduced N-gram sequences are extracted. The malware can be detected based on the detection model. Furthermore, the information about malware family can be obtained based on the classification model.

The threats mentioned in Section 3 are almost the hardest part of all malicious detection methods. Our solution mitigates some threats as follows:

Repacking and Reflection. Our study has shown that reflection APIs are more frequently used by malware set than in the benign set, which makes them part of the feature vector for the classification.

Malware in MultiDEX. In the feature extraction process, we analyze all the DEX files to extract the opcodes and generate the N-gram sequences.

Run-Based Malware. This is a difficult point of malicious detection and cannot be detected by static analysis. The method in this paper may be insufficient for such threats.

Native Code. Since our detection tool only works at Dalvik bytecode level, it is not able to detect any dangerous methods invoked. However, the use of invoking calls such as *invoke-direct* is also used as a feature by our model.

4.2. Feature Extraction and Simplification. Dalvik instructions contained 230 instructions such as method call instructions, data manipulation instructions, and return instructions [25]. For malicious detection, the full instruction set is too complicated. In addition, there are some instructions of the same semantics. Condensing these instructions would not affect the test results but also improve the efficiency of detection. For example, Hang et al. [26] classified all 218 Dalvik instructions into a simplified instruction set with only 13 types of instructions, which is called SDIL.

Thus, we have designed a more simplified symbolic instruction set that is suitable for more efficient machine learning. The detailed steps are as follows:

(1) We analyze many smali source code files and found that many instructions have multiple versions due to the different parameters. For example, consider two Dalvik move instructions (i.e., *move-object/from16 vAA, vBBBB; move-object/16 vAAAA, vBBBB*), the only difference between them is how many bits they use to represent registers indices (*vAA* and *vAAAA* require 8 and 16 bits). According to the above rules, finally, we identified 107 representative Dalvik instructions which have the core semantics and high frequency of occurrence.

(2) Then, we assign them into 10 types of symbols according to the semantics of each instruction. For example, *G* stands for all the jump instructions, it could be *goto*, *goto16*, or *goto/32*, because they have the almost the same semantics. Table 1 shows the full symbolic instruction set with different semantics we proposed.

(3) We use Apktool to decompile apk for the sake of getting smali source code. Then, we extract instructions according to the reduced instruction set. Finally, we replace ordered instructions with symbols.

N-gram [27] is widely used for natural language processing (NLP) and also used for the analysis of malware [16, 28]. When the N-gram sequences of Dalvik instructions are generated, they can be easily analyzed for the detection of Android malware. N-gram is a type of probabilistic model, and it assumes that the appearance of the nth word only correlates with the previous $n-1$ words. For example,

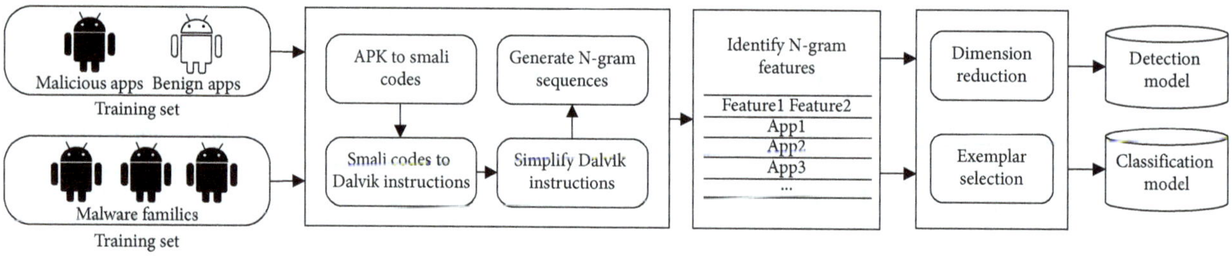

FIGURE 1: The structure of the proposed detection and classification model.

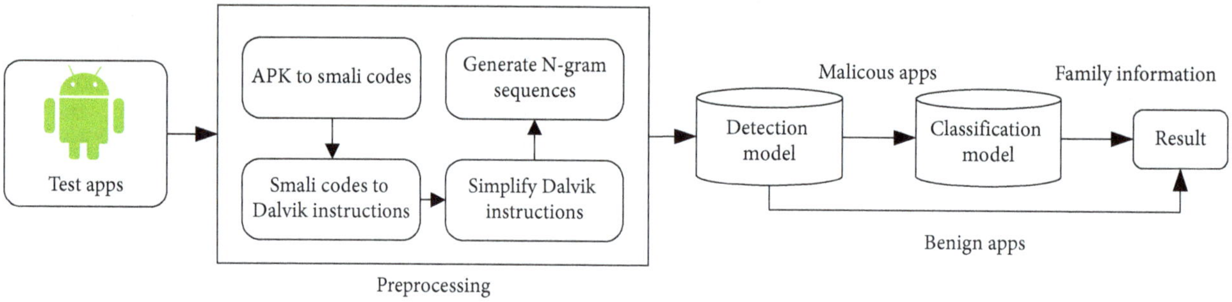

FIGURE 2: The process of predicting Android malware.

suppose that we have a sequence of symbolic instructions "MRGITPV," the 3-gram features are [{MRG}, {RGI}, {GIT}, {ITP}, and {TPV}]. Table 2 shows the occurrence rate of several 3-gram features in the selected APKs.

Since the number of N-gram features is excessive, so we use information gain (IG) for feature reduction as it has been proven to be effective [29]. The information gain of a given feature X with respect to the class attribute Y is the reduction in uncertainty about the value of Y, after observing values of X. It is denoted as $IG(Y|X)$. The uncertainty about the value of Y is measured by its entropy defined as

$$H(Y) = -\sum_i P(y_i)\log_2(P(y_i)), \quad (1)$$

where $P(y_i)$ is the prior probabilities for all values of Y. The uncertainty about the value of Y after observing values of X is given by the conditional entropy of Y given X defined as

$$H(Y|X) = -\sum_j P(x_j)\sum_i P\left(y_i|x_j\right)\log_2\left(P\left(y_i|x_j\right)\right), \quad (2)$$

where $P(y_i|x_j)$ is the posterior probabilities of Y given the values of X. The information gain is thus defined as $IG(Y|X) = H(Y) - H(Y|X)$. According to this measure, a feature X is regarded more correlated to class Y than feature Z, if $IG(Y|X) > IG(Y|Z)$. By calculating information gain, we can rank the correlations of each feature to the class and select key features based on this ranking.

4.3. Exemplar Selection and Optimization. Recent researches show that the clustering algorithm affinity propagation [30] can be used to generate good representative data for intrusion detection training [31]. The intrusion detection training is very similar to the malware detection training, so we choose this clustering algorithm in our paper. Affinity propagation is employed as a reduction method to select a smaller set composed of representative exemplars from the original large training data. When the size of the training dataset becomes smaller, the detection model efficiency would be improved after the sample optimization.

Affinity propagation (AP) clusters instances by passing messages between data points iteratively. Unlike clustering algorithms such as k-means, AP does not require the number of clusters to be determined or estimated before running the algorithm, and the only parameter that should be set is the *preference*. The number of clusters generated by AP is decided by its *preference*, so we choose the preferences between the minimum and the maximum of the similarities to generate the expected number of clusters that are separately distributed.

The time complexity of AP is $O(N^2T)$, where N is the number of samples and T is the number of iterations until convergence. Further, the memory complexity is $O(N^2)$ if a dense similarity matrix is used, but reducible if a sparse similarity matrix is used.

5. Evaluation

5.1. Experimental Setup. Classification algorithm is a typical method of machine learning. The most common evaluation metrics include true positive (TP), false positive (FP), true negative (TN), and false negative (FN) [32]. These four metrics can make up a confusion matrix as shown in Table 3.

Depending on these basic metrics, a series of common evaluation metrics can also be generated as follows:

TABLE 1: Symbolic instruction set definition.

Symbols	Semantics	Quantity	Representative opcodes prefixes
C	Comparison	5	cmpl-float|cmpg-float|cmpl-double|cmpg-double|cmp-long
D	Definition	11	const|const/4|const/16|const-wide|const/high|const-string
M	Manipulation	13	move|move-wide|move-object|move-result|move-exception
R	Return	4	return|return-void|return-wide|return-object
L	Monitor	2	monitor-enter|monitor-exit
G	Jump	3	goto|goto/16|goto/32
I	Judgment	12	if-eq|if-ne|if-lt|if-ge|if-gt|if-le|if-eqz|if-nez|if-ltz|if-gez|if-gtz|if-lez
T	Reading	21	aget|iget|sget|aget-wide|aget-object|aget-boolea|aget-byte|aget-char
P	Writing	21	aput|iput|sput|aput-wide|aput-object|aput-boolea|aput-byte|aput-char
V	Method call	15	invoke-virtual|invoke-super|invoke-direct|invoke-static

TABLE 2: Illustration for the frequency statistics of 3-gram features.

	CCC	CCD	CCM	CCR	...	VVV
DroidKungFu.apk	0.032016	0.010601	0.019507	0.036822	...	0.071454
BaseBridge.apk	0.059207	0.011801	0.018162	0.032164	...	0.047486
Plankton.apk	0.030508	0.006391	0.017376	0.038064	...	0.053952
...	...	...	...	...	...	...

TABLE 3: Confusion matrix.

Prediction	Malicious	Benign
Malicious	TP	FN
Benign	FP	TN

$$\begin{aligned} \text{TP rate} &= \text{recall} = \frac{\text{TP}}{\text{TP}+\text{FN}}, \\ \text{FP rate} &= \frac{\text{FP}}{\text{FP}+\text{TN}}, \\ \text{precision} &= \frac{\text{TP}}{\text{TP}+\text{FP}}, \\ \text{accuracy} &= \frac{\text{TP}+\text{TN}}{\text{TP}+\text{TN}+\text{FP}+\text{FN}}, \\ \text{F}-\text{measure} &= \frac{2\times\text{recall}\times\text{precision}}{\text{recall}+\text{precision}}. \end{aligned} \tag{3}$$

True-positive rate (TPR) is sometimes called detection rate or recall rate. False-positive rate (FPR) is sometimes called false alarm rate. Receiver operating characteristic (ROC) is a graphical plot that illustrates the performance of a binary classifier system as its discrimination threshold is varied. The curve is created by plotting the TPR against FPR at various threshold settings. The area under the curve (AUC) is the area under the ROC curve. The closer that this value is to 1, the better the performance is.

We used a dataset of real Android applications and malware to conduct the experiments. The dataset includes all Android malware from Drebin [33]. Additionally, all the benign applications are randomly collected from Google Play [34]. Previous work [35] has pointed out that only 2 in 52208 applications were malware from Google Play, so the applications can be considered as benign applications.

We utilize Waikato Environment for Knowledge Analysis (Weka) to train data and create the model. Weka is an open-source software, and it integrates a lot of machine learning algorithms, so we choose it as our experimental tool.

5.2. Malware Detection Performance. We randomly choose 300 malware and benign samples, respectively, as experimental data. 2-gram sequences, 3-gram sequences, and 4-gram sequences are generated, respectively. These sequences are respectively fed into Random Forest, Support Vector Machine (SVM), *k*-Nearest Neighbor (kNN), and Naive Bayes to build the classifier. The time complexity is shown in Table 4. 10-fold cross-validation is selected to evaluate our method. The results are shown in Tables 5–7.

AUC, TPR, FPR, precision, recall, and F-Measure are selected as performance measurement indicators. In particular, AUC is an important indicator because it can comprehensively reflect TPR and FPR. As shown in Table 5, Random Forest has the best performance on AUC and FPR, and its TPR is a little less than kNN. As shown in Table 6, every indicator of Random Forest is the best. As shown in Table 7, Random Forest performs best on the two indicators of F-Measure and AUC. Through the above analysis, we can see Random Forest is the optimal algorithm among 2-gram sequences, 3-gram sequences, and 4-gram sequences.

Next, we will find out the optimal N-gram sequences based on Random Forest. By analyzing Tables 5–7, we can easily find that Random Forest's AUC from good to bad: 4-gram, 3-gram, and 2-gram; Random Forest's FPR from good to bad: 3-gram, 4-gram, and 2-gram. The performance of 2-gram sequences is the worst, so it is no longer to be considered. 4-gram sequences' AUC is 2%, which is higher than 3-gram sequences', but 4-gram sequences' TPR is 30% higher than 3-gram sequences'. Through the above analysis, 3-gram sequences based on Random Forest is the optimal method.

Table 4: Classifiers' time complexity.

Classifier	Time complexity	Parameter
Random Forest	$O(\text{ndlogn})$	n is the number of samples and d is the number of features
SVM	$O(\max(n,d), \min((n,d)^2))$	n is the number of points and d is the number of features
kNN	$O(\text{ndk})$	n is the number of samples and d is the number of features
Naive Bayes	$O(\text{nd})$	n is the number of samples and d is the number of features

Table 5: The detection results based on 2-gram sequences.

Classifier	TPR	FPR	Precision	Recall	*F*-Measure	AUC
Random Forest	0.915	0.106	0.876	0.915	0.895	0.97
SVM	0.852	0.173	0.801	0.852	0.826	0.84
kNN	0.93	0.136	0.848	0.93	0.887	0.897
Naive Bayes	0.87	0.221	0.763	0.87	0.813	0.888

Table 6: The detection results based on 3-gram sequences.

Classifier	TPR	FPR	Precision	Recall	*F*-Measure	AUC
Random Forest	0.956	0.079	0.908	0.956	0.931	0.982
SVM	0.922	0.13	0.853	0.922	0.886	0.896
kNN	0.952	0.148	0.84	0.952	0.892	0.903
Naive Bayes	0.867	0.17	0.807	0.867	0.836	0.903

Table 7: The detection results based on 4-gram sequences.

Classifier	TPR	FPR	Precision	Recall	*F*-Measure	AUC
Random Forest	0.956	0.103	0.884	0.956	0.918	0.984
SVM	0.944	0.106	0.879	0.944	0.911	0.919
kNN	0.904	0.088	0.894	0.904	0.899	0.909
Naive Bayes	0.9	0.103	0.877	0.9	0.888	0.917

Table 8: The results of different size of samples.

Number of samples	TPR	FPR	Precision	AUC
600 samples	0.956	0.086	0.908	0.982
1200 samples	0.980	0.079	0.917	0.992
2400 samples	0.988	0.077	0.921	0.994

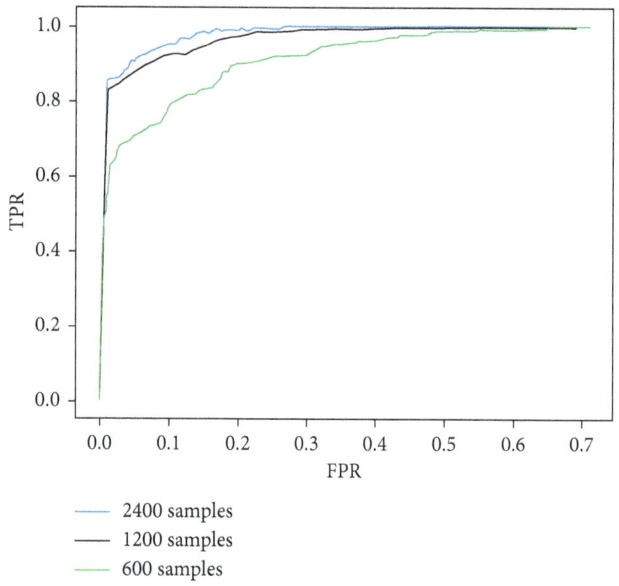

Figure 3: ROC curves with different size of samples.

Table 9: The time and performance effectiveness with different size of exemplars selected.

Training data	Training time	TPR	FPR
2400 samples	2.9 s	0.988	0.077
834 exemplars (after AP)	0.86 s	0.964	0.083
516 exemplars (after AP)	0.56 s	0.959	0.091

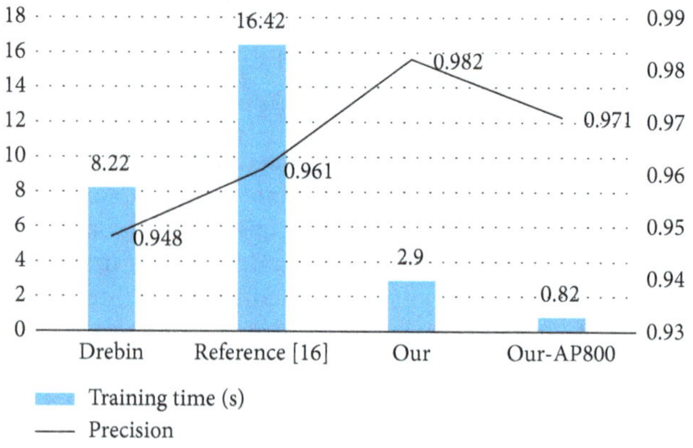

Figure 4: Comparison of training time.

Different size of sample dataset would affect the evaluation results. Three different sizes of sample dataset are randomly generated: 600 samples, 1200 samples, and 2400 samples. The 3-gram sequences based on Random Forest method is used for the experiment. The final model is tested by 10-fold cross-validation. As shown in Table 8 and Figure 3, the results show that performance becomes better when the number of samples increases.

5.3. Comparison of Training Time. In order to further improve malware detection performance, especially for a large volume of training data, affinity propagation is used to generate the excellent representative data for malware detection training. We randomly choose 2400 samples as training data. Then, the training data are processed by AP algorithm; 834 exemplars and 516 exemplars are generated, respectively. The representative exemplars are trained to create a detection model and the model is tested by 10-fold cross-validation. As shown in Table 9, TPR and FPR show little change, but the time cost has declined dramatically. The time performance is improved after exemplar selection.

Then, we compare the proposed method with several baselines. Drebin [23] is a lightweight method of Android

TABLE 10: Detection rates of TinyDroid and antivirus scanners.

Num.	TinyDroid	Avira	Dr.Web	AVG	Kaspersky	ESET	GDATA	Rising	MSE	Avast
2000 samples	95.3%	100.0%	97.0%	94.0%	91.5%	91.0%	66.0%	49.5%	15.5%	11.0%
4000 samples	98.6%	99.1%	92.7%	92.3%	78.7%	90.0%	42.9%	45.0%	7.5%	10.2%

TABLE 11: False-positive rates of TinyDroid and antivirus scanners.

Num.	TinyDroid	Avira	Dr.Web	AVG	Kaspersky	ESET	GDATA	Rising	MSE	Avast
2000 samples	5.0%	0.5%	0.5%	0%	0%	0%	0.5%	0.5%	0%	0%
4000 samples	1.4%	1.6%	0.9%	0%	0%	0%	0%	1.2%	0%	0%

malware detection. They apply linear-time static analysis and learning techniques for efficiency. Canfora et al. [16] proposed a method using frequencies of N-grams of opcodes to detect Android malware effectively. We extract the corresponding number of features as much as possible for experiment and use the same dataset (randomly choose 2400 samples) to train three models, respectively. The results are shown in Figure 4.

Here, we divide our method into two parts. In the first part, we only use opcode instruction symbolization and N-gram techniques to reduce feature dimensions (Our in Figure 4). In the second part, on the basis of the first part, we use the AP clustering algorithm for sample selection (Our-AP800 in Figure 4).

We can see that the time spent to perform model training in our method (first part) is about 2.9 s, which is better than the other two baselines. Subsequently, we performed AP clustering algorithm processing on the samples and selected 800 center samples to retrain. We can see that the accuracy of the model has decreased a little, but the time cost has declined dramatically. And our method still has higher accuracy than other methods. The experimental results clearly show that *TinyDroid* is better than the state-of-the-art methods.

5.4. Comparison with Real World AV Scanners. In this part, two different sizes of sample dataset are chosen to evaluate the detection rate of *TinyDroid*. For this experiment, we randomly split the dataset into a training set (60%) and a testing set (40%). To compete with common antivirus products in practice, we send each sample to the VirusBook service [36] and get the output from 9 common antivirus scanners (Avira, Dr.Web, AVG, Kaspersky, ESET, GDATA, Rising, MSE, and Avast). Finally, we obtain the detection rates and false-positive rates by the statistics of output.

The results of the experiments are shown in Table 10 and Table 11. While most scanners detect over 90% of the malware, the detection rates for some scanners are below 50%. Obviously, these antivirus scanners may not be specialized in detecting mobile applications. On the 2000 samples' dataset, *TinyDroid* provides the third best performance with detection of 95.3% and outperforms 7 out of the 9 scanners. On the 4000 samples' dataset, *TinyDroid* provides the second best performance with detection of 98.6% and outperforms 8 out of the 9 scanners. This shows that *TinyDroid* achieves better detection rates when the number of samples increases. Also, these samples have been public for a longer period, so almost all antivirus scanners provide proper signatures for this dataset. The machine learning method has much more advantage than the traditional technique when the samples are unknown malware.

TABLE 12: The results on the classification of malware family using TinyDroid.

Malware family	TPR	FPR	Precision	Recall	*F*-Measure	AUC
Plankton	0.949	0.003	0.974	0.949	0.961	0.999
DroidKungF	0.833	0.003	0.972	0.833	0.897	0.996
GinMaster	1	0.037	0.755	1	0.86	0.997
FakeDoc	0.976	0	1	0.976	0.988	1
FakeInstalle	0.865	0.006	0.941	0.865	0.901	0.978
Opfake	1	0.019	0.882	1	0.938	0.997
BaseBridge	0.816	0	1	0.816	0.899	0.995
K_{min}	1	0	1	1	1	1
Iconosys	1	0	1	1	1	1

The false-positive rates of the antivirus scanners range from 0% to 1% on our dataset of benign applications. Although the false-positive rate of *TinyDroid* is higher than the antivirus scanners, Table 11 shows that the false-positive rate of *TinyDroid* has seen a decline when the number of samples increases.

5.5. Malware Classification Performance. In addition to the detection performance, we need to evaluate the malware classification performance. In this part, we select 9 top malware families as the experimental data. The experimental data contains a total of 900 samples and is separated into a training set (60%) and a testing set (40%). The results of the experiments are shown in Table 12. All the values of AUC are higher than 0.97 and the values of FPR are lower than 0.04. *TinyDroid* shows a good performance on malware classification.

6. Conclusion

This paper proposes a novel and lightweight static detection system *TinyDroid*. We firstly use the reverse engineering to get the Dalvik instructions from DEX files and simplify the instructions into a small symbol set. Then, the detection model integrating N-gram, exemplar selection method, and machine learning algorithm is set up based on those above reduced symbolic sequences. We compare *TinyDroid* against the antivirus scanners in real world. The experimental results show that *TinyDroid* outperforms the state-of-the-art tools on detection and classification.

There are still some deficiencies and improvements to be found in this study: The samples used in this paper mainly originate from the academic sample set of the organization or platform and lack metamorphic malware samples. Metamorphic malware could escape the proposed method. It is clearly vulnerable to obfuscation and packing.

In the next step, according to the limitation, dynamic analysis methods can be studied and combined to extract more useful features. Meanwhile, other optimization methods can also be further studied to keep the lightweight and improve the detection efficiency.

Acknowledgments

This paper was partially supported by the Key National Natural Science Foundation of China with Grant nos. 61772026, U1509214, and 61602412.

References

[1] "Gartner Report," http://www.gartner.com/newsroom/id/3215217.
[2] A. Shabtai, Y. Fledel, U. Kanonov et al., "Google android: a state-of-the-art review of security mechanisms," 2009, https://arxiv.org/abs/0912.5101.
[3] A. P. Felt, M. Finifter, E. Chin, S. Hanna, and D. Wagner, "A survey of mobile malware in the wild," in *Proceedings of 1st ACM Workshop on Security and Privacy in Smartphones and Mobile Devices (SPSM'11)*, pp. 3–14, ACM, Chicago, IL, USA, 2011.
[4] Alibaba Security, http://jaq.alibaba.com/.
[5] H. Kang, J. W. Jang, A. Mohaisen, and H. K. Kim, "Detecting and classifying android malware using static analysis along with creator information," *International Journal of Distributed Sensor Networks*, vol. 11, no. 6, Article ID 479174, 2015.
[6] J. W. Jang and H. K. Kim, "Function-oriented mobile malware analysis as first aid," *Mobile Information Systems*, vol. 2016, Article ID 6707524, 11 pages, 2016.
[7] H. J. Zhu, Z. H. You, Z. X. Zhu, W. L. Shi, X. Chen, and L. Cheng, "DroidDet: effective and robust detection of Android malware using static analysis along with rotation forest model," *Neurocomputing*, vol. 272, pp. 638–646, 2018.
[8] J. Li, L. Sun, Q. Yan, Z. Li, W. Srisa-an, and H. Ye, "Significant permission identification for machine learning based Android malware detection," *IEEE Transactions on Industrial Informatics*, vol. 15, no. 7, pp. 3216–3225, 2018.
[9] C. Wang, Z. Li, X. Mo, H. Yang, and Y. Zhao, "An Android malware dynamic detection method based on service call co-occurrence matrices," *Annals of Telecommunications*, vol. 72, no. 9-10, pp. 607–615, 2017.
[10] K. Tam, A. Feizollah, N. B. Anuar, R. Salleh, and L. Cavallaro, "The evolution of android malware and Android analysis techniques," *ACM Computing Surveys*, vol. 49, no. 4, pp. 1–49, 2017.
[11] K. O. Elish, X. Shu, D. D. Yao, B. G. Ryder, and X. Jiang, "Profiling user-trigger dependence for Android malware detection," *Computers & Security*, vol. 49, pp. 255–273, 2015.
[12] Y. Liu, K. Guo, X. D. Huang, Z. Zhou, and Y. Zhang, "Detecting android malwares with high-efficient hybrid analyzing methods," *Mobile Information Systems*, vol. 2018, Article ID 1649703, 12 pages, 2018.
[13] X. Wang, D. F. Zhang, X. Su, and W. J. Li, "Mlifdect: Android malware detection based on parallel machine learning and information fusion," *Security and Communication Networks*, vol. 2017, article 6451260, 14 pages, 2017.
[14] A. Feizollah, N. B. Anuar, R. Salleh, and A. W. A. Wahab, "A review on feature selection in mobile malware detection," *Digital Investigation*, vol. 13, pp. 22–37, 2015.
[15] B. B. Rad, M. Masrom, and S. Ibrahim, "Opcodes histogram for classifying metamorphic portable executables malware," in *Proceedings of International Conference on e-Learning and e-Technologies in Education (ICEEE)*, pp. 209–213, IEEE, Lodz, Poland, September 2012.
[16] G. Canfora, A. De Lorenzo, E. Medvet, F. Mercaldo, and C. A. Visaggio, "Effectiveness of opcode ngrams for detection of multi family Android malware," in *Proceedings of 10th International Conference on Availability, Reliability and Security (ARES)*, pp. 333–340, IEEE, Toulouse, France, August 2015.
[17] B. Kang, S. Y. Yerima, K. McLaughlin, and S. Sezer, "N-opcode analysis for Android malware classification and categorization," in *Proceedings of International Conference On Cyber Security And Protection Of Digital Services (Cyber Security)*, pp. 1–7, IEEE, London, UK, June 2016.
[18] P. O'Kane, S. Sezer, and K. McLaughlin, "N-gram density based malware detection," in *Proceedings of World Symposium on Computer Applications & Research (WSCAR)*, pp. 1–6, IEEE, Sousse, Tunisia, January 2014.
[19] B. Wolfe, K. Elish, and D. Yao, "High precision screening for Android malware with dimensionality reduction," in *Proceedings of 13th International Conference on Machine Learning and Applications (ICMLA)*, pp. 21–28, IEEE, Detroit, MI, USA, December 2014.
[20] M. Zhao, T. Zhang, F. B. Ge, and Z. J. Yuan, "RobotDroid: a lightweight malware detection framework on smartphones," *Journal of Networks*, vol. 7, no. 4, pp. 715–722, 2012.
[21] I. Santos, F. Brezo, X. Ugarte-Pedrero, and P. G. Bringas, "Opcode sequences as representation of executables for data-mining-based unknown malware detection," *Information Sciences*, vol. 231, pp. 64–82, 2013.
[22] D. Arp, M. Spreitzenbarth, M. Hübner, H. Gascon, and K. Rieck, "DREBIN: effective and explainable detection of android malware in your pocket," in *Proceedings of 21st Annual Network and Distributed System Security Symposium (NDSS'14)*, pp. 1–15, San Diego, CA, USA, February 2014.
[23] G. Suarez-Tangil, S. K. Dash, M. Ahmadi, J. Kinder, G. Giacinto, and L. Lorenzo Cavallaro, "DroidSieve: fast and accurate classification of obfuscated android malware," in *Proceedings of Seventh ACM on Conference on Data and Application Security and Privacy*, pp. 309–320, ACM, Scottsdale, AZ, USA, March 2017.
[24] Apktool, https://github.com/iBotPeaches/Apktool/.
[25] Dalvik opcodes, http://pallergabor.uw.hu/androidblog/dalvik_opcodes.html.
[26] H. Dong, N.-Q. He, G. Hu, Q. Li, and M. Zhang, "Malware detection method of android application based on simplification instructions," *Journal of China Universities of Posts and Telecommunications*, vol. 21, pp. 94–100, 2014.
[27] P. F. Brown, P. V. Desouza, R. L. Mercer, V. J.D. Pietra, and J. C. Lai, "Class-based n-gram models of natural language," *Computational linguistics*, vol. 18, no. 4, pp. 467–479, 1992.
[28] E. Raff, R. Zak, R. Cox et al., "An investigation of byte n-gram features for malware classification," *Journal of Computer Virology and Hacking Techniques*, vol. 14, no. 1, pp. 1–20, 2018.

[29] W. Wang, S. Gombault, and T. Guyet, "Towards fast detecting intrusions: using key attributes of network traffic," in *Proceedings of Third International Conference on Internet Monitoring and Protection (ICIMP'08)*, pp. 86–91, IEEE, Bucharest, Romania, June-July 2008.

[30] B. J. Frey and D. Dueck, "Clustering by passing messages between data points," *Science*, vol. 315, no. 5814, pp. 972–976, 2007.

[31] T. M. Chen, X. Zhang, S. C. Jin, and O. Kim, "Efficient classification using parallel and scalable compressed model and its application on intrusion detection," *Expert Systems with Applications*, vol. 41, no. 13, pp. 5972–5983, 2014.

[32] J. W. Han, J. Pei, and M. Kamber, *Data Mining: Concepts and Techniques*, Elsevier, New York, NY, USA, 2011.

[33] The Drebin Dataset, http://user.informatik.uni-goettingen.de/~darp/drebin/.

[34] Google Play, https://play.google.com/.

[35] M. Grace, Y. Zhou, Q. Zhang, S. Zou, and X. Jiang, "Riskranker: scalable and accurate zero-day android malware detection," in *Proceedings of 10th International Conference on Mobile Systems, Applications, and Services*, pp. 281–294, ACM, Ambleside, UK, June 2012.

[36] VirusBook, https://www.virusbook.cn/.

5

Asynchronous Channel-Hopping Scheme under Jamming Attacks

Yongchul Kim,[1] Young-Hyun Oh,[2] and Jungho Kang[1]

[1]*Korea Military Academy, Seoul, Republic of Korea*
[2]*Department of Computer Science, North Carolina State University, Raleigh, NC, USA*

Correspondence should be addressed to Yongchul Kim; kyc6454@gmail.com

Academic Editor: Francesco Gringoli

Cognitive radio networks (CRNs) are considered an attractive technology to mitigate inefficiency in the usage of licensed spectrum. CRNs allow the secondary users (SUs) to access the unused licensed spectrum and use a blind rendezvous process to establish communication links between SUs. In particular, quorum-based channel-hopping (CH) schemes have been studied recently to provide guaranteed blind rendezvous in decentralized CRNs without using global time synchronization. However, these schemes remain vulnerable to jamming attacks. In this paper, we first analyze the limitations of quorum-based rendezvous schemes called asynchronous channel hopping (ACH). Then, we introduce a novel sequence sensing jamming attack (SSJA) model in which a sophisticated jammer can dramatically reduce the rendezvous success rates of ACH schemes. In addition, we propose a fast and robust asynchronous rendezvous scheme (FRARS) that can significantly enhance robustness under jamming attacks. Our numerical results demonstrate that the performance of the proposed scheme vastly outperforms the ACH scheme when there are security concerns about a sequence sensing jammer.

1. Introduction

The unlicensed spectrum band has become overcrowded as the demand for smart phones and portable devices has increased, while the licensed spectrum band is always underutilized (e.g., the occupancy of the licensed spectrum is less than 6% [1]). To solve the spectrum scarcity issue in wireless communications, the Federal Communications Commission (FCC) allows the unlicensed users (i.e., secondary users) to make use of the licensed spectrum as long as they do not interfere with the licensed users (i.e., primary users) [2, 3]. This promising paradigm introduces the use of cognitive radio networks (CRNs) as a key technology for opportunistically exploiting the spectrum. In CRNs, secondary users (SUs) must establish a link before communicating with each other. In other words, two SUs should meet on a common channel, which is not occupied by primary users (PUs), to exchange handshake information. This is often referred to as a rendezvous process. The best known approach to solving the rendezvous problem is using a common control channel (CCC) [4, 5]. The advantages in using this approach are the simple implementation and management of a rendezvous. However, maintaining a CCC in CRNs is not feasible since the availability of the spectrum will change dynamically over time. Moreover, this approach may result in a bottleneck, and it potentially creates a single point of failure. Therefore, the rendezvous process in a distributed manner without having any centralized controller or dedicated CCCs is preferable in a more practical scenario. This process is often referred to as a blind rendezvous. To achieve blind rendezvous, a channel-hopping (CH) technique is used as a fundamental strategy to visit the available channels. Most research regarding CH algorithms consider that the time is divided into slots; thus, a successful rendezvous can be achieved when two SUs hop on the same channel in the same timeslot. The number of timeslots that are required until the successful rendezvous after all SUs have begun their CH sequences is defined as time to rendezvous (TTR). According to the geographical locations of the SUs, the number of available channels sensed by each SU might not be the same. If all SUs have the same number of available channels, we call this the symmetric system. All others are asymmetric systems. For reliable performance in CRNs, most CH schemes must guarantee rendezvous in more than one channel within a sequence period, and the TTR value must be bounded and small. To evaluate the performance of proposed CH schemes, two critical metrics are commonly used: the maximum TTR (MTTR) and the expected TTR (ETTR).

There has been a great deal of research directed toward providing guaranteed rendezvous as well as minimizing MTTR. In particular, quorum-based rendezvous algorithms are well-known schemes for providing some level of security under hostile jamming attack scenarios with low time latency. However, these schemes remain vulnerable to sophisticated jamming attacks. In this paper, we present a noble sequence sensing jamming attack (SSJA) in which the jammer can estimate the SU's channel-hopping sequences quickly and begin jamming the rest of the timeslots. We then examine the limitations of those quorum-based rendezvous schemes including frequency quorum rendezvous (FQR) [6] and asynchronous channel-hopping (ACH) [7] schemes under a sophisticated jamming attack. Since FQR requires time synchronization between SUs, we focus more on ACH rendezvous scheme for evaluating the effectiveness of SSJA. Moreover, we introduce our proposed fast and robust asynchronous rendezvous scheme (FRARS) that can reduce TTR as well as increase robustness significantly against jamming attacks.

The contributions of this paper are twofold: first, we analyze the limitations of the well-known quorum-based rendezvous scheme under a sophisticated jamming attack by introducing a novel SSJA model. In the SSJA, a jammer can detect the sender's entire CH sequence of ACH schemes on $N/2$ time frames, where N is the number of available channels. Thus, the rendezvous success rates of ACH schemes will be dramatically decreased under SSJAs. Second, we present our proposed FRARS in order to evaluate the performance under an SSJA. We include a theoretical analysis in addition to extensive numerical analysis under SSJAs. Our numerical results demonstrate that our proposed scheme outperforms others that are recently proposed. The balance of this paper is organized as follows. In Section 2, we review related work in CRNs. In Section 3, we introduce SSJA model and present several CH schemes including FQR and ACH as well as our proposed FRARS. The numerical analysis of our jamming attack for both ACH and FRARS and the results are discussed in Section 4. Section 5 concludes this paper.

2. Related Work

Due to the drawbacks of using a centralized controller or dedicated CCC, many studies have focused on blind rendezvous systems. To solve the blind rendezvous problem in CRNs, a purely random [8] or an improved random algorithm [9] provides a trivial CH algorithm where each SU hops from one channel to another among available channels in a purely random way. That is, when two SUs hop on the same channel at the same time by chance, rendezvous occurs. These schemes can be applied to almost any system but cannot guarantee a bounded TTR between any two SUs.

2.1. Synchronous Rendezvous Algorithms. To achieve guaranteed rendezvous in finite time, several algorithms have been proposed with the assumption of global time synchronization [10–12]. Bahl et al. [10] proposed a link-layer protocol called Slotted Seeded Channel Hopping (SSCH) that increases the capacity of an IEEE 802.11 network by utilizing frequency diversity. Krishnamurthy et al. [11] proposed a two-phase autoconfiguration algorithm that enables SUs to dynamically compute the globally common channel set in a distributed manner. Bian et al. [12] introduced quorum-based two CH schemes, namely, M-QCH and L-QCH: the first design ensures ETTR by minimizing the MTTR and the second design guarantees the even distribution of the rendezvous points in terms of both time and frequency. However, these synchronous systems may not be feasible in certain types of networks, for example, ad hoc networks. Moreover, under the assumption of synchronization, the impact of a jamming attack can be significant.

2.2. Asynchronous Rendezvous Algorithms. To overcome these challenges, many asynchronous CH algorithms (i.e., one that does not require clock synchronization) have been proposed recently in the literature [13–21]. DaSilva and Guerreiro [13] proposed a sequence-based rendezvous and was later referred to as the generated orthogonal sequence (GOS) algorithm in which all SUs use the identical predefined CH sequences. This algorithm is used with interspersed permutation channels to guarantee rendezvous even when SUs are not synchronized. Theis et al. [14] showed better performance than that of GOS in terms of ETTR by presenting modular clock (MC) and modified modular clock (MMC) algorithms. In the MC system, the SUs can rendezvous any time although they independently generate their CH sequences by using prime number and rate (forward-hop). Lin et al. [15] proposed a jump-stay (JS) algorithm in which each SU has a jump and stay pattern to find a common channel. Intuitively, SUs jump on available channels during the jump pattern and stay on a specific channel during the stay pattern. The enhanced jump-stay (EJS) algorithm [16] was also proposed by the same authors to improve the MTTR and ETTR performances for asymmetric systems. Liu et al. [17] introduced the ring walk (RW) algorithm to guarantee asynchronous rendezvous by using the concept of velocity. The SUs in this scheme walk on the ring by visiting vertices of channels with different velocities. The higher velocity SU will eventually catch the lower velocity SU. Chuang et al. [18] presents a new alternate HOP-and-WAIT channel-hopping method (E-AHW) to minimize the time to rendezvous (TTR) than existing methods. In this method, each SU has a unique alternating sequence of HOP and WAIT. Most recently, Salehkaleybar et al. [22] proposed periodic jump rendezvous (PJR) and modified version of PJR (mPJR) algorithms for role-based and nonrole-based cases, respectively, in order to reduce TTR. Pu et al. [23] studied the dynamic rendezvous problem in CRNs where the status of the licensed channels varies over time by introducing the available channel probabilities. This work aims to propose more realistic models under adversaries as a future work. Thus, the vulnerability against a sophisticated jamming attack is not addressed. The deterministic rendezvous sequence (DRSEQ) and channel rendezvous sequence (CRSEQ) algorithms proposed in [19, 20] provide fast asynchronous rendezvous under symmetric and asymmetric models, respectively. The upper bounds of MTTR of those algorithms are significantly small as shown in [24]. Yadav and Misra [21] develop an algorithm that generates a deterministic CH sequence, which guarantees rendezvous even faster than the DRSEQ algorithm. However, they are not applicable to jamming attack scenarios due to deterministic CH sequences.

2.3. Jamming Attack Scenarios. Most of the aforementioned algorithms focus on minimizing MTTR without using time synchronization in either symmetric or asymmetric scenarios. None of them are considered to be robust in jamming attack environments. In wireless communications, an adversary (enemy and jammer) is a malicious entity that can easily disrupt legitimate communications by intentionally injecting noise-like signals (or dummy packets) into the wireless medium. To alleviate this vulnerability problem in CRNs, quorum-based rendezvous schemes are proposed in [6, 7]. The FQR algorithm [6] exploits a quorum system where each SU independently constructs a random sequence by scrambling the sender's frequency quorum sequences for every frame as will be addressed in the next section. This makes jamming difficult and inefficient. An enhanced version of FQR, advanced FQR (AFQR) algorithm [6], adds more timeslots mapping random frequency channels at arbitrary positions within each frame to improve the rendezvous probability while it might degrade time overhead with the increased number of timeslots in a period. However, FQR only supports synchronous systems where any two SUs must start their sequences at the same time in order to rendezvous. Abdel-Rahman and Krunz [25] studied the rendezvous problem in the presence of an insider attack using a game-theoretic framework. This work showed that the rendezvous performance improves if the receiver and jammer are time synchronized and both have a common guess about the transmitter's strategy. However, the jammer model in this work is not a smart jammer but an insider jammer. Thus, the vulnerability against a sophisticated jamming attack is not addressed. Bian and Park [7] proposed a quorum-based ACH algorithm to ensure that the TTR is upper bounded even if the SU's clocks are asynchronous, and it maximizes the rendezvous probability between any pair of SUs by enabling rendezvous on every available channel. The sender and the receiver in an ACH algorithm independently generate their own sequences and rendezvous within a favorable amount of time compared to other schemes. Nevertheless, the ACH algorithm is significantly vulnerable to a sophisticated jamming attack. A survey paper [26] on jamming and antijamming techniques shows the classification of jammers. As a reactive channel-hopping jammer, we introduce an SSJA model in this paper to show how effectively it attacks the ACH system by adding more sophisticated capabilities such as estimating the SU's CH sequence within a short time. To overcome this vulnerability against SSJA, we proposed a FRARS algorithm that employs randomized permutation in every period. Due to the random features of FRARS, it is unfeasible for the SSJA to estimate the CH sequences. Thus, the effectiveness of the SSJA is negligible for the FRARS. Our proposed scheme is comparable to the ACH algorithm since both the sender and the receiver in FRARS independently generate their own sequences like those in ACH. The performance results of FRARS as compared with ACH will be addressed in Section 4.

3. Channel-Hopping Schemes

In this section, we present two well-known quorum-based rendezvous schemes, FQR and ACH algorithms, with a brief definition of a quorum system. Then, we introduce the SSJA model to show how effectively it attacks quorum-based rendezvous schemes. We also present our FRARS algorithm to enhance robustness against jamming attack and compare the effectiveness of the SSJA on ACH and FRARS.

3.1. The Quorum System. A quorum system has two fundamental properties, that is, the intersection property and the rotation closure property [27]. All quorum systems hold the intersection property, but the rotation closure property may not be present in some cases. The FQR exploits a cyclic quorum system [28] to design a set of hopping sequences. We provide those definitions in this subsection, and we borrow all the terminologies defined in [6, 7, 27, 28].

Definition 1. Given a finite universal set $U = Z_N = \{0, 1, \ldots, N-1\}$ of N elements, a quorum system Q under U is a collection of nonempty subsets of U, which satisfies the intersection property:

$$p \cap q \neq \varnothing, \forall p, q \in Q. \tag{1}$$

Each p or $q \in Q$ is called a quorum, and Z_N denotes the set of nonnegative integers less than n. Given a nonnegative integer k and a quorum q in a quorum system Q under the universal set U, we define

$$\text{rotate}(q, k) = \{(i+k) \bmod n \mid i \in q\}. \tag{2}$$

Definition 2. A quorum system Q under U has the rotation closure property if the following holds:

$$\forall p, q \in Q, \forall k \in [0, \ldots, N-1], \text{rotate}(p, k) \cap q \neq \varnothing. \tag{3}$$

Definition 3. A set $D = \{a_1, \ldots, a_k\} \subset Z_N$, $a_i \in \{0, \ldots, N-1\}$, and $k \leq N$ is called a cyclic (N, k) difference set if for every $d \not\equiv 0 \bmod N$ there exists at least one ordered pair (a_i, a_j), where $a_i, a_j \in D$, such that $a_i - a_j \equiv d \bmod N$, that is, d is a difference value between two elements of D.

Definition 4. Given a (N, k) difference set $D = \{a_1, \ldots, a_k\} \subset Z_N$, a cyclic quorum system constructed by D is $Q = \{C_0, \ldots, C_{N-1}\}$, where $C_i = \{a_1 + i, a_2 + i, \ldots, a_k + i\} \bmod N, i = 0, \ldots, N-1$. For a (7, 3) difference set $D = \{1, 2, 4\} \subset Z_7$, for example, the cyclic quorum system is $Q = \{C_0, \ldots, C_6\}$, where $C_0 = \{1, 2, 4\}, C_1 = \{2, 3, 5\}, C_2 = \{3, 4, 6\}$, $C_3 = \{4, 5, 0\}, C_4 = \{5, 6, 1\}, C_5 = \{6, 0, 2\}$, and $C_6 = \{0, 1, 3\}$.

3.2. FQR Base System. The authors in [25] apply the abovementioned properties of quorum systems to design a FQR system. The frequency-hopping sequence of an SU in FQR is constructed by assigning frequencies to t timeslots in one period X, which is denoted by $X = \{x_0, \ldots, x_{t-1}\} = \{(0, f_0), \ldots, (t-1, f_{t-1})\}$, where $x_i \in X$ contains a tuple of *(timeslot index, frequency index)* and $f_i \in \{0, \ldots, N-1\}$ represents the frequency index at timeslot t in a period. In FQR, an SU generates two different hopping sequences: a sending sequence and a receiving sequence. If an SU has data to transmit, it follows a sending sequence, otherwise, a receiving sequence. For example, consider that a sender and a receiver use a (7, 3) different set, that is, $N = 7$ and $k = 3$. And, a cyclic quorum system $Q = \{C_0, \ldots, C_6\}$ is constructed from $D = \{1, 2, 4\}$. Thus, the sender and the receiver can select a random number and obtain a quorum $S = C_i$ and $R = C_j$, respectively (e.g., $S = C_0 = \{1, 2, 4\}, R = C_4 = \{5, 6, 1\}$). A sender node constructs a sending sequence X by assigning a frequency index to the timeslot i using $C_0 = \{c_0, c_1, c_2\} = \{1, 2, 4\} : x_i = (i, c_m)$, where $0 \le i \le k^2 - 1$ and $m = i \bmod k$. Then, it obtains $X = \{(0, 1), (1, 2), (2, 4), (3, 1), (4, 2), (5, 4), (6, 1), (7, 2), (7, 4)\}$. A receiver node constructs a receiving sequence Y by assigning frequency index to the timeslot i using $C_4 = \{c_0, c_1, c_2\} = \{5, 6, 1\} : y_i = (i, c_n)$, where $0 \le i \le k^2 - 1$ and $n = (i - (i \bmod k))/k$. Then, it obtains $Y = \{(0, 5), (1, 5), (2, 5), (3, 6), (4, 6), (5, 6), (6, 1), (7, 1), (7, 1)\}$. Additionally, permuting frequency indexes in each frame of X is done as a last step for constructing FQR sequences.

Figure 1(a) shows FQR sequences of both sender and receiver, respectively. It also shows that the sender and receiver rendezvous on frequency 1 at time T_6. The upper bound of TTR of FQR system is only k^2 timeslots, which is approximately, equal to N. In AFQR, one more timeslot is added into each frame of the X sequence, and it is assigned a random frequency f_i such that $f_i \notin C_k$, where C_k is a quorum selected by the sender. That is, the selected frequency f_i is inserted into a random timeslot in each frame as shown in Figure 1(c). Although this will increase the length of the time period (i.e., $k(k+1)$ timeslots), the expected number of rendezvous within $k(k+1)$ timeslots also increases to $1 + (k - 1/N - k)$. The selected frequency index 5 is added to the sequence in Figure 1(c); thus, one more rendezvous on frequency 5 at time T_0 is provided. The techniques used in FQR and AFQR such as scrambling the hopping sequence and inserting additional timeslots make it difficult for a jammer to predict the hopping sequences. However, these schemes are not appropriate in an emergency or tactical scenario where time synchronization between randomly meeting nodes cannot be assumed. Figures 1(b) and 1(d) show that a clock shift nullifies the guaranteed upper bound of rendezvous.

3.3. ACH Base System. In ACH system, the TTR between any pair of CH sequences is upper bounded without requiring clock synchronization by exploiting two properties of the array-based quorum systems: the intersection property and the rotation closure property. Let u denote the CH sequence of an SU and T denote a period of the CH sequence: $u = \{u_0, u_1, \ldots, u_i, \ldots, u_{T-1}\}$, where $u_i \in [0, N-1]$ is the channel index of sequence u in the ith timeslot of a CH period. The SUs in ACH generate their CH sequences by using an $N \times N$ array-based quorum system and assigning channel index values to quorums' columns or spans. The sender and receiver construct different CH sequences for rendezvous in ACH. The sender assigns N channel indexes to the N columns of an $N \times N$ array, $S[\cdot][\cdot]$, such that each column has different channel indexes and all array elements in the same column are assigned the same channel index. If we denote h_j the channel index value assigned to the jth column, where $j \in [0, N-1]$, the sequence of the sender will be constructed by the following way: $u_{i \cdot N+j} = h_j$, where $i, j \in [0, N-1]$. The receiver assigns N channel indexes to the N spans of an $N \times N$ array, $R[\cdot][\cdot]$, such that each span has a different channel index and all array elements in the same span are assigned the same channel index. Let s_k denote the kth span and h_k denote the channel index value assigned to the s_k, where $k \in [0, N-1]$. The sequence of the receiver will be constructed in the following way: $v_{i \cdot N+j} = h_k$ if $R[i][j] \in s_k$, where $i, j \in [0, N-1]$. Therefore, the sender's sequence u and receiver's sequence v are generated by using the two $N \times N$ arrays, and the period of the sequences will be N^2. Figure 2 shows an example of the ACH system when $N = 3$.

Without loss of generality, the sender's sequence u starts first and then i timeslots later, the receiver's sequence v starts. The operation rotate (u, i) yields a new CH sequence u^*, where all the elements in the same column are assigned the same channel index. Thus, u^* and v have N distinct rendezvous channels within a sequence period $T = N^2$. It is obvious that the asynchronous rendezvous problem is solved in ACH system since the sender and receiver CH sequences satisfy the rotation closure property.

3.4. The Sequence Sensing Jamming Attacks. To evaluate the performance of the ACH system under a sophisticated jamming attack, we present a noble sequence sensing jamming attack (SSJA) model in which a jammer can estimate the SU's CH sequences. Since one period of the sender's CH sequence consists of N repeated N timeslots ($T = N^2$), a jammer can effectively nullify the whole ACH system by estimating only N channels (one subperiod). We call this subperiod a frame, that is, one period of the ACH sequence consists of N frames. We assume that the number of available N channels does not change during a sequence period T to clearly illustrate the characteristics of the SSJA. We also assume that the jammer resides in the network

	One time period								
	Frame 1			Frame 2			Frame 3		
Time slot	T_0	T_1	T_2	T_3	T_4	T_5	T_6	T_7	T_8
X (quorum C_0)	2	1	4	4	1	2	1	4	2
Y (quorum C_4)	5	5	5	6	6	6	1	1	1

(a)

Time slot	T_0	T_1	T_2	T_3	T_4	T_5	T_6	T_7	T_8	T_0
Sender X	2	1	4	4	1	2	1	4	2	
Receiver X		5	5	5	6	6	6	1	1	1

(b)

	One time period											
	Frame 1				Frame 2				Frame 3			
Time slot	T_0	T_1	T_2	T_3	T_4	T_5	T_6	T_7	T_8	T_9	T_{10}	T_{11}
X (quorum C_0)	5	2	1	4	4	5	1	2	1	4	2	5
Y (quorum C_4)	5	5	5	5	6	6	6	6	1	1	1	1

(c)

Time slot	T_0	T_1	T_2	T_3	T_4	T_5	T_6	T_7	T_8	T_9	T_{10}	T_{11}	T_{12}
Sender X	5	2	1	4	4	5	1	2	1	4	2	5	
Receiver X		5	5	5	5	6	6	6	6	1	1	1	1

(d)

FIGURE 1: (a) An illustration of rendezvous in FQR, (b) a simple clock shift can nullify FQR rendezvous, (c) an illustration of rendezvous in AFQR, and (d) a simple clock shift can nullify AFQR rendezvous.

before the communication starts, that is, the jammer waiting for the sender starts its sequence and the sender does not know of the existence of the jammer. We also assume that the jammer has capabilities similar to those of the normal user with one transmitting channel and one listening channel. Thus, the jammer knows the number of N available channels whenever it changes due to the occupation of PUs. The jammer randomly selects one distinct listening channel and waits until the sender hops to that channel. As soon as the jammer detects one channel, it chooses another channel (not previously tried) for detecting the next channel. Therefore, on average, the jammer can detect two channels within a frame (N timeslots) of the sequence. For all detected channels, the jammer immediately jams those channels for the rest of the ACH sequence. Moreover, the average time for the jammer to compute the entire ACH sequence is $N/2$ frames. When $N-1$ channels are detected among N timeslots, the last channel is automatically detected. In other words, after an average of $(N/2)\times N$ timeslots, the jammer can completely jam the remaining channels. Therefore, our SSJA model can significantly decrease the rendezvous probability of the ACH scheme.

3.5. Proposed System. In this subsection, we introduce our proposed FRARS system that can enhance robustness significantly against jamming attacks by generating new CH sequences in every period. The sender and receiver construct different CH sequences for rendezvous in a way similar to ACH. When an SU has data to send, it follows a sending CH sequence. Otherwise, it follows a receiving CH sequence. The length of an SU's CH sequence period is only $2N-1$, thus $u=\{u_0,u_1,\ldots,u_i,\ldots,u_{2N-2}\}$, where $u_i\in[0,N-1]$ is the channel index of the sequence u in the ith timeslot of a CH period. Unlike the ACH scheme, the CH sequence will change in every period. Let I denote the slot number and u and v denote two CH sequences in FRARS. When there is k slot time difference between u and v as shown in Figure 3, we say that two SUs rendezvous in the ith timeslot on channel c when $u_{\{i=I \bmod 2N-1\}}=v_{\{j=(I-k) \bmod 2N-1\}}=c$, where $c\in[0,N-1]$.

The sending CH sequence employs a random permutation channel in the first N timeslots. And then, the reverse order of the random permutation excluding the last channel will be used for the next $N-1$ timeslots. The first N timeslots and the next $N-1$ timeslots are referred to as *permutation*

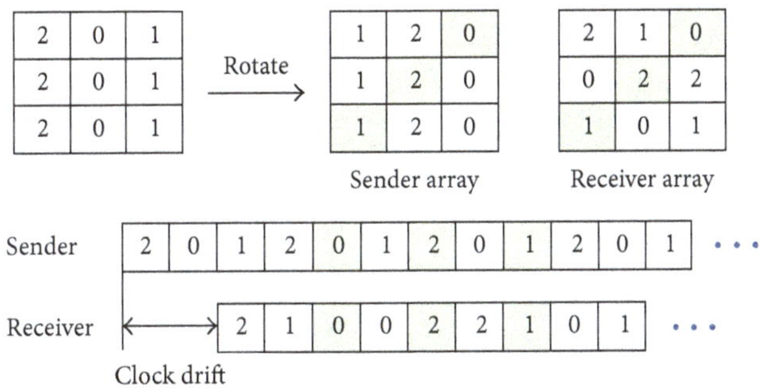

FIGURE 2: An example of ACH system when $N = 3$.

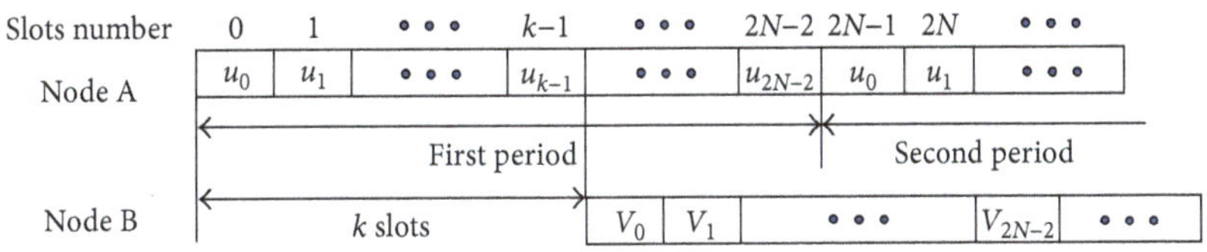

FIGURE 3: FRARS system model structure.

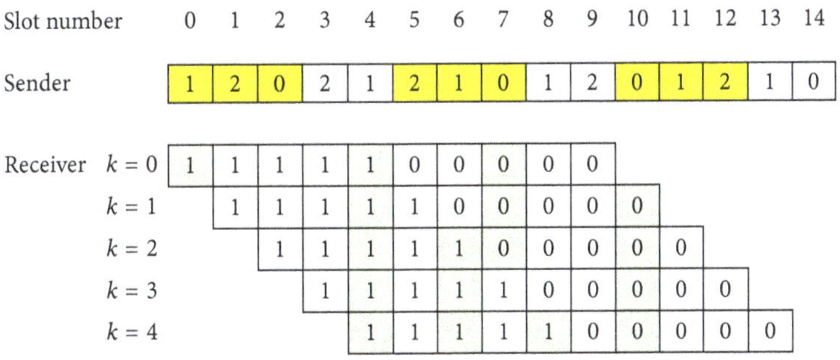

FIGURE 4: Illustration of FRARS when the number of available channel is 3.

parts and *reversed repetition parts*, respectively. If we represent the permutation part as $R = \{r_0, r_1, \ldots, r_i, \ldots, r_{N-1}\}$, where $r_i \in [0, N-1]$, the generation of sending CH sequence with N available channels can be expressed as

$$u_i = \begin{cases} r_i & \text{for } 0 \leq i \leq N-1, \\ r_{2N-2-i} & \text{for } N \leq i \leq 2N-2. \end{cases} \quad (4)$$

Figure 4 shows an illustration of rendezvous in FRARS when the number of available channels is 3. The sending sequence shows three periods, and the receiving sequence shows two periods. The shaded timeslots in the sending sequence represent the permutation parts, and the rest are the reversed repetition parts. For example, the sender visits channel 2 in slot number 3, which is equal to the one in slot number 1, and the sender visits channel 1 in slot number 4, which is equal to the one in slot number 0. The receiving sequence selects one random channel from N and stays on that channel during one period. The shaded timeslots in the receiving sequence indicate rendezvous. For all k, FRARS guarantees that any pair of sender and receiver nodes rendezvous within at most $2N - 1$ slots. Since the sender's sequence is generated by using a randomized permutation of N channels for every period, it is not feasible for the sophisticated jammer to estimate when the sender and receiver might rendezvous.

4. Performance Evaluation

In this paper, we implemented the sequence sensing jamming attack (SSJA) model to evaluate its effectiveness against the asynchronous channel-hopping (ACH) [7] schemes. For the sake of simplicity, we did not consider multiple SU scenario, since it is more difficult to analyze the performance of rendezvous algorithms due to collision problem between SUs. Therefore, we focused on the symmetric scenario of ACH scheme in which two SUs have the same number of available channels. The two SUs do not know each other's existence, and they are not time synchronized. When they want to communicate with each other (i.e., one of them has data to send), they go through a rendezvous algorithm. During a rendezvous process, each SU has to hop every available channel according to their CH sequence until they successfully rendezvous on the same channel. In our implementation, the sender starts the communication first and the receiver can start at any timeslot within the N^2

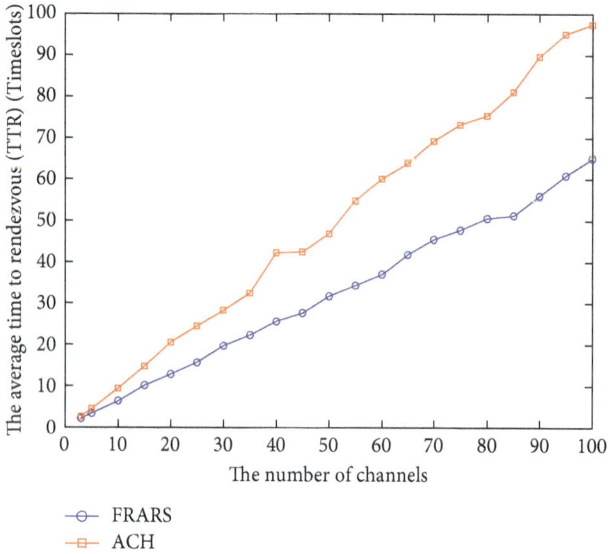

FIGURE 5: The average time to rendezvous (TTR) for ACH and FRARS when the available channels are $[3, \ldots, 100]$ with no jamming attacks.

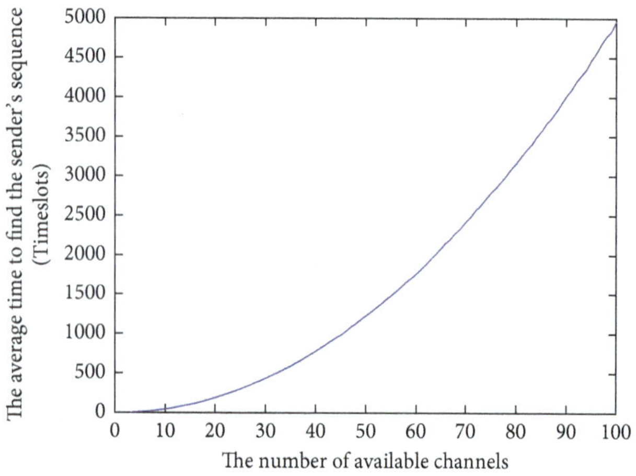

FIGURE 6: The average time to find the entire sender's CH sequence for ACH.

timeslots of the sender because there is no time synchronization. We also make a typical assumption that the SSJA jammer resides in the network and is listening on one channel before the communication starts. This is a normal situation since the jammer is trying to disrupt the communication and is invisible to the sender. Then, we implement our FRARS scheme to demonstrate that it is more robust against jamming attacks compared to the ACH scheme. In the jamming attack for the FRARS scheme, a jammer randomly selects one channel for each timeslot and jams the channel until the MTTR. Through our extensive numerical analysis under SSJA, we demonstrate that our proposed FRARS scheme outperforms the quorum-based state-of-the-art ACH scheme.

First, we compare the average time to rendezvous (TTR) for both ACH and FRARS schemes. Figure 5 gives the average TTR for ACH and FRARS schemes where the number of available channels N varies from 3 to 100 with no jamming attacks. We repeated the process 1000 times for each available channel N and calculated the average TTR for them. This figure shows that the average TTR for both ACH and FRARS schemes increases steadily as the number of available channels increases. Our numerical results for the ACH's TTR are close to its theoretical TTR $O(N)$ discussed in [7]. Overall, the average TTR for the FRARS system is lower than ACH system's TTR for all N.

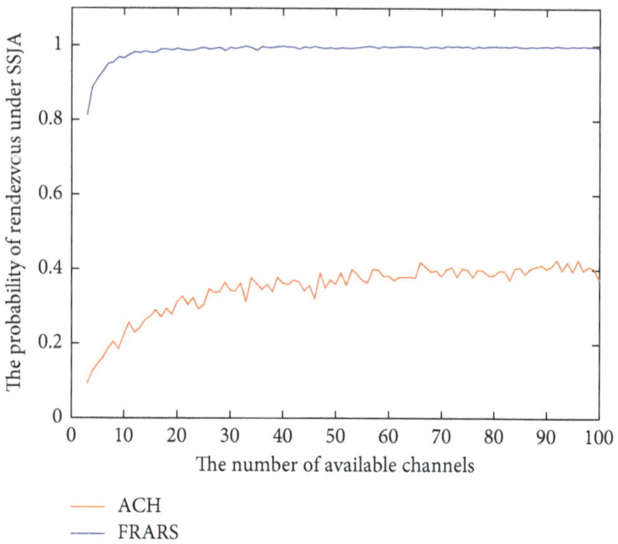

FIGURE 7: The probability of rendezvous for both ACH and FRARS under SSJA.

Second, we implemented the expected time to find the entire sender's CH sequence for ACH where the number of available channels M varies from 3 to 100. We repeated the process 1000 times for each number of available channels and calculated the average time for finding the sender's CH sequence. Our numerical results show that the average time for finding the entire sender's CH sequence is close to an average of $\lfloor (N/2) \times N \rfloor$ timeslots as we discussed in Section 3.4. This means that there are no rendezvous after $\lfloor (N/2) \times N \rfloor$ timeslots so that the rendezvous success rates of ACH schemes will be dramatically decreased under our SSJA attack as we can see the average time to find the entire sender's CH sequence for ACH in (Figure 6).

Lastly, we compare the rendezvous probability for the ACH and FRARS schemes under SSJA. Figure 7 shows the probability of rendezvous for both ACH and FRARS systems. The rendezvous probability for the ACH scheme is less than 40% for almost all available channels under SSJA. Since there is no rendezvous after an average $\lfloor (N/2) \times N \rfloor$ timeslots, the rendezvous probability of the ACH scheme is dramatically decreased under the SSJA. However, the rendezvous probability for the FRARS scheme is almost steady and more than 80% for all the available channels N under jamming attacks.

5. Conclusion

In this paper, we examined the drawbacks of quorum-based state-of-the-art schemes such as FQR and ACH. In particular, the SSJA model is used to show how effectively it attacks the ACH system. The sender's CH sequence in ACH is fully detected by the jammer within an average $(N/2) \times N$ timeslots, after that the jammer can completely jam the remaining channels. To remedy this jamming problem, we proposed a new FRARS algorithm that can significantly enhance robustness by allowing randomized permutation in every period of the sender's CH sequence. Our numerical results demonstrate that the rendezvous probability of the ACH under SSJA is dramatically decreased from 100% to less than 40%. On the other hand, the rendezvous probability of our FRARS is almost steady under jamming attacks and close to 100% as the number of available channels increases. Therefore, our FRARS vastly outperforms other recently proposed schemes under a sophisticated jamming attack. As a future work, we will continue to analyze the performance of the proposed scheme under multiple SUs and PUs by taking into account collision problems between SUs.

Acknowledgments

This work was supported by Hwarang-Dae Research Institute of Korea Military Academy under Grant no. 20130401.

References

[1] FCC, *Notice of Proposed Rulemaking: In the Matter of Unlicensed Operation in the TV Broadcast Bands, FCC 04-113*, FCC, Washington, DC, USA, 2004.

[2] I. F. Akyildiz, W.-Y. Lee, M. C. Vuran, and S. Mohanty, "Next generation/dynamic spectrum access/cognitive radio wireless networks: a survey," *Computer Networks*, vol. 50, no. 13, pp. 2127–2159, 2006.

[3] S. Bhattarai, J. M. J. Park, B. Gao, K. Bian, and W. Lehr, "An overview of dynamic spectrum sharing: ongoing initiatives, challenges, and a roadmap for future research," *IEEE Transactions on Cognitive Communications and Networking*, vol. 2, no. 2, pp. 110–128, 2016.

[4] Z. Htike, C. S. Hong, and S. Lee, "The life cycle of the rendezvous problem of cognitive radio ad hoc networks," *Journal of Computing Science and Engineering*, vol. 7, no. 2, pp. 81–88, 2013.

[5] C. Cordeiro, K. Challapali, D. Birru, and S. Shankar, "IEEE 802.22: the first worldwide wireless standard based on cognitive radios," *Journal of Communications*, vol. 1, no. 1, pp. 38–47, 2006.

[6] E.-K. Lee, S. Y. Oh, and M. Gerla, "Frequency quorum rendezvous for fast and resilient key establishment under jamming attack," *ACM SIGMOBILE-Mobile Computing and Communications Review*, vol. 14, no. 4, pp. 1–3, 2010.

[7] K. Bian and J. M. Park, "Asynchronous channel hopping for establishing rendezvous in cognitive radio networks," in *2011 Proceedings IEEE INFOCOM*, pp. 236–240, Shanghai, China, April 2011.

[8] M. D. Silvius, F. Ge, A. Young, A. B. MacKenzie, and C. W. Bostian, "Smart radio: spectrum access for first responders," in *Proceedings of the Wireless Sensing and Processing III*, pp. 698008–698012, Orlando, FL, USA, 2008.

[9] C. Cormio and K. R. Chowdhury, "Common control channel design for cognitive radio wireless ad hoc networks using adaptive frequency hopping," *Ad Hoc Networks*, vol. 8, no. 4, pp. 430–438, 2010.

[10] P. Bahl, R. Chandra, and J. Dunagan, "Ssch: slotted seeded channel hopping for capacity improvement in ieee 802.11 adhoc wireless networks," in *Proceedings of the 10th Annual International Conference on Mobile Computing and Networking, ser. MobiCom'04*, pp. 216–230, ACM, New York, NY, USA, 2004.

[11] S. Krishnamurthy, M. Thoppian, S. Kuppa et al., "Time-efficient distributed layer-2 auto-configuration for cognitive radio networks," *Computer Networks*, vol. 52, no. 4, pp. 831–849, 2008.

[12] K. Bian, J.-M. Park, and R. Chen, "A quorum-based framework for establishing control channels in dynamic spectrum access networks," in *Proceedings of the 15th Annual International Conference on Mobile Computing and Networking, ser. MobiCom'09*, pp. 25–36, ACM, New York, NY, USA, 2009.

[13] L. A. DaSilva and I. Guerreiro, "Sequence-based rendezvous for dynamic spectrum access," in *2008 3rd IEEE Symposium on New Frontiers in Dynamic Spectrum Access Networks*, pp. 1–7, Chicago, IL, USA, October 2008.

[14] N. C. Theis, R. W. Thomas, and L. A. DaSilva, "Rendezvous for cognitive radios," *IEEE Transactions on Mobile Computing*, vol. 10, no. 2, pp. 216–227, 2011.

[15] Z. Lin, H. Liu, X. Chu, and Y. W. Leung, "Jump-stay based channel-hopping algorithm with guaranteed rendezvous for cognitive radio networks," in *2011 Proceedings IEEE INFOCOM*, pp. 2444–2452, Shanghai, China, April 2011.

[16] Z. Lin, H. Liu, X. Chu, and Y.-W. Leung, "Enhanced jump-stay rendezvous algorithm for cognitive radio networks," *IEEE Communications Letters*, vol. 17, no. 9, pp. 1742–1745, 2013.

[17] H. Liu, Z. Lin, X. Chu, and Y. W. Leung, "Ring-walk based channel-hopping algorithms with guaranteed rendezvous for cognitive radio networks," in *Green Computing and Communications (GreenCom), 2010 IEEE/ACM Int'l Conference on Int'l Conference on Cyber, Physical and Social Computing (CPSCom)*, pp. 755–760, Hangzhou, China, December 2010.

[18] I. H. Chuang, H. Y. Wu, and Y. H. Kuo, "A fast blind rendezvous method by alternate hop-and-wait channel hopping in cognitive radio networks," *IEEE Transactions on Mobile Computing*, vol. 13, no. 10, pp. 2171–2184, 2014.

[19] D. Yang, J. Shin, and C. Kim, "Deterministic rendezvous scheme in multichannel access networks," *Electronics Letters*, vol. 46, no. 20, pp. 1402–1404, 2010.

[20] J. Shin, D. Yang, and C. Kim, "A channel rendezvous scheme for cognitive radio networks," *IEEE Communications Letters*, vol. 14, no. 10, pp. 954–956, 2010.

[21] R. N. Yadav and R. Misra, "Periodic channel-hopping sequence for rendezvous in cognitive radio networks," in *2015 International Conference on Advances in Computing, Communications and Informatics (ICACCI)*, pp. 1787–1792, Kochi, India, August 2015.

[22] S. Salehkaleybar and M. R. Pakravan, "A periodic jump-based rendezvous algorithm in cognitive radio networks," *Computer Communications*, vol. 79, pp. 66–77, 2016.

[23] H. Pu, Z. Gu, X. Lin, Q. S. Hua, and H. Jin, "Dynamic rendezvous algorithms for cognitive radio networks," in *Proceedings of the 2016 IEEE International Conference on Communications (ICC)*, pp. 1–6, Kuala Lumpur, Malaysia, May 2016.

[24] H. Liu, Z. Lin, X. Chu, and Y. W. Leung, "Taxonomy and challenges of rendezvous algorithms in cognitive radio networks," in *Proceedings of the 2012 International Conference on Computing, Networking and Communications (ICNC)*, pp. 645–649, Maui, HI, USA, January–February 2012.

[25] M. J. Abdel-Rahman and M. Krunz, "Game-theoretic quorum-based frequency hopping for anti-jamming rendezvous in dsa

networks," in *Proceedings of the 2014 IEEE International Symposium on Dynamic Spectrum Access Networks (DYSPAN)*, pp. 248–258, Tysons Corner, VA, USA, April 2014.
[26] K. Grover, A. Lim, and Q. Yang, "Jamming and anti-jamming techniques in wireless networks: a survey," *International Journal of Ad Hoc and Ubiquitous Computing*, vol. 17, no. 4, pp. 197–215, 2014.
[27] J.-R. Jing, Y.-C. Tseng, C.-S. Hsu, and T.-H. Lai, "Quorum-based asynchronous power-saving protocols for ieee 802.11 ad hoc networks," in *Proceedings of the 2003 International Conference on Parallel Processing*, pp. 257–264, Montreal, Canada, October 2003.
[28] W.-S. Luk and T.-T. Wong, "Two new quorum based algorithms for distributed mutual exclusion," in *Proceedings of 17th International Conference on Distributed Computing Systems*, pp. 100–106, Baltimore, MD, USA, May 1997.

6

Demystifying Authentication Concepts in Smartphones: Ways and Types to Secure Access

Sandeep Gupta, Attaullah Buriro, and Bruno Crispo

Department of Information Engineering and Computer Science, University of Trento, Trento, Italy

Correspondence should be addressed to Sandeep Gupta; sandeep.gupta@unitn.it

Academic Editor: Fabio Gasparetti

Smartphones are the most popular and widespread personal devices. Apart from their conventional use, that is, calling and texting, they have also been used to perform multiple security sensitive activities, such as online banking and shopping, social networking, taking pictures, and e-mailing. On a positive side, smartphones have improved the quality of life by providing multiple services that users desire, for example, anytime-anywhere computing. However, on the other side, they also pose security and privacy threats to the users' stored data. User authentication is the first line of defense to prevent unauthorized access to the smartphone. Several authentication schemes have been proposed over the years; however, their presentation might be perplexing to the new researchers to this domain, under the shade of several buzzwords, for example, active, continuous, implicit, static, and transparent, being introduced in academic papers without comprehensive description. Moreover, most of the reported authentication solutions were evaluated mainly in terms of accuracy, overlooking a very important aspect—the usability. This paper surveys various types and ways of authentication, designed and developed primarily to secure the access to smartphones and attempts to clarify correlated buzzwords, with the motivation to assist new researchers in understanding the gist behind those concepts. We also present the assessment of existing user authentication schemes exhibiting their security and usability issues.

1. Introduction

The birth of smartphones can be traced back to 1973, when Motorola launched their first phone—the Dynatac 8000X [1]. In the last 40 years, mobile device manufacturers have invested heavily in the innovation of mobile phones, transforming a device invented merely for calling and short text messaging into the personal, portable and powerful device of nowadays, equipped with many advanced software and hardware features.

Smartphones, undoubtedly, bring rich digital experiences to the users by offering personalized services, for example, chatting, e-mailing, GPS-navigation, net banking, online shopping, social networking, and video conferencing. Most of these services collect and store a large amount of the user's personal data on the device; thus, any unauthorized access to the user's data could have unfavorable consequences. Hence, it becomes extremely important to prevent any unauthorized access to the smartphone. Typically, access to modern smartphones is secured by enabling different authentication solutions, such as PINs/passwords, face recognition, and fingerprint.

By and large multiple terminologies in the field of authentication are being used by researchers not always with clear definitions, which is obviously disconcerting for students and new researchers. Triandopoulos et al. [2] described one-time authentication as "one-time passcodes" or "one-time password" (OTP) as the second authentication factor, although OTP is a more widely accepted term. Crouse et al. [3] described continuous authentication as a periodical composition of one-shot authentication. However, Feng et al. [4] mentioned periodic authentication as equivalent to automatic logouts due to user's inactivity. Patel et al. [5] considered continuous authentication and active authentication systems as the same. Similarly, Dutt et al. [6] suggested the use of transparent modalities in conjunction with explicit authentication methods, such as passwords, PINs, or secret patterns for authenticating users, whereas the study by De Luca et al. [7] considered the use of a transparent modality with or without other schemes and termed it *implicit authentication*. That modality could be used as standalone or to complement the explicit authentication schemes to enhance their usability [8, 9]. More specifically the concept of *transparent authentication* is explained as implicitly fingerprinting the user's device interaction logs to authenticate the user [10].

Causey [11] considered *risk-based authentication* similar to an *adaptive authentication* scheme. Traore et al. [12] described *risk based authentication* on the basis of contextual and historical information, extracted from their activities, to build users' risk profiles, for making later the authentication and authorization decisions. Ayed [13] patented the idea for *adaptive authentication* in mobile phones by specifying that *adaptive authentication* uses different authentication methods and different data protection methods depending on the user's location, availability of the network, and the importance of the data. It is pretty much evident from the above discussion that these definitions are correlated, but there is need to relate them to each other by trying to provide consistent definitions for all these terms.

We start this paper by explaining the prevalent ways to authenticate humans along with different types of authentication mechanisms, in the context of smartphones. Then, we try to homogenize different terminologies used in the context of user authentication with the vision that it will benefit the new researchers in understanding existing approaches. Our contribution can help new researchers to get acquainted with different user authentication concepts along with the assessment of their solutions on the basis of modalities, usability, and security.

The rest of work is organized as follows: Section 2 presents the different ways and types of authentication mechanisms. Ways refer to the common factors used to authenticate humans, while types refer to different authentication mechanisms, for example, one-shot, multifactor, continuous, and multimodal, utilizing these factors. Also, we discuss design goals for usable authentication systems and usability evaluation methods. Section 3 surveys the different state-of-the-art solutions proposed over the years for user authentication on smartphones. The related work on the ways and types of user authentication concepts available for smartphones is evaluated on the basis of their usability and security. Finally, Section 4 concludes the paper.

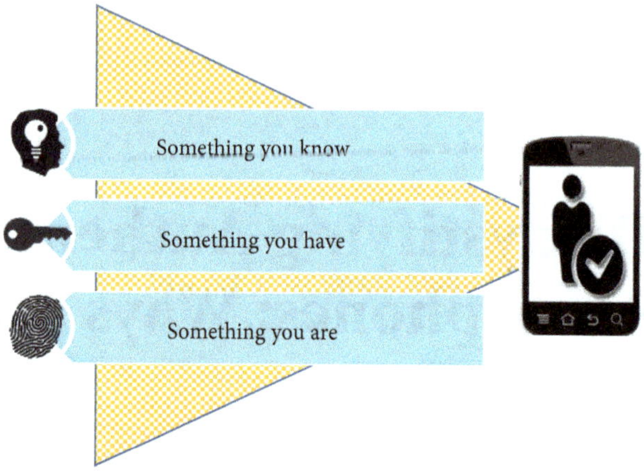

FIGURE 1: Ways to authenticate humans.

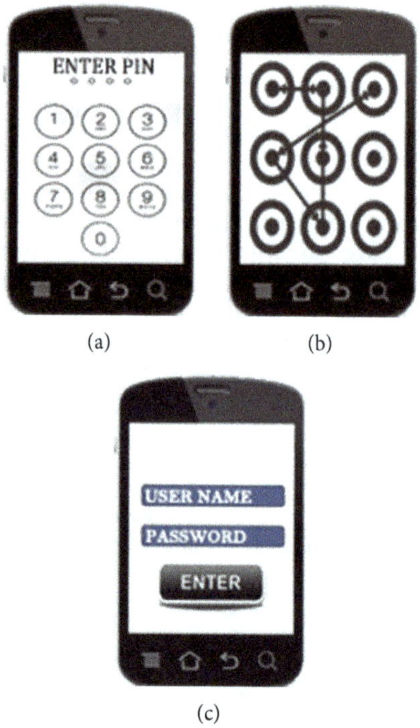

FIGURE 2: (a) PIN, (b) graphical pattern, and (c) password.

2. Comprehensive Study

In this section, we explain the ways to authenticate the users and the types of authentication mechanisms developed using them, in the context of smartphones.

2.1. Ways to Authenticate Users. The ways in which humans can be authenticated are broadly categorized in three categories [14], that is, "Something you know," "Something you have," and "Something you are," as depicted in Figure 1.

2.1.1. Something You Know. Knowledge-based authentication (KBA) schemes, that is, PINs (Figure 2(a)), graphical passwords (Figure 2(b)), and password (Figure 2(c)), are the most widely used schemes on the smartphones. KBA is based on some sort of a secret knowledge that user sets up earlier during the enrollment and needs to remember as long as he or she continues using the scheme.

2.1.2. Something You Have. This mechanism is also referred as token-based authentication. Many service providers and financial institutions are offering sensitive services, such as net banking, e-wallet, and e-commerce, adopting 2-factor authentication, that is, one-time passcodes (OTPs) along with usual username/password for authentication purpose. Service providers usually supply a small security device to each of their users for generating the one-time passcodes.

OTP schemes can be easily implemented on smartphones (Figure 3(a)) which could be sent either via SMS on the registered number or user could generate this OTP offline (Figure 3(b)) on the mobile apps provided by service

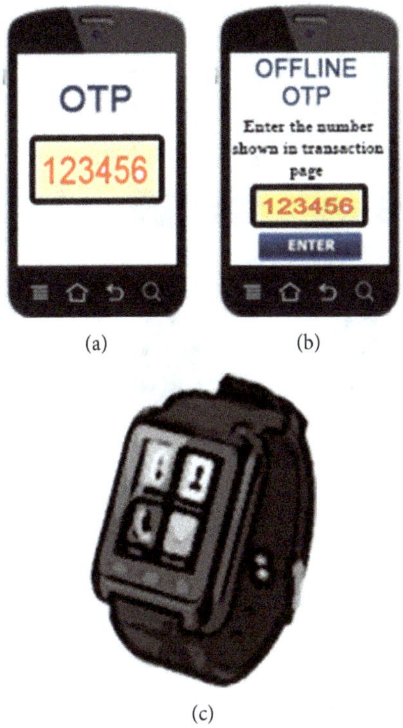

FIGURE 3: (a) One-time passcode (OTP) via SMS, (b) offline OTP using app, and (c) paired devices.

providers. Additionally, wearable devices (Figure 3(c)) could be used for receiving the OTPs via SMS.

2.1.3. Something You Are. This authentication mechanism relies on the measurement of biometric characteristics of users and is further classified as physiological and behavioral biometrics. Figure 4 illustrates the commonly available authentication ways for smartphone users under this category.

On smartphones, physical traits, that is, ear and face, can be collected using the built-in hardware, that is, camera; however, fingerprint and iris recognition require additional dedicated hardware. Similarly, behavioral biometric modalities, such as gait, grip, swipe, pickup, touch, and voice, can be profiled unobtrusively, using various built-in sensors [15], namely, accelerometer, gyroscope, magnetometer, proximity sensor, touch screens, and microphone. Touch-based solutions authenticate users based on their unique interactions with the device, while they perform a specific task. Additionally, behavioral biometric-based authentication is cost-effective; they generally do not require any special hardware and are considered lightweight in implementation [8].

2.2. Types of Authentication Mechanisms. Researchers have been investigating the utilization of different ways, that is, PIN, passwords, OTP, face, touch, and so on, to design and develop the different types of authentication solutions. These types are briefly explained below:

2.2.1. One-Shot Authentication. One-shot authentication is a type of authentication mechanism in which users' credentials are verified at the beginning of the session [16–18]. This is simply a process where a user claims his or her identity by providing the correct credentials or fulfilling the challenges in order to gain the access to a device. For example, PINs, passwords, graphical patterns, fingerprints, face, and iris are some of the commonly used modalities on the smartphones, for authenticating users. If the verification is successful (e.g., right password is entered), the access is granted; otherwise, the access is denied. Session remains valid until the user signs off or closes the session.

2.2.2. Periodic Authentication. Periodic authentication is simply the variant of "one-shot authentication" in which idle timeout duration is set, for closing the session, automatically [4, 19]. If a user remains inactive for more than the idle timeout duration, the device locks itself.

2.2.3. Single Sign-On (SSO) Authentication. Single sign-on (SSO) is a long-term or persistent authentication type in which a user remains signed on till the time he or she revokes or terminates the session. In case, if the system observes any discrepancy with respect to fix set of attributes, for example, change in location, network connection, and anomaly in usage pattern, the session is terminated or the user is asked for reauthentication [20–22]. VMware identity manager provides APIs to implement mobile sign-on authentication for airwatch-managed Android devices [23]. Similarly, Google offers G Suite apps for single sign-on for Android devices which can be done by pairing smartphones with smartwatches [24].

2.2.4. Multifactor Authentication. Multifactor authentication utilizes the concept of combining 2 or more authentication ways, that is, e-mail verification, OTP via SMS, phone call to the predefined numbers, push notification to the paired device, smart tokens, and so on, along with the usual method of authentication [25–27]. A very common practice is registering ones mobile number with service providers, and whenever the corresponding user accesses that service for sensitive operation, for example, online banking, service provider sends the one-time passcodes (OTPs) via SMS, getting assured that a legitimate user has requested access to that service.

2.2.5. Static and Dynamic Authentication. The static authentication mechanism presents the fixed set of challenges to the users, whereas dynamic authentication mechanism capitalizes the concept in which diverse set of prestored challenges are presented every time users unlock their smartphones [28, 29].

2.2.6. Continuous Authentication. As the name implies, continuous authentication mechanisms are developed to authenticate a legitimate owner throughout their entire session. If any anomaly is detected by the device, the access to the device is stopped, immediately, and the device asks for explicit reauthentication [4, 29, 30]. In other words, the users are passively and periodically monitored throughout their interactive session with any device or system [5]. This concept

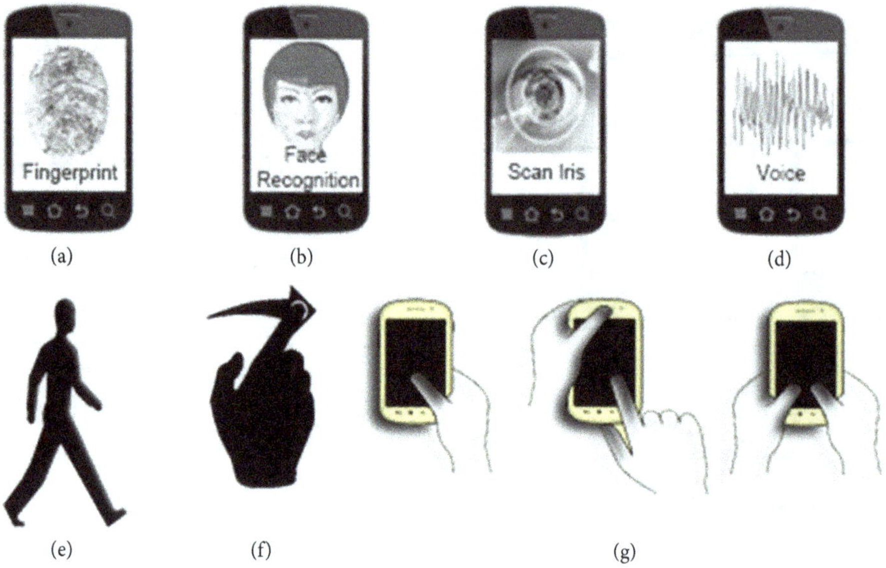

FIGURE 4: (a) Fingerprint, (b) face, (c) iris, (d) voice, (e) gait, (f) swipe, and (g) touch.

seems to promise higher security as compared to the other authentication mechanisms, such as *one-shot* authentication, *one-time* authentication, and *periodic* authentication, but at the same time much more complex to implement. Additionally, it is desirable that a *continuous* authentication system should not interrupt the user's normal activity and be lightweight, that is, on battery consumption.

2.2.7. Transparent Authentication. This concept stresses more on the procedure of collecting and analyzing user authentication identifiers [4, 10]. More specifically, if the system performs authentication steps in background (without requiring explicitly user cooperation) [10, 31], they are termed as *implicit, transparent, or unobtrusive* authentication systems. However, various authentication types (one-shot, risk-based, or continuous) could collect input transparently.

2.2.8. Risk-Based Authentication. Risk-based authentication schemes are mostly based on nonstatic authentication decision engine, where the decision to accept or reject authentication is based on a risk score computed in real-time, which is compared with the stored risk profiles of the users, and then the system challenges the users for authentication [32], accordingly. For instance, if a user is checking a bank account balance from a verified secure location (home or workplace), verification of identity should not be required. While in case of nonverified location, for example, the service requires additional evidence about the identity of the user thus asking for the authentication credentials. Nowadays, risk-based authentication schemes tend to offer frictionless authentication providing user experience, that could be tailored as per threats observed by the service providers [11, 12, 33, 34].

2.2.9. Adaptive Authentication. Adaptive user authentication boasts the concept having ability to change and to prepare for different conditions and situations, while securing any unauthorized access [13, 35, 36]. It entails for multifactor user authentication mechanisms which should be readily configurable and deployable.

2.2.10. Unimodal and Multimodal Authentication. This term is typically used for biometric authentication schemes. The literal meaning of modality (https://dictionary.cambridge.org/dictionary/english/modality) is a particular way of doing or experiencing something. This concept is based on the number of modalities or traits being used in the authentication systems [37–39]. Unimodal authentication systems leverage only a single biometric modality or trait, whereas multimodal systems are developed by combining two or more modalities. Multimodal authentication systems demonstrate several advantages, such as higher recognition rate, accuracy, and universality [39].

2.3. Usable Authentication System Design Goals. Usability along with security plays a pivotal role in evaluating user authentication schemes. This leads to an important question—how to trade-off between security and usability [40]? We present the guidelines described by Yee for usable security designs [41]. Yee's work focused on addressing valid and nontrivial concerns specific to usable security. We explain below the design goals from usability perspective as suggested in [41]:

(i) *Appropriate boundaries*: this goal is based on *the principle of boundaries* [42]. In order to distinguish among objects and actions along the boundaries, which are relevant to users, system should expose the boundaries and must acknowledge the users. For example, in the context of mobile devices, popular Operating Systems (OS), such as Android (Ver. 6 onwards) and iOS, allow users to grant permissions to the applications and services accessing resources while installing them. Here, the object could be assumed as the apps or services for the devices and actions could be defined as the indicators that the apps or services demand from users to serve them and to use the system's resources. However, boundaries are the thin line that defines the users' decisions affecting the security of system due to human factors.

(ii) *Path of least resistance*: choosing the most natural method in granting the authority is the most secure way.

(iii) *Explicit authorization*: any authorization to other actors must only be granted in accordance with user actions which should be well understood by a user while acknowledging the consent.

(iv) *Visibility*: a user should be aware of others' active authority affecting any security-relevant decisions.

(v) *Revocability*: a user should be able to revoke others' authority to access the system.

(vi) *Self-awareness*: maintain accurate awareness of the user's own authority to control the system.

(vii) *Trusted path*: protect the user's channels to any entity that manipulate authority on the user's behalf.

(viii) *Identifiability*: any specific objects and specific actions must be clearly identifiable and apparent to the user.

(ix) *Expressiveness*: enable the user to express safe security policies in terms that fit the user's goals.

(x) *Clarity*: notify the consequences of any security-relevant decisions precisely that the user is most likely to perform.

2.4. Usability Evaluation. System usability scale (SUS) questionnaire [43] is utilized to gather subjective assessments about the usability of the proposed systems [8]. The questionnaire consists of 10 questions or statements. The response to each question/statement is measured on a 5-point scale ranging from "strongly disagree" to "strongly agree." The final SUS score ranges between 0 and 100, where a higher value indicates a more usable system. The system usability scale (SUS) template for questionnaire and scoring is available online [44].

3. Literature Review and Analysis

In this section, we review the recent literature emphasizing on the types of authentication mechanisms and the ways on which they are developed and analyze them from security and usability point of view. More specifically, we present the assessment of commonly used user authentication mechanisms on smartphones, focusing on the security and usability issues.

3.1. Ways of Authentication. The usability of authentication mechanisms is one of the dominant attributes that influence users' acceptance of a particular authentication scheme [45]. The ISO standard:13407 defines usability as "*the extent to which a product can be used by specified users to achieve specified goals with effectiveness, efficiency, and satisfaction, in a specified context of use*" [46]. Further, the study [47] suggests that the usability can be done on the basis of three criteria: task performance, user satisfaction, and user cost.

Conventional authentication schemes, that is, PIN, passwords, and graphical patterns, are no more considered secure and convenient [48] because they are not able to distinguish between the users, rather they authorize everyone (regardless of whether that person is the legitimate owner of the device or not) who enter the correct credentials. Physiological biometric-based solutions are considered more secure because it is assumed that human body traits cannot be shared, copied, lost or stolen. Moreover, they genuinely authenticate their users by forcing them to present themselves physically to the system. However, they are less preferable on smartphones due to their inherent usability issues [49]. As such, security experts are focusing on developing the usable authentication systems because they believe that behavioral biometrics will restructure the authentication landscape in the next 5–8 years [50].

In each subsections, we have included tables presenting the synopsis of each authentication ways being used as different authentication types along with the references that either indicating usability pros and cons or reporting security solutions and concerns.

3.1.1. Something You Know. As per the web report [51], average smartphone users get themselves engaged in 76 separate phone sessions, while heavy users (the top 10%) peaked to 132 sessions per day. PIN/passwords, and graphical patterns, require users to memorize their text, they had set earlier, to unlock their devices, every time they need to initiate the session (76 times a day). The capacity of the human brain to process the information varies from person to person [52]. Zhang et al. [53] found that users faced problems in remembering their passwords and more especially, to memorize and correctly recall numerous passwords. This encouraged users going for an easy or simple password which is quick to remember [54], but this opens plenty of opportunities for attackers to guess or crack their passwords, easily [55]. When the system enforces stringent password policies, users due to memorability issues [56], allow their browsers or password managers to save their username/password information to make future logins easier. However, users trusting their browsers or password managers are more likely to be a victim of a wide variety of attacks [57, 58]. Overall, 82% of end users are frustrated with managing passwords [59]. Clearly, this indicates the lack of usability, and a result, nearly, 75 million smartphones users in the US do not use any of PIN, pattern, or passwords because they consider them annoying and an obstacle in quick access to their smartphones [60].

From security perspective, PINs and passwords are vulnerable to various attacks, for example, guessing [61], because users choose date of births [57], easier digits (1111, 2222, etc.,) [62] to set up their PIN. Alternatively, Android

Table 1: Synopsis of knowledge-based schemes.

Modalities	Authentication types	Usability pros and cons indicated	Security solutions or concerns reported
PIN [57, 60, 62]; password [53, 60]; pattern [60, 64]	One-shot; static; periodic; single sign-on; unimodal	[48, 52–57, 60, 62, 71, 72]	[58, 61–70]

Table 2: Synopsis of token-based schemes.

Modalities	Authentication types	Usability pros and cons indicated	Security solutions or concerns reported
OTP [75]; device pairing [77, 79]	Multifactor; adaptive; dynamic; risk-based	[40, 79, 80–82]	[73–78]

users (40% of them) prefer graphical patterns for device unlocking. But this approach too requires users to remember them; hence users choose simple and less secure patterns, that is, if a user connects at least four dots without repeating any of them in their patterns, the maximum number of combinations are 389,112 which could be easily cracked by brute force [63]. Ye at al. [64] managed to crack 95% of 120 unique patterns collected from 215 independent users within just five attempts by recording their smartphone screen, remotely, while they were unlocking their devices. In addition, these schemes are more vulnerable to shoulder surfing than textual passwords [65].

Knowledge-based authentication schemes are generally used as one-shot, static, or unimodal authentication types (refer Table 1) due to usability issues they are prone to several attacks, such as smudge attacks [66], shoulder surfing or observation attacks [61, 67], dictionary-based attacks, or rainbow table password attacks [68]. Recently, Mehrnezhad et al. [69] demonstrated the recovery of entered PIN or password from the sensory data collected, while the users were entering their secrets. They installed PINlogger.js—a JavaScript-based side-channel attack, capable of recording motion and orientation sensor streams without requiring any user permission from the user. The attack resulted in 94% accuracy in recovering the correct PIN number in just three rounds of tries. Similarly, Sarkisyan et al. [70] demonstrated an approach to exploit smartwatch motion sensors to recover the entered PINs. They infested smartwatches with malware to get access to the smartwatch motion sensors and inferred user activities and PINs. In a controlled scenario, authors obtained PIN numbers within 5 guesses with an accuracy of at least 41% using random forest classifier over a dataset of 21 users.

3.1.2. Something You Have. As defined in Section 2.1.2, smartphones are being utilized for authentication purposes in several sensitive operations by the means of OTP via SMS, offline OTP using Apps, or pairing the wearable devices, for example, smartwatches, smartglasses, and smartcards. However, this idea of enhancing security with multifactor authentication, that is, topping *knowledge-based authentication* with *token-based authentication* (one-time passcode), eventually perishes too due to side-channel attacks, for example, MITM (man-in-the-middle) and MITPC/Phone (man-in-the-PC/phone) [73]. Software-based OTP solutions also do not guarantee the confidentiality of the generated passwords or the seeds as the mobile OS could be compromised, at the same time, could also suffer from denial-of-service attacks on the account of mobile OS crashes [74].

The adversaries by the means of real-time phishing or intercept attacks could reveal the users' secret information and valid OTP by breaking into their smartphones [75]. As per the Verizon Data Breach Investigations Report [76], NIST stopped recommending the users for two-factor authentication via SMS, as malicious code infesting mobile endpoints could surreptitiously capture second factors delivered by SMS or offline OTP generated using apps. Secure device pairing schemes allow access to the smartphones by pairing it with a trusted Bluetooth device like a smartwatch and use the same to unlock the phone. This concept from the usability point of view is a very elegant solution but not safe from insider attacks or sniffing attacks [77, 78].

Token-based authentication (TBA) schemes are used in multifactor, adaptive, dynamic, and risk-based authentication types (Table 2). Unfortunately, they could not add too much to the usability because the users are required to manage always an additional hardware for the sole purpose of authentication. As a result, Braz and Robert [40] gave usability rating 3 (out of 5) to one-time generator acquisition devices. Additionally, Belk mentioned that token-based authentication mechanism incurred more cost to users and are comparatively slower [79]. According to a study by Zink and Waldvogel [82], 83.3% users considered that SMS-based transaction authentication number is not a usable solution. Another in-depth usability study by Krol et al. [81] evaluated 2-factor authentication on 21 online banking customers (16 among 21 were having multiple accounts with more than one bank). Total 90 separate login sessions of all the participants were collected meticulously, over the period of 11 days. Their analysis showed approximately 13.3% faced problems due to mistyped credentials, misplaced token, forgotten credentials and so on.

3.1.3. Something You Have: Insertable Biometrics. Insertable biometrics [83–85] (Table 3) including implantable medical devices (IMDs) [86] and emerging technologies such as Bespoke devices [87, 88], neodymium magnets [89], NFC or RFID chips [90, 91], smart piercings [92, 93], and smart tattoos [93] are the newer addition to biometrics that

Table 3: Synopsis of insertable biometrics.

Modalities	Authentication types	Usability pros and cons indicated	Security solutions or concerns reported
Bespoke devices [87, 88]; neodymium magnets [89]; NFC or RFID chips [90, 91]; smart piercings [92, 93]; smart tattoos [93]	Continuous; multimodal; transparent	[94]	Data not available

Table 4: Synopsis of physiological biometrics.

Modalities	Authentication types	Usability pros and cons indicated	Security solutions or concerns reported
Face [99, 100]; eyes [10, 101]; iris [102]; fingerprint [103, 104]	One-shot; multifactor; multimodal	[49, 105–109]	[17, 95–97, 99, 100, 102–104, 110]

potentially can be used to provide increased usability over the existing solutions [94]. Researches are exploring the further possibilities of insertable biometrics as go-to solution for improving digital security and usability in smartphones.

3.1.4. Something You Are: Physiological Biometrics. Mobile device manufacturers have started embedding biometric sensors in their flagship smartphones for reliable and convenient user authentication with the intuition that biometric approaches are better than their conventional authentication schemes. For example, Apple, Huawei, Lenovo (Motorola), Microsoft (Nokia), Samsung, and many other leading manufacturers have integrated fingerprint sensors, iris scanners, and face recognition algorithms, in some of their high-end devices. These advancements are akin to replacing a hay castle with a glass house to ward off attacks from sophisticated cyber pirates.

Physiological biometrics, for example, face, fingerprint, iris, and eyes, are commonly used as one-shot or multifactor/multimodel (combining with other modalities) authentication schemes for smartphones (Table 4). Unexpectedly, biometric systems have shown to be exposed to different types of attacks, for example, impersonation, replay, spoofing, and hill climbing [95], exposing their security loopholes. These schemes suffer from their data leakage; that is, a user's face can be easily found on social media websites, or his or her fingerprints can be extracted from the photos from their gestures, to mount a presentation attack [96] against them. Additionally, these solutions also suffer from lack of secrecy [97] and vulnerability to various spoofing attacks [98].

Recent research has shown that these schemes can be hacked very easily with almost negligible investment and efforts. For example, iPhone X face ID was hacked with 3D-printed mask costing just $150 approximately [100], while Samsung S8 facial recognition technology [99] was simply fooled with a photo of the owner. Similarly, German Chaos Computer Club cracked the Samsung Galaxy S8 iris scanner [102] with a dummy eye made from pictures of the iris, taken by a digital camera in a night mode, and covered it with a contact lens to match the curvature of the eye, within a month of S8 launch. The same club earlier cracked the iPhone 5S fingerprint sensor protection within two days after the device went on sale worldwide [103]. Their hacking team photographed the glass surface containing the fingerprint of a user and created a "fake fingerprint" using a thin film to unlock the phone. Japan's National Institute of Informatics (NII) researcher Isao Echizen [104] demonstrated that fingerprints can easily be recreated from photos, taken just from three meters distance, without the use of any sophisticated process and warned casually making a peace sign in front of a camera, which could lead to fingerprint theft.

From the usability perspective, smartphone users have not shown optimistic inclination to physiological biometric-based authentication schemes. For example, De Luca et al. [49] determined smartphone users felt like as if they are taking selfies all day to authenticate themselves. Additionally, the performance of these schemes is affected by several exogenous factors, such as accessories, camera movement, capturing distance, clothing, illumination, interoperability of the sensors, noise, occlusion, operators, postures, and training, which makes the authentication process more challenging and less usable to the user [106–109].

3.1.5. Something You Are: Behavioral Biometrics. Behavioral biometrics [111] is described as the future of user authentication. Thus, the focus of the research has been shifted to develop newer behavioral biometric-based solutions. For example, applications like e-wallet, m-commerce, and mobile banking are some of the sensitive domains, where behavioral biometric-based solutions have shown to be handy in authenticating the customers on their smartphones.

Although the behavioral modalities are not considered to be unique enough for identification purposes, they have proved to be sufficiently unique for user authentication [112, 113]. One or more modalities can be combined to increase their accuracy and enhance their usability. These schemes could be stitched to the existing user authentication mechanisms as an additional transparent authentication layer [8, 9, 114] enhancing the reliability of whole authentication process without affecting the usability. Behavioral biometric techniques could be deployed as adaptive, continuous, multimodal, risk-based, transparent authentication (Table 5).

Gait recognition is a process of identifying or verifying individuals on the basis of their walking style. In clinical applications, human gait was already getting utilized for the studies related to the health of a person, and nearly 25 key patterns from gait were detected using different techniques like

Table 5: Synopsis of behavioral biometrics.

Modalities	Authentication types	Usability pros and cons indicated	Security solutions or concerns reported
Touch [9, 113]; keystroke [115]; hold [8]; gait [116–118]; behavior profiling [119]	Adaptive; continuous; multimodal; risk-based; transparent	[3, 5, 10, 113, 119–121]	[8, 12, 29, 112, 113, 115–117, 122–127]

image processing, floor sensors, and sensors placed on the body [118]. Recently, smartphones and wearable devices have also started utilizing it for authentication purposes [128]. As users are not required to perform any explicit interaction with their devices, gait modality can be collected unobtrusively, and this leads to making it convenient for a user-friendly access system [116]. Muaaz and Mayrhofer [116] evaluated the security strength of a smartphone-based gait recognition system against zero-effort and live-minimal-effort impersonation attacks under realistic scenarios and achieved an equal error rate (EER) of 13% on a dataset of 35 participants. However, more testing is required to check the robustness against impersonation attacks. Hestbek et al. [117] introduced a method using wearable sensors and noncyclic feature extraction and achieved 18.92% half total error rate (HTER) on a dataset of 36 users. Similarly, the grip is another natural way to authenticate users. It is robust too as the finger movements and pressure applied while gripping the mobile device are visibly unseen and difficult to be replicated or imitated by the impostor. Murao et al. [124] proposed a grip-based authentication solution, which profiles grip gestures using pressure sensors mounted on the lateral and back sides of a smartphone and achieved a 2% ERR, which is equivalent to face recognition-based authentication.

Keystroke or touch dynamics refers to the typing characteristics (due to the timing differences) of individuals to fingerprint their identity. Researchers have proved its effectiveness in both fixed text and text independent scenarios. Since designing such systems does not require any additional dedicated hardware and data can be collected, unobtrusively, they have been widely tested and evaluated [9, 114]. Zheng et al. [115] proposed authentication mechanism based on tapping; they collected tapping data from over 80 users; and their system achieved high accuracy with averaged 3.65% EER. Another bimodal authentication scheme developed using client-server architecture for online financial environments achieved 96% true acceptance rate (TAR) and 0.01% false acceptance rate (FAR) using 15 training samples on a dataset of 95 users [9]. This scheme used motion-based touch-types biometrics, that is, touch typing and phone movements by users and collected data, transparently, while users entering their credentials to sign in to their banking apps using 8-digit PIN/password [9], while the "touchstroke" scheme used 4-digit PIN/password [114]. Buriro et al. [8], proposed, implemented, and evaluated the "Hold and Sign" scheme on commercially available smartphones and achieved 95% TAR on a dataset of 30 volunteers. This was a bimodal behavioral biometric based on user's smartphone holding style, by examining the hand and finger micromovements of users, while the users were signing on device's touchscreen. In an another approach, Buriro et al. [113] proposed multimodal behavioral biometrics (swipe, pickup movement, and voice) for user authentication on smartphones and reported 7.57% HTER in an experiment involving 26 participants.

Brunet et al. [123] experimented on voice modality for user authentication on a public database (Sphinx Database of the Carnegie Mellon University [129]). They digitized the user's voice and extracted Mel Frequency Cepstral Coefficients (MFCCs) features and computed the Euclidean distance to authenticate the user and reported an EER of 4.52%. Behavior profiling techniques were based on the applications, and the services utilized in past for generating a user profile and compared it against the current activity of a user in real-time [5]. If any significant variation is observed, the system could take action for a possible intrusion. Sultana et al. [119] combined social behavioral information of individuals that was extracted from the online social networks to fuse with traditional face and ear biometrics, to enhance the performance of the traditional biometric systems.

Studies suggest that no single biometric trait can ideally fit all the scenarios; however, by trying multimodal biometric approaches, most of the limitation of unimodal systems can be addressed [121, 122, 125]. The selection of proper modalities and combining them, systematically, most of the times increase the accuracy, usability, and security. In a study conducted by Saevanee et al. [126], the unimodal systems, namely, behavior profiling, keystroke dynamics, and linguistic profiling, were proved less accurate; they yielded an EER of 20%, 20%, and 22%, respectively. However, by applying matching-level fusion, the error rate was decreased, significantly (EER 8%). Additionally, the use of users' transparent characteristics for data collection and classification also increases the usability of the system. Thus, in order to furnish users with an adequate security, a better usability is also required to design the authentication solutions for smartphones.

3.2. Authentication Types

3.2.1. One-Shot Authentication. One-shot authentication schemes are designed to authenticate a user at the initiation of a session (subject's identity is verified only once, just before allowing access to the resources) [16, 18]. Roth et al. [18] also discussed the limitations of one-shot authentication, such as short sensing time, inability to rectify decisions, and enabling the access for potentially unlimited periods of time. Meng et al. [17] introduced the term one-off authentication for one-shot authentication. They also concluded that authenticating just once leaves the possibilities for impostors to gain the access to the current session and retrieve sensitive information from mobile phones.

3.2.2. Periodic Authentication. Bertino et al. [19] defined *periodic authorization* with a mathematical expression "{[begin,

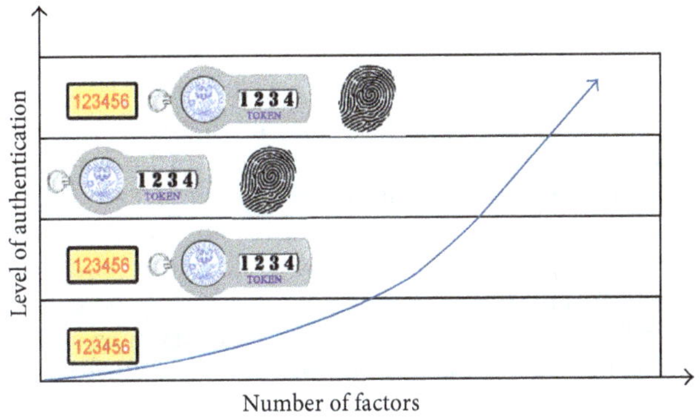

Figure 5: Factors of authentication [27].

end], P, auth}" holding of 3 prime attributes, where "begin" is authorization start date, "end" is either the constant ∞, or a deauthorization date after the start date, "P" is the duration of a session, and "auth" is an authorization function. Feng et al. [4] determined that periodic authentication or automatic logouts are more detrimental while one-shot authentication solutions are prone to a wide variety of attacks. Typing an error-free username and/or password on smartphone's keyboard is really a tedious task, especially when an average user initiates 76 phone sessions a day [51]. *Single sign-on* (SSO) has been seen as the solution to the problem.

3.2.3. Single Sign-On. *Single sign-on* (SSO) enables users to sign in to an app using a single or federated identity, for example, Facebook, Twitter, and Google+. But this concept is severely risky for mobile devices as they are more likely to be misplaced or could be inadvertently shared with someone. In an SSO system, the user is authenticated to a single identity provider (IDP) which acts as a trusted party between the user and multiple service providers (SPs), and on the demand of the user, IDP generates an authentication token for a specific SP asserting the users' identity; in turn, SP allows the user to access the services [20]. Users can access different applications using SSO, once they are authenticated to the system. SSO is further divided into two categories, that is, Enterprise Single Sign-ON (ESSO) and Reduced Sign-ON (RSSO) [21]. ESSO enables a user to enter the same id and password to sign into multiple applications within an enterprise domain. The system is considered the least secure because there could be potential curious adversary which can try to spoof and consequently resulting in an identity theft. Therefore, it is also known as RSSO.

3.2.4. Multifactor Authentication. Security experts also suggest the use of *multifactor authentication* by processing multiple factors, simultaneously, for the verification purposes [27]. In multifactor authentication, generally, a PIN or password is the baseline authentication standard, while more factors can be augmented from a wide variety of available sources to verify users (Figure 5). It could be observed in Figure 5 that as the number of factors increases, the level of authentication also increases. For an instance, if only PIN is used, the authentication level is minimum, but when other factors like tokens and

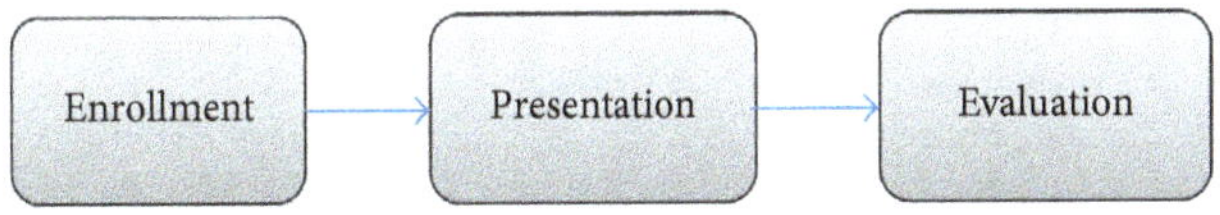

Figure 6: Static authentication process [29].

fingerprints are added, the authentication level tends to increase proportionally.

The most common authentication mechanism is the secondary code that can be delivered either via SMS to the registered mobile number or can be obtained directly from a secure authenticator mobile app. Other forms of multifactor authentication involve the use of a smart card or smart token entitled to the user, biometrics like the face or fingerprint scans, or a dedicated code generator linked to user's account [25]. This concept is mainly influenced by the notions that not all the authentication factors could be hacked at the same time. Stanislav [26] in his paper explained various technical methods by which two-factor authentication can be implemented.

3.2.5. Static versus Dynamic Authentication. *Static authentication* process, like other authentication types, mainly consists of three steps: enrollment, presentation, and evaluation as illustrated in Figure 6, and the outcome of the evaluation is a binary decision [29]. In the enrollment step, system generates a feature template by processing the information gathered from the user, profiles the feature vectors with the label of the user, and saves it for the evaluation or matching. During the presentation step, system asks the user to confirm his or her credentials. In the final step, that is, evaluation, information given by the user is compared with the stored templates of the claimed identity. Conclusively, the access is granted or denied as per the match result.

Static authentication verifies the individual's identity only at the start of a session like one-shot authentication does, whereas in *dynamic authentication* the user is presented with a varying set of challenges to enable the dynamic scaling of access controls. Ren and Wu [28] explained dynamic authentication as a scheme that utilizes one-time password derived from the user's password, the authenticating time, and a unique attribute only known to the user.

3.2.6. Continuous Authentication. *Continuous authentication* is a mechanism to repeatedly verify the identity of a user for the entire duration of an authorized session as illustrated in Figure 7 [29]. More specifically a continuous authentication is an approach that constantly verifies a user's identity and locks the system once the change in users' identity is observed [29]. Continuous authentication process dynamically iterates in between the three steps involved (Figure 6) throughout the session. However, these iterations can be event-based or can be adjusted at fix intervals (periodically) or randomly [29]. A continuous authentication is an approach that constantly verifies a user's identity and locks the system once the change in user identity is observed. Thus, overcoming the limitations of one-shot authentication, where authentication happens only at the time of login, and any future changes in user identity go undetected [130]. Behavioral biometric-based

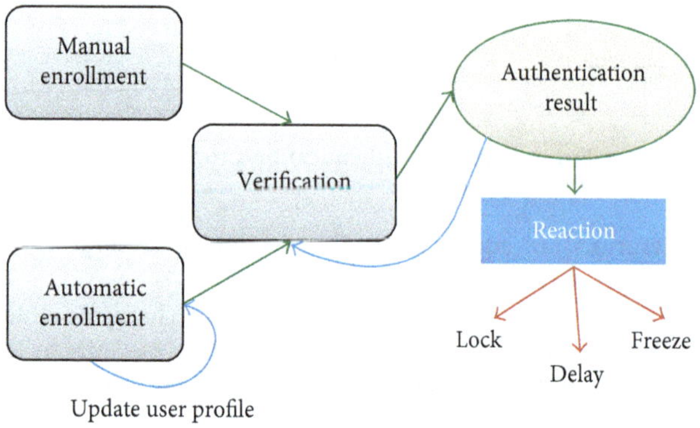

FIGURE 7: Continuous authentication process [29].

continuous authentication solutions have shown to be more attractive to the researchers of the domain because these behavioral modalities can be collected and utilized, unobtrusively, for authentication purposes [30].

However, continuous authentication, active authentication, implicit authentication, and transparent authentication have been interchangeably used in many papers [10, 120, 131, 132]. Patel et al. [5] considered continuous authentication and active authentication systems as similar and explained it as continuous monitoring of the user activities after the initial access to the mobile device. Active authentication, as defined by Stolerman et al. [132], is the process of continuously verifying users based on their on-going interaction with the device. The Defense Advanced Research Projects Agency (DARPA) started Active Authentication program [133] in order to seek solutions by shifting the focus during authentication from the password to people themselves. The first phase of their Active Authentication program focused on the behavioral traits, that is, cognitive fingerprint, which could be processed without the need for additional sensors.

According to Fridman et al. [134], active authentication is the problem of continuously verifying the identity of an individual. They conducted an experiment using Android mobile devices and collected several biometric modalities, namely, text entered via soft keyboard, applications used, websites visited, physical location of the device as determined from GPS (when outdoors) or WiFi (when indoors), and stylometry, of 200 volunteers approximately for a period of at least 30 days. Their authentication system achieved an ERR of 0.05 (5%) after 1 minute of user interaction with the device, and an EER of 0.01 (1%) after 30 minutes in identifying a legitimate user. In another stylometric-based continuous authentication, an EER of 12.42% for message blocks of 500 characters is achieved using support vector machine (SVM) for classification [135]. However, stylometry-based authentication schemes must improve the accuracy, delays, and forgery.

Khan et al. [120] mentioned that *implicit authentication* employs behavioral biometrics in a continuous and transparent manner to recognize and validate smartphone users' identity and conducted a field study on implicit authentication usability and security perceptions with 37 participants. Their experiment indicated that 91% of participants found implicit authentication to be convenient and 81% perceived defined the protection level to be satisfactory.

3.2.7. Transparent Authentication. *Transparent authentication* [10] was suggested as an alternative authentication mechanism with minimal or no noticeable involvement of users. Transparent authentication implicitly authenticates the users on the basis of their unique interactions with the device and creates a logic for authentication decisions. Feng et al. [4] utilized the term transparent and continuous for their Finger-gestures Authentication System using Touchscreen (FAST) to protect the mobile system. The approach transparently captures the touch data without intervening to user's normal user-device interactions. After the user's login, FAST continues to authenticate the mobile user in the background using intercepted touch data from their normal user-smartphone interactions.

3.2.8. Risk-Based Authentication. ClearLogin [136] defines *risk-based user authentication* as a method which adapts authentication levels based on the apparent risks, to mitigate the potential intrusion, before they happen. Existing *risk-based user authentication* schemes generate a risk profile to determine the complexity of challenge to authenticate a user during a session, that is, higher-risk profiles lead to stronger authentication, whereas usual authentication scheme should be sufficient in normal scenarios [137]. Identity Automation [138] considers *risk-based user authentication* similar to *adaptive authentication* because they adapt to the stringency of authentication processes based on the likelihood that access to a given system could result in its compromise.

Earlier risk-based user authentication mechanisms were mainly based on contextual or historical user information or both [139]. Furthermore, these systems use ad hoc or simplistic risk management models based on some rule-based techniques, which are proved to be ineffective due to human factors [140]. However, nowadays as NuData Security [34] mentioned risk-based authentication schemes are getting fueled by behavior piercing technology that gives maximum security with minimal interruption to the user experience. Risk-based user authentication can be applied from two different perspectives: proactive or re-active [12]. When applied proactively, risk-based authentication actively anticipates the genesis of potential attacks, failures, or any kind of security issues and takes prompt action. In contrast, re-active risk-based authentication accepts some of the risks until the risk score goes beyond the permissible threshold level, and consequently, reauthentication is required.

3.2.9. Adaptive Authentication. *Adaptive authentication* [141] is a way by which two- or multifactor authentication can be configured and deployed by doing risk assessment. Thus, it is a method for selecting the appropriate authentication factors accustomed to the situation accordingly to the user risk profile and tendencies. It can be deployed as follows:

(i) By setting static policies based on risk levels for different factors, such as user role, resource importance, location, time of day, or day of the week

(ii) By learning day-to-day activities of users based on their habits to generate dynamic policies

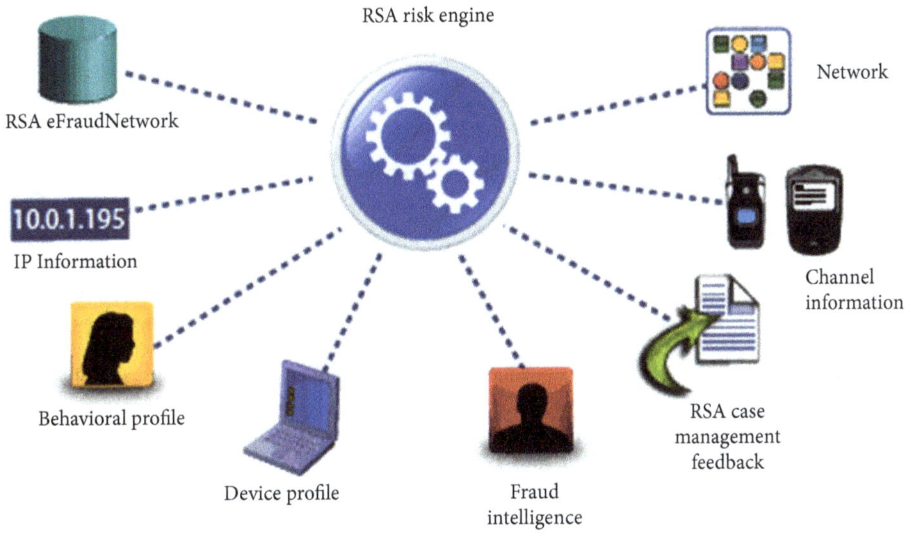

FIGURE 8: RSA adaptive authentication [36]. The RSA Risk Engine measures over one hundred indicators and assigns a unique risk score to each activity.

(iii) Lastly, by combing of both static and dynamic policies

Hulsebosch et al. [35] exploited the ability to sense and use context information to augment or replace the traditional static security measures by making them more adaptable to a given context and thereby less intrusive to derive context sensitive adaptive authentication. RSA Risk Engine [36] used self-learning risk model and adapts itself on the basis of received feedback. The feedback loop includes case resolution and genuine or failed authentication results as well as chargeback files for *adaptive authentication* for e-commerce (Figure 8).

3.2.10. Unimodal and Multimodal Authentication Systems. *Unimodal authentication systems* use single modality for establishing user identity, whereas *multimodal authentication systems* include multiple modalities (sources of information) [39]. Unimodal and multimodal terms are more associated with biometric systems where person recognition is based on distinctive personal traits or characteristics [37]. Unimodal physiological biometric based on face, fingerprint, and iris are already deployed on the smartphones; however, multimodal systems are yet to be deployed. Behavioral biometric-based solutions based on touch-stroke dynamics, voice, gait, and so on have been widely tested and evaluated by researchers; however, their deployment to the smartphones is still awaited.

Jain et al. [38] showed that multimodal biometric systems driven by multiple biometric sources perform, generally, better recognition performance as compared to unimodal systems. As per the type of multiple modalities being used, multimodal biometric systems can be further divided into three categories: (1) multiphysiological, (2) multibehavioral, and (3) hybrid multimodal systems [142]. The multiphysiological category includes multimodal biometric systems, where only physiological traits, such as face, fingerprint, and iris, are fused at different levels, whereas the multibehavioral system combines data from keyboard, mouse, and graphical user interface interactions. Hybrid multimodal system [143] fused face, ear, and signature with social network analysis at the decision level to enhance the biometric recognition performance.

Researchers have been actively working on combining different modalities to develop multimodal solutions; however, these systems have yet to appear on the real products.

4. Conclusion

In this paper, we presented the gist of ways and types of user authentication concepts in the context of smartphones. We surveyed the different state-of-the-art solutions proposed over the years and attempted to homogenize correlated buzzwords used in this field, with the motivation to assist new researchers in understanding these concepts. Then, we evaluated the related work on the ways and types of user authentication mechanisms available for smartphones, on the basis of their usability and security. Also, we discussed design goals for usable authentication systems and usability evaluation methods.

Acknowledgments

This project has received funding from the European Union's Horizon 2020 research and innovation programme under the Marie Sklodowska-Curie grant agreement no. 675320.

References

[1] Motorola, "Timely achievements," 2017, https://www.motorola.com/us/about/motorola-history-milestones.

[2] N. Triandopoulos, A. Juels, R. L. Rivest, and J. Brainard, "Multi-server one-time passcode verification on respective high order and low order passcode portions," US Patent 9,454,654, 2016.

[3] D. Crouse, H. Han, D. Chandra, B. Barbello, and A. K. Jain, "Continuous authentication of mobile user: fusion of face image and inertial measurement unit data," in *Proceedings of the International Conference on Biometrics (ICB)*, pp. 135–142, IEEE, Phuket, Thailand, May 2015.

[4] T. Feng, Z. Liu, K.-A. Kwon et al., "Continuous mobile authentication using touchscreen gestures," in *Proceedings of the IEEE Conference on Technologies for Homeland Security (HST)*, pp. 451–456, IEEE, Waltham, MA, USA, November 2012.

[5] V. M. Patel, R. Chellappa, D. Chandra, and B. Barbello, "Continuous user authentication on mobile devices: recent progress and remaining challenges," *IEEE Signal Processing Magazine*, vol. 33, no. 4, pp. 49–61, 2016.

[6] D. C. Dutt, A. B. Somayaji, and M. J. K. Bingham, "System and method for behavioural biometric authentication using program modelling," US Patent App. 15/059,692, 2016.

[7] A. De Luca, A. Hang, F. Brudy, C. Lindner, and H. Hussmann, "Touch me once and i know it's you!: implicit authentication based on touch screen patterns," in *Proceedings of the SIGCHI Conference on Human Factors in Computing Systems*, pp. 987–996, ACM, New York, NY, USA, 2012.

[8] A. Buriro, B. Crispo, F. Del Frari, and K. Wrona, "Hold and sign: a novel behavioral biometrics for smartphone user authentication," in *Proceedings of the 2016 IEEE Security and Privacy Workshops (SPW)*, pp. 276–285, IEEE, San Jose, CA, USA, May 2016.

[9] A. Buriro, S. Gupta, and B. Crispo, "Evaluation of motion-based touch-typing biometrics in online financial environments," in *Proceedings of the 16th International Conference of the Biometrics Special Interest Group (BIOSIG)*, Darmstadt, Germany, September 2017.

[10] H. Crawford and K. Renaud, "Understanding user perceptions of transparent authentication on a mobile device," *Journal of Trust Management*, vol. 1, no. 1, p. 7, 2014.

[11] B. Causey, "Adaptive authentication: an introduction to risk-based authentication," 2013, http://searchsecurity.techtarget.com/tip/Adaptive-authentication-An-introduction-to-risk-based-authentication.

[12] I. Traore, I. Woungang, M. S. Obaidat, Y. Nakkabi, and I. Lai, "Online risk-based authentication using behavioral biometrics," *Multimedia tools and applications*, vol. 71, no. 2, pp. 575–605, 2014.

[13] M. B. Ayed, "Method for adaptive authentication using a mobile device," US Patent 8,646,060, 2014.

[14] F. B. Schneider, "Something you know, have, or are," 2005, https://www.cs.cornell.edu/courses/cs513/2005fa/NNLauthPeople.html.

[15] N. Forsblom, "Were you aware of all these sensors in your smartphone?," 2015, https://blog.adtile.me/2015/11/12/were-you-aware-of-all-these-sensors-in-your-smartphone/.

[16] A. Buriro, "Behavioral biometrics for smartphone user authentication," Ph.D. thesis, University of Trento, Trento, Italy, 2017.

[17] W. Meng, D. S. Wong, S. Furnell, and J. Zhou, "Surveying the development of biometric user authentication on mobile phones," *IEEE Communications Surveys & Tutorials*, vol. 17, no. 3, pp. 1268–1293, 2015.

[18] J. Roth, X. Liu, A. Ross, and D. Metaxas, "Biometric authentication via keystroke sound," in *Proceedings of the International Conference on Biometrics (ICB)*, pp. 1–8, IEEE, Madrid, Spain, June 2013.

[19] E. Bertino, C. Bettini, E. Ferrari, and P. Samarati, "An access control model supporting periodicity constraints and temporal reasoning," *ACM Transactions on Database Systems*, vol. 23, no. 3, pp. 231–285, 1998.

[20] F. Feldmann, *Binding Credentials: Securing (SSO) Authentication*, Ruhr University Bochum, Bochum, Germany, 2016.

[21] HuntingtonVentures, "Single sign on: the business of authentication," 2006, https://archive.is/20140315095827/.

[22] A. Salazar, "SSO vs. centralized authentication," 2014, https://stormpath.com/blog/sso-vs-centralized-auth.

[23] VMware, "VMware identity manager documentation center," 2017, https://pubs.vmware.com/vidm/index.jsp?topic=%2Fcom.vmware.wsair-administration%2FGUID-1E5128A5-1394-4A50-8098-947780E38166.html.

[24] Google, "G suite: single sign-on on an android device," 2016, https://support.google.com/a/users/answer/2758865?hl=en.

[25] R. Ritchie, D. Rubino, K. Michaluk, and P. Nickinson, "The future of authentication: Biometrics, multi-factor, and co-dependency," 2013, https://www.androidcentral.com/talk-mobile/future-authentication-biometrics-multi-factor-and-co-dependency-talk-mobile.

[26] M. Stanislav, *Two-Factor Authentication*, IT Governance Ltd., Ely, UK, 2015.

[27] D. G. Warnock and C. C. Peck, "A roadmap for biomarker qualification," *Nature biotechnology*, vol. 28, no. 5, pp. 444-445, 2010.

[28] X. Ren and X.-W. Wu, "A novel dynamic user authentication scheme," in *Proceedings of the International Symposium on Communications and Information Technologies (ISCIT)*, pp. 713–717, IEEE, Gold Coast, QLD, Australia, October 2012.

[29] I. Traoré and A. A. E. Ahmed, *Introduction to Continuous Authentication, in Continuous Authentication Using Biometrics: Data, Models, and Metrics: Data, Models, and Metrics*, University of Victoria, Victoria, BC Canada, 2011.

[30] M. Frank, R. Biedert, E. Ma, I. Martinovic, and D. Song, "Touchalytics: on the applicability of touchscreen input as a behavioral biometric for continuous authentication," *IEEE Transactions on Information Forensics and Security*, vol. 8, no. 1, pp. 136–148, 2013.

[31] A. Buriro, B. Crispo, and Y. Zhauniarovich, "Please hold on: unobtrusive user authentication using smartphone's built-in sensors," in *Proceedings of the IEEE International Conference on Identity, Security and Behavior Analysis (ISBA)*, pp. 1–8, IEEE, New Delhi, India, February 2017.

[32] Wikipedia, "Risk-based authentication," 2017, https://en.wikipedia.org/wiki/Risk-based_authentication.

[33] A. J. Harris and D. C. Yen, "Biometric authentication: assuring access to information," *Information Management & Computer Security*, vol. 10, no. 1, pp. 12–19, 2002.

[34] NuData Security, "What is risk based authentication?," 2017, https://nudatasecurity.com/blog/ecommerce/what-is-risk-based-authentication/.

[35] R. Hulsebosch, M. S. Bargh, G. Lenzini, P. Ebben, and S. M. Iacob, "Context sensitive adaptive authentication," in *Proceedings of the European Conference on Smart Sensing and Context*, pp. 93–109, Springer, Kendal, UK, October 2007.

[36] RSA, "RSA adaptive authentication system," 2017, https://www.rsa.com/content/dam/rsa/PDF/h9096-rsa-risk-engine-sb-11-2.pdf.

[37] A. K. Jain, S. Pankanti, S. Prabhakar, L. Hong, and A. Ross, "Biometrics: a grand challenge," in *Proceedings of the 17th International Conference on Pattern Recognition (ICPR)*, vol. 2, pp. 935–942, IEEE, Cambridge, UK, August 2004.

[38] A. Jain, K. Nandakumar, and A. Ross, "Score normalization in multimodal biometric systems," *Pattern Recognition*, vol. 38, no. 12, pp. 2270–2285, 2005.

[39] A. Ross and A. Jain, "Information fusion in biometrics," *Pattern recognition letters*, vol. 24, no. 13, pp. 2115–2125, 2003.

[40] C. Braz and J.-M. Robert, "Security and usability: the case of the user authentication methods," in *Proceedings of the 18th Conference on l'Interaction Homme-Machine*, pp. 199–203, ACM, New York, NY, USA, 2006.

[41] K.-P. Yee, "User interaction design for secure systems," in *Proceedings of the 4th International Conference on Information and Communications Security*, pp. 278–290, Singapore, December 2002.

[42] Microsoft, "Key principles of software architecture," 2018, https://msdn.microsoft.com/en-us/library/ee658124.aspx.

[43] J. Brooke, *SUS-A Quick and Dirty Usability Scale: Usability Evaluation in Industry*, Taylor and Francis, Oxford, UK, 1996.

[44] Usability, "System usability scale (sus)," 2017, https://www.usability.gov/how-to-and-tools/methods/system-usability-scale.html.

[45] M. F. Theofanos, R. J. Micheals, and B. C. Stanton, "Biometrics systems include users," *IEEE Systems Journal*, vol. 3, no. 4, pp. 461–468, 2009.

[46] ISO, *Human-Centred Design Processes for Interactive Systems*, International Organization for Standardization, Geneva, Switzerland, 1999.

[47] M. A. Sasse, "Red-eye blink, bendy shuffle, and the yuck factor: a user experience of biometric airport systems," *IEEE Security & Privacy*, vol. 5, no. 3, pp. 78–81, 2007.

[48] M. Harbach, E. Von Zezschwitz, A. Fichtner, A. De Luca, and M. Smith, "It's a hard lock life: a field study of smartphone (un) locking behavior and risk perception," in *Proceedings of the Symposium on Usable Privacy and Security (SOUPS)*, pp. 9–11, Menlo Park, CA, USA, July 2014.

[49] A. De Luca, A. Hang, E. Von Zezschwitz, and H. Hussmann, "I feel like I'm taking selfies all day!: towards understanding biometric authentication on smartphones," in *Proceedings of the 33rd Annual ACM Conference on Human Factors in Computing Systems*, pp. 1411–1414, ACM, New York, NY, USA, 2015.

[50] T. Sloane, "Behavioral biometrics: the restructuring of the authentication landscape," 2017, https://www.mercatoradvisorygroup.com/Webinars/Behavioral_Biometrics__The_Restructuring_of_the_Authentication_Landscape/.

[51] M. Winnick, "Putting a finger on our phone obsession," 2016, https://blog.dscout.com/mobile-touches.

[52] N. Cowan, C. C. Morey, Z. Chen, A. L. Gilchrist, and J. S. Saults, "Theory and measurement of working memory capacity limits," *Psychology of Learning and Motivation*, vol. 49, pp. 49–104, 2008.

[53] J. Zhang, X. Luo, S. Akkaladevi, and J. Ziegelmayer, "Improving multiple-password recall: an empirical study," *European Journal of Information Systems*, vol. 18, no. 2, pp. 165–176, 2009.

[54] A. Adams and M. A. Sasse, "Users are not the enemy," *Communications of the ACM*, vol. 42, no. 12, pp. 40–46, 1999.

[55] M. Awad, Z. Al-Qudah, S. Idwan, and A. H. Jallad, "Password security: password behavior analysis at a small university," in *Proceedings of the 5th International Conference on Electronic Devices, Systems and Applications (ICEDSA)*, pp. 1–4, IEEE, Ras Al Khaimah, UAE, December 2016.

[56] S. Komanduri, R. Shay, P. G. Kelley et al., "Of passwords and people: measuring the effect of password-composition policies," in *Proceedings of the SIGCHI Conference on Human Factors in Computing Systems*, pp. 2595–2604, ACM, New York, NY, USA, 2011.

[57] J. Bonneau, S. Preibusch, and R. J. Anderson, "A birthday present every eleven wallets? The security of customer-chosen banking pins," in *Proceedings of the 16th International Conference Financial Cryptography and Data Security*, vol. 7397, pp. 25–40, Springer, Kralendijk, Bonaire, February-March 2012.

[58] D. Silver, S. Jana, D. Boneh, E. Y. Chen, and C. Jackson, "Password managers: attacks and defenses," in *Proceedings of the 23rd USENIX Security Symposium*, pp. 449–464, San Diego, CA, USA, August 2014.

[59] T. Bhattasali, K. Saeed, N. Chaki, and R. Chaki, "A survey of security and privacy issues for biometrics based remote authentication in cloud," in *Proceedings of the IFIP International Conference on Computer Information Systems and Industrial Management*, pp. 112–121, Springer, Ho Chi Minh City, Vietnam, November 2014.

[60] PandaSecurities, "No password? You're asking to be hacked," 2016, https://www.pandasecurity.com/mediacenter/tips/smartphone-risk-dont-use-password.

[61] C. Katsini, M. Belk, C. Fidas, N. Avouris, and G. Samaras, "Security and usability in knowledge-based user authentication: a review," in *Proceedings of the 20th Pan-Hellenic Conference on Informatics*, p. 63, ACM, Patras, Greece, November 2016.

[62] J. K. Thorpe, *On the Predictability and Security of User Choice in Passwords*, Carleton University, Ottawa, ON, Canada, 2008.

[63] Secure Group, "Lock pattern, pin, or password: What is the most reliable way to lock a phone," 2017, https://blog.securegroup.com/lock-pattern-pin-or-password-what-is-the-most-reliable-way-to-lock-a-phone.

[64] G. Ye, Z. Tang, D. Fang et al., "Cracking android pattern lock in five attempts," in *Proceedings of the Network and Distributed System Security Symposium*, San Diego, CA, USA, February-March 2017.

[65] F. Tari, A. Ozok, and S. H. Holden, "A comparison of perceived and real shoulder-surfing risks between alphanumeric and graphical passwords," in *Proceedings of the Second Symposium on Usable Privacy and Security*, pp. 56–66, ACM, Pittsburgh, PA, USA, July 2006.

[66] A. J. Aviv, K. L. Gibson, E. Mossop, M. Blaze, and J. M. Smith, *Smudge Attacks on Smartphone Touch Screens*, Vol. 10, Woot, Carrollton, TX, USA, 2010.

[67] X. Suo, Y. Zhu, and G. S. Owen, "Graphical passwords: a survey," in *Proceedings of the 21st Annual Computer Security Applications Conference*, p. 10, IEEE, Tucson, AZ, USA, December 2005.

[68] CAPEC-Release1.6, "Common attack pattern enumeration and classification," 2016, http://capec.mitre.org.

[69] M. Mehrnezhad, E. Toreini, S. F. Shahandashti, and F. Hao, "Stealing pins via mobile sensors: actual risk versus user perception," *International Journal of Information Security*, pp. 1–23, 2016.

[70] A. Sarkisyan, R. Debbiny, and A. Nahapetian, "Wristsnoop: smartphone pins prediction using smartwatch motion sensors," in *IEEE International Workshop on Information Forensics and Security (WIFS)*, pp. 1–6, IEEE, Abu Dhabi, UAE, December 2015.

[71] F. Breitinger and C. Nickel, "User survey on phone security and usage," in *Proceedings of the Biometrics and Electronic Signatures (BIOSIG)*, pp. 139–144, Darmstadt, Germany, September 2010.

[72] M. Meeker and L. Wu, "Kleiner Perkins Caufield and Byers (KPCB): internet trends," 2017, http://www.kpcb.com/internet-trends.

[73] H. Choi, H. Kwon, and J. Hur, "A secure OTP algorithm using a smartphone application," in *Proceedings of the Seventh International Conference on Ubiquitous and Future Networks (ICUFN)*, pp. 476–481, IEEE, Sapporo, Japan, July 2015.

[74] H. Sun, K. Sun, Y. Wang, and J. Jing, "TrustOTP: transforming smartphones into secure one-time password tokens," in *Proceedings of the 22nd ACM SIGSAC Conference on Computer and Communications Security*, pp. 976–988, ACM, Denver, CO, USA, October 2015.

[75] B. Cha, N. Kim, and J. Kim, "Prototype analysis of OTP key-generation based on mobile device using voice characteristics," in *Proceedings of the International Conference on Information Science and Applications (ICISA)*, pp. 1–5, IEEE, Jeju, Korea, April 2011.

[76] Verizon, "How long since you took a hard look at your cybersecurity?," 2017, http://www.verizonenterprise.com/verizon-insights-lab/dbir/2017.

[77] I. Agadakos, C.-Y. Chen, M. Campanelli et al., *Jumping the Air Gap: Modeling Cyber-Physical a Ack Paths in the Internet-of-Things*, ACM, New York, NY, USA, 2017.

[78] M. Fomichev, F. Alvarez, D. Steinmetzer, P. Gardner-Stephen, and M. Hollick, "Survey and systematization of secure device pairing," *IEEE Communications Surveys & Tutorials*, vol. 20, no. 1, pp. 517–550, 2018.

[79] M. Belk, P. Germanakos, C. Fidas, and G. Samaras, "A personalization method based on human factors for improving usability of user authentication tasks," in *Proceedings of International Conference on User Modeling, Adaptation, and Personalization*, pp. 13–24, Springer, Aalborg, Denmark, July 2014.

[80] M.-C. Chuang and M. C. Chen, "An anonymous multi-server authenticated key agreement scheme based on trust computing using smart cards and biometrics," *Expert Systems with Applications*, vol. 41, no. 4, pp. 1411–1418, 2014.

[81] K. Krol, E. Philippou, E. De Cristofaro, and M. A. Sasse, ""They brought in the horrible key ring thing!" Analysing the usability of two-factor authentication in UK online banking," arXiv preprint arXiv:1501.04434, 2015.

[82] T. Zink and M. Waldvogel, "X. 509 user certificate-based two-factor authentication for web applications," in *Proceedings of the 10. DFN-Forum Kommunikationstechnologien*, pp. 51–61, Berlin, Germany, May 2017.

[83] K. J. Heffernan, F. Vetere, L. M. Britton, B. Semaan, and T. Schiphorst, "Insertable digital devices: voluntarily under the skin," in *Proceedings of the 2016 ACM Conference Companion Publication on Designing Interactive Systems*, pp. 85–88, ACM, Brisbane, QLD, Australia, June 2016.

[84] P. Strohmeier, C. Honnet, and S. Von Cyborg, "Developing an ecosystem for interactive electronic implants," in *Proceedings of the Conference on Biomimetic and Biohybrid Systems*, pp. 518–525, Springer, Edinburgh, UK, July 2016.

[85] M. Janiak, C. Schaub, D. Lynam, B. Howe, G. Wachter, and G. Krueger, "Biometric authentication device for use with a personal digital assistant," US Patent App. 09/854,078, 2001.

[86] K. J. Heffernan, "Insertables workshop," 2016, https://insertables.wordpress.com/.

[87] TechTarget, "Bespoke," 2017, http://whatis.techtarget.com/definition/bespoke.

[88] M. Gupta, C. Holloway, B. M. Heravi, and S. Hailes, "A comparison between smartphone sensors and bespoke sensor devices for wheelchair accessibility studies," in *Proceedings of the IEEE 10th International Conference on Intelligent Sensors, Sensor Networks and Information Processing (ISSNIP)*, pp. 1–6, IEEE, Singapore, April 2015.

[89] First4magnets, "Use of neodymium magnets," 2016, https://www.first4magnets.com/tech-centre-i61/information-and-articles-i70/neodymium-magnet-information-i82/common-applications-of-neodymium-magnets-i88.

[90] P. Urien and S. Piramuthu, "Framework and authentication protocols for smartphone, NFC, and RFID in retail transactions," in *Proceedings of the IEEE 8th International Conference on Intelligent Sensors, Sensor Networks and Information Processing*, pp. 77–82, IEEE, Melbourne, VIC, Australia, April 2013.

[91] I. J. Forster and A. N. Farr, "Method for preventing unauthorized diversion of nfc tags," US Patent App. 15/659,941, 2017.

[92] H. Hassan, S. Wacquant, and H. B. Seifert, "Vehicle driver monitoring system," US Patent App. 15/463,293, 2017.

[93] M. Cuff, "Smart digital tattoos," 2014, http://www.stylus.com/scckpj.

[94] K. J. Heffernan, F. Vetere, and S. Chang, "Towards insertables: devices inside the human body," *First Monday*, vol. 22, no. 3, 2017.

[95] R. M. Bolle, J. H. Connell, S. Pankanti, N. K. Ratha, and A. W. Senior, *Guide To Biometrics*, Springer Science & Business Media, Berlin, Germany, 2013.

[96] R. Ramachandra and C. Busch, "Presentation attack detection methods for face recognition systems: a comprehensive survey," *ACM Computing Surveys*, vol. 50, no. 1, pp. 1–37, 2017.

[97] A. K. Jain, K. Nandakumar, and A. Ross, "50 years of biometric research: accomplishments, challenges, and opportunities," *Pattern Recognition Letters*, vol. 79, pp. 80–105, 2016.

[98] A. Hadid, N. Evans, S. Marcel, and J. Fierrez, "Biometrics systems under spoofing attack: an evaluation methodology and lessons learned," *IEEE Signal Processing Magazine*, vol. 32, no. 5, pp. 20–30, 2015.

[99] S. Kovach, "Business insider-Samsung's Galaxy S8 facial recognition feature can be fooled with a photo," 2017, http://www.businessinsider.com/samsung-galaxy-s8-facial-recognition-tricked-with-a-photo-2017-3?IR=T.

[100] J. Titcomb, "Hackers claim to beat iPhone X's face id in one week with 115 mask," 2017, http://www.telegraph.co.uk/technology/2017/11/13/hackers-beat-iphone-xs-face-one-week-115-mask/.

[101] H. Saevanee, N. Clarke, S. Furnell, and V. Biscione, "Continuous user authentication using multi-modal biometrics," *Computers & Security*, vol. 53, pp. 234–246, 2015.

[102] A. Hern, "The guardian-Samsung Galaxy S8 iris scanner fooled by German hackers," 2017, https://www.theguardian.com/technology/2017/may/23/samsung-galaxy-s8-iris-scanner-german-hackers-biometric-security.

[103] A. Charles, "The guardian-iPhone 5S fingerprint sensor hacked by Germany's Chaos Computer Club," 2013, https://www.theguardian.com/technology/2013/sep/22/apple-iphone-fingerprint-scanner-hacked.

[104] C. McGoogan and D. Demetriou, "The telegraph-peace sign selfies could let hackers copy your fingerprints," 2017, http://www.telegraph.co.uk/technology/2017/01/12/peace-sign-selfies-could-let-hackers-copy-fingerprints.

[105] C. Bhagavatula, B. Ur, K. Iacovino, S. M. Kywe, L. F. Cranor, and M. Savvides, "Biometric authentication on iPhone and Android: usability, perceptions, and influences on adoption," in *Proceedings of the Workshop on Usable Security (USEC)*, pp. 1-2, San Diego, CA, USA, January 2015.

[106] A. Jain, R. Bolle, and S. Pankanti, *Biometrics: Personal Identification in Networked Society*, Vol. 479, Springer Science & Business Media, Berlin, Germany, 2006.

[107] M. Kumar, A. Insan, N. Stoll, K. Thurow, and R. Stoll, "Stochastic fuzzy modeling for ear imaging based child identification," *IEEE Transactions on Systems, Man, and Cybernetics: Systems*, vol. 46, no. 9, pp. 1265–1278, 2016.
[108] M. Sultana, M. Gavrilova, and S. Yanushkevich, "Multi-resolution fusion of DTCWT and DCT for shift invariant face recognition," in *Proceedings of the IEEE International Conference on Systems, Man and Cybernetics (SMC)*, pp. 80–85, IEEE, Hong Kong, China, October 2014.
[109] Y. Xu, X. Fang, X. Li et al., "Data uncertainty in face recognition," *IEEE transactions on cybernetics*, vol. 44, no. 10, pp. 1950–1961, 2014.
[110] C. Song, A. Wang, K. Ren, and W. Xu, "EyeVeri: a secure and usable approach for smartphone user authentication," in *Proceedings of the 35th Annual IEEE International Conference on Computer Communications, IEEE INFOCOM 2016*, pp. 1–9, IEEE, San Francisco, CA, USA, April 2016.
[111] IBIA, "Behavioral biometrics," 2017, https://www.ibia.org/biometrics-and-identity/biometric-technologies/behavioral-biometrics.
[112] L. M. Mayron, "Behavioral biometrics for universal access and authentication," in *Proceedings of the International Conference on Universal Access in Human-Computer Interaction*, pp. 330–339, Springer, Los Angeles, CA, USA, August 2015.
[113] A. Buriro, B. Crispo, F. Del Frari, J. Klardie, and K. Wrona, "ITSME: multi-modal and unobtrusive behavioural user authentication for smartphones," in *Proceedings of the International Conference on Passwords*, pp. 45–61, Springer, Cambridge, UK, December 2015.
[114] A. Buriro, B. Crispo, F. Del Frari, and K. Wrona, "Touchstroke: smartphone user authentication based on touch-typing biometrics," in *Proceedings of the International Conference on Image Analysis and Processing*, pp. 27–34, Springer, Genoa, Italy, September 2015.
[115] N. Zheng, K. Bai, H. Huang, and H. Wang, "You are how you touch: user verification on smartphones via tapping behaviors," in *Proceedings of the IEEE 22nd International Conference on Network Protocols (ICNP)*, pp. 221–232, IEEE, Raleigh, NC, USA, October 2014.
[116] M. Muaaz and R. Mayrhofer, "Smartphone-based gait recognition: from authentication to imitation," *IEEE Transactions on Mobile Computing*, vol. 16, no. 11, pp. 3209–3221, 2017.
[117] M. R. Hestbek, C. Nickel, and C. Busch, "Biometric gait recognition for mobile devices using wavelet transform and support vector machines," in *Proceedings of the 19th International Conference on Systems, Signals and Image Processing (IWSSIP)*, pp. 205–210, IEEE, Vienna, Austria, April 2012.
[118] A. Muro-De-La-Herran, B. Garcia-Zapirain, and A. Mendez-Zorrilla, "Gait analysis methods: an overview of wearable and non-wearable systems, highlighting clinical applications," *Sensors*, vol. 14, no. 12, pp. 3362–3394, 2014.
[119] M. Sultana, P. P. Paul, and M. L. Gavrilova, "Social behavioral information fusion in multimodal biometrics," *IEEE Transactions on Systems, Man, and Cybernetics: Systems*, 2017.
[120] H. Khan, U. Hengartner, and D. Vogel, "Usability and security perceptions of implicit authentication: convenient, secure, sometimes annoying," in *Proceedings of the Symposium on Usable Privacy and Security (SOUPS 2015)*, pp. 225–239, Santa Clara, CA, USA, July 2015.
[121] N. Poh and J. Korczak, "Hybrid biometric person authentication using face and voice features," in *Proceedings of the 3rd International Conference Audio- and Video-Based Biometric Person Authentication (AVBPA)*, vol. 1, pp. 348–353, Springer, Halmstad, Sweden, June 2001.
[122] R. Brunelli and D. Falavigna, "Person identification using multiple cues," *IEEE Transactions on Pattern Analysis and Machine Intelligence*, vol. 17, no. 10, pp. 955–966, 1995.
[123] K. Brunet, K. Taam, E. Cherrier, N. Faye, and C. Rosenberger, "Speaker recognition for mobile user authentication: an Android solution," in *8ème Conférence sur la Sécurité des Architectures Réseaux et Systèmes d'Information (SAR SSI)*, p. 10, Mont-de-Marsan, France, September 2013.
[124] K. Murao, H. Tobise, T. Terada, T. Iso, M. Tsukamoto, and T. Horikoshi, "Mobile phone user authentication with grip gestures using pressure sensors," *International Journal of Pervasive Computing and Communications*, vol. 11, no. 3, pp. 288–301, 2015.
[125] J. Kittler, J. Matas, K. Jonsson, and M. R. Sánchez, "Combining evidence in personal identity verification systems," *Pattern Recognition Letters*, vol. 18, no. 9, pp. 845–852, 1997.
[126] H. Saevanee, N. Clarke, and S. Furnell, "Multi-modal behavioural biometric authentication for mobile devices," in *Information Security and Privacy Research*, pp. 465–474, Springer, Heidelberg, Germany, 2012.
[127] Y. Tang, N. Hidenori, and Y. Urano, "User authentication on smart phones using a data mining method," in *Proceedings of the International Conference on Information Society (i-Society)*, pp. 173–178, IEEE, London, UK, June 2010.
[128] S. M. Welten, *Sensing with Smartphones*, 2013.
[129] Y. Obuchi, "PDA speech database," 2006, http://www.speech.cs.cmu.edu/databases/pda/index.html.
[130] S. Eberz, K. B. Rasmussen, V. Lenders, and I. Martinovic, "Evaluating behavioral biometrics for continuous authentication: challenges and metrics," in *Proceedings of the 2017 ACM on Asia Conference on Computer and Communications Security*, pp. 386–399, ACM, Abu Dhabi, UAE, April 2017.
[131] U. Mahbub, S. Sarkar, V. M. Patel, and R. Chellappa, "Active user authentication for smartphones: a challenge data set and benchmark results," in *Proceedings of the IEEE 8th International Conference on Biometrics Theory, Applications and Systems (BTAS 2016)*, pp. 1–8, IEEE, Buffalo, NY, USA, September 2016.
[132] A. Stolerman, A. Fridman, R. Greenstadt, P. Brennan, and P. Juola, "Active linguistic authentication revisited: real-time stylometric evaluation towards multi-modal decision fusion," in *Proceedings of the Tenth Annual IFIP WG 11.9 International Conference on Digital Forensics*, vol. 11, pp. 1–11, Vienna, Austria, January 2014.
[133] R. P. Guidorizzi, "Security: active authentication," *IT Professional*, vol. 15, no. 4, pp. 4–7, 2013.
[134] L. Fridman, S. Weber, R. Greenstadt, and M. Kam, "Active authentication on mobile devices via stylometry, application usage, web browsing, and GPS location," *IEEE Systems Journal*, vol. 11, no. 2, pp. 513–521, 2017.
[135] M. L. Brocardo, I. Traore, and I. Woungang, "Toward a framework for continuous authentication using stylometry," in *Proceedings of the IEEE 28th International Conference on Advanced Information Networking and Applications (AINA 2014)*, pp. 106–115, IEEE, Victoria, Canada, May 2014.
[136] ClearLogin, "Risk-based authentication," 2017, http://www.clearlogin.com/glossary/risk-based-authentication/.
[137] B. Schneier, "Risk-based authentication," 2013, https://www.schneier.com/blog/archives/2013/11/risk-based_auth.html.
[138] Identity Automation, "Risk-based authentication," 2017, https://www.identityautomation.com/iam-platform/rapididentityidentity-access-management/multi-factor-authentication/risk-based-authentication/.

[139] D. Hintze, E. Koch, S. Scholz, and R. Mayrhofer, "Location-based risk assessment for mobile authentication," in *Proceedings of the 2016 ACM International Joint Conference on Pervasive and Ubiquitous Computing: Adjunct*, pp. 85–88, ACM, Heidelberg, Germany, September 2016.

[140] Y. Y. Haimes, *Risk Modeling, Assessment, and Management*, John Wiley & Sons, Hoboken, NJ, USA, 2015.

[141] IdentityAutomation, "What is adaptive authentication?," 2017, http://blog.identityautomation.com/what-is-adaptive-authentication.

[142] A. Jain and A. Kumar, "Biometric recognition: an overview," in *Second Generation Biometrics: The Ethical, Legal and Social Context*, E. Mordini and D. Tzovaras, Eds., pp. 49–79, Springer, Dordrecht, Netherlands, 2012.

[143] P. P. Paul, M. L. Gavrilova, and R. Alhajj, "Decision fusion for multimodal biometrics using social network analysis," *IEEE Transactions on Systems, Man, and Cybernetics: Systems*, vol. 44, no. 11, pp. 1522–1533, 2014.

7

The Dangers of Rooting: Data Leakage Detection in Android Applications

Luca Casati and Andrea Visconti

Department of Computer Science, Università degli Studi di Milano, Via Comelico 39/41, 20135 Milano, Italy

Correspondence should be addressed to Andrea Visconti; andrea.visconti@unimi.it

Academic Editor: Jinglan Zhang

Mobile devices are widely spread all over the world, and Android is the most popular operative system in use. According to Kaspersky Lab's threat statistic (June 2017), many users are tempted to root their mobile devices to get an unrestricted access to the file system, to install different versions of the operating system, to improve performance, and so on. The result is that unintended data leakage flaws may exist. In this paper, we (i) analyze the security issues of several applications considered relevant in terms of handling user sensitive information, for example, financial, social, and communication applications, showing that 51.6% of the tested applications suffer at least of an issue and (ii) show how an attacker might retrieve a user access token stored inside the device thus exposing users to a possible identity violation. Notice that such a token, and a number of other sensitive information, can be stolen by malicious users through a man-in-the-middle (MITM) attack.

1. Introduction

In everyday routine, smartphones, laptops, tablets or, more in general, mobile devices have become an essential need for everyone. They are widely used to read e-mails, carry out financial transactions, browse maps, chat with other people, and so on. Mobile devices have to face a number of issues due to the resource constraints (performance issue [1, 2], e.g.) and also security issues (data leakage [3, 4], privacy concern [5, 6], etc.). In particular, the latter may be affected by the applications installed. Usually users choose such applications focusing on the number of total downloads [7], the reviews provided by users [8, 9], and so on. A typical environment where ratings can be easily found is *Google Play Store*, the largest app store which counts over 3 million applications available [10] split into two major categories: *Apps* and *Games*—with 2.5 million and 500 thousand apps, respectively [11]. However, it often happens that people who provide ratings evaluate the appearance, functionality, usability, and performances of an application without focusing on security aspects. In addition, as reported in Kaspersky Lab's threat statistic (June 2017) [12] summarized in Table 1, security issues are further amplified by users when they root their phones. Notice that users obtain superuser access privileges to change the current Android version, to get access to the file system without restrictions, to install modified apps and gain more privileges, to improve performance, and so on. However, these access privileges may affect the security of installed applications [12, 13, 15], providing an access door to many sensitive information [16–18]. In this scenario, unintended data leakage flaws may exist.

In order to identify such flaws, in this paper, we extend and improve our previous work [19]. In particular, we improve our testing activities by analyzing not only the security issues of Android password managers but also those applications that are considered particularly relevant in terms of handling user sensitive information, such as financial, social, and communication applications. Notice that we do not describe innovative techniques but rather we measure the impact of a well-known technique (e.g., Xposed framework) on a rooted device, executing an extensive testing activities and observing that several applications do not implement the minimum security requirements.

TABLE 1: The top 10 (out of 100) countries where Android devices are rooted most frequently and where mobile devices are attacked most often by a malware [12].

Country	Rooted devices	Place in top 100 countries attacked
Bangladesh	13%	2
Indonesia	12%	3
Nepal	12%	5
Algeria	19%	7
Nigeria	13%	9
Ghana	12%	10
Venezuela	26%	13
Moldova	15%	22
Ecuador	11%	25
Italy	12%	66

In addition, we show the possibility to retrieve an access token, exposing users to a possible identity violation. Finally, we show that the same token (and many other sensitive information) can be retrieved through a man-in-the-middle (MITM) attack because several applications do not implement adequate cryptographic techniques for data protection or do not implement them at all.

The remainder of the paper is organized as follows. In Section 2, we describe a number of approaches that can be used to analyze applications. In Section 3, we show the solution adopted to retrieve sensitive information from Android applications. Particular attention is paid to describe hooking techniques. In Section 4, we present our testing activities, showing how malicious users might retrieve sensitive information. Finally, conclusions are drawn in Section 5.

2. Different Approaches to Analyze Applications

When an application lands on the market, it becomes suddenly available to be used by everyone. This means that it can be tested and analyzed under all possible conditions. Every internal element of an app should share the necessary information to perform a specific task without any data leakage. Unfortunately, this does not always happen.

In order to recognize possible data leakages, two well-known approaches can be used: *static* and *dynamic analysis*.

(i) Static analysis is based on the examination of an application without the execution of it [20]. Its radius of action is quite limited because many applications adopt obfuscation [21, 22] and dynamic code loading [23] to restrict access to internal information. However, it may be interesting to understand if the application's associated files, such as database, backup, or log files, are encrypted. In this case, entropic techniques are very useful [24].

(ii) Dynamic analysis, instead, relies on the execution of the applications [25, 26]. The main idea is to collect (at runtime) the values that gradually come out from the called instructions. The advantage of this approach is to be less susceptible to code obfuscation. In general, Android applications can assume many behaviors; thus it is necessary to monitor their activities, for example, through interface or automatic event injectors [27–29].

But there is also a third approach, situated halfway between the previous: the *hybrid analysis* [30, 31]. To work well, a system which adopts this technique must be designed in such a way that if the first was lacking, the second would take place, covering the gap [30].

In mobile device analysis, there is no a standard approach (static or dynamic) to collect data optimally. More precisely, we collect data via static analysis, and then we employ them in a dynamic scanning. This was accomplished through hooking techniques (hooking means to intercept methods with a known signature called by an application, acquiring its complete control), setting up the scenario shown in Figure 1. Taking into account a Java class named *Signature*, we notice that (a) the method *initSign* is invoked, (b) *initSign* receives a *PrivateKey* object, (c) *initSign* pass the object itself to another method, that is, *engineInitSign* of Figure 1, and (d) *Hooker* could take control of the method call, spying, or replacing its contents.

To better understand how this mechanism works, we explain in detail the hooking techniques—*Xposed* framework [32]—in Section 3.

3. How to Retrieve Sensitive Information

A generic Android application is a single compressed archive which includes essential information about the app [33]. Among all this information, we focus on the DEX file (Figure 2) because it provides interesting features related to the target application [34, 35].

We developed a tool, called *Apk2Method*, which

(i) opens the APK of the target application;

(ii) identifies the *classes.dex* file that looks for a specific marker, that is, 6465 780a 3033 3500 in *Hex*;

(iii) reads all methods invoked related to cryptographic field;

(iv) finally, outputs a text file where all gathered data are stored in a convenient format for a subsequent parsing. For the sake of simplicity, we call such a file *file.txt*.

Then, we developed an Android application which

(v) inputs data previously stored in *file.txt* and parses such a file using Java reflections and regular expressions;

(vi) runs inside a module of the Xposed framework, called *Prober*, which is able to select the target application.

More precisely, *Prober* represents the real execution engine of the hooking technique, implemented by *Xposed*. The Xposed framework, in turn, takes control of each method called by the target application, spying, or replacing each passed argument. Doing so, the control flow of an application can be changed, providing us the ability to execute our own code enriched with specific security tests.

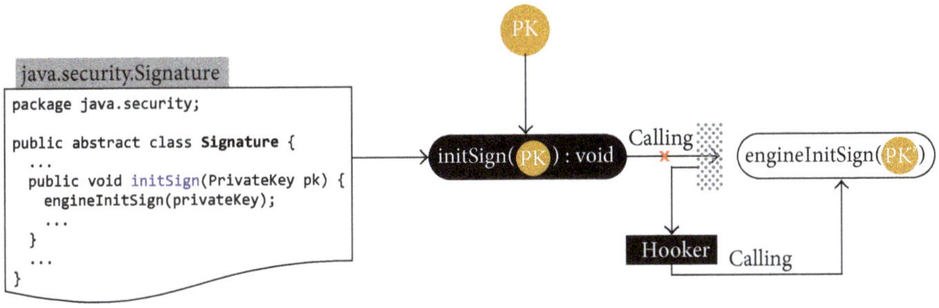

FIGURE 1: An example of the hooking technique in action, specialized in spying.

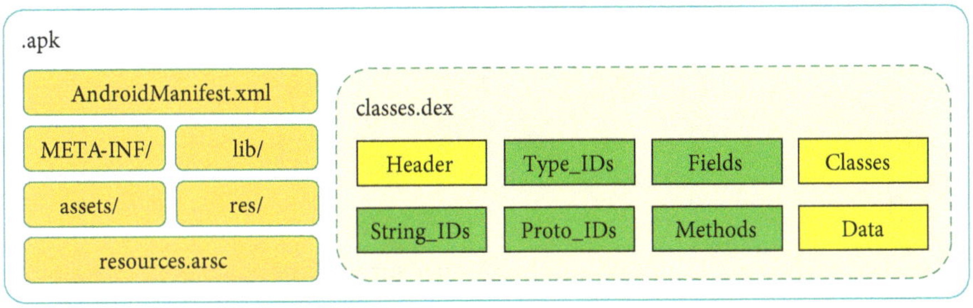

FIGURE 2: A compact view of an APK file, pointing out the DEX file components.

Notice that it may happen that a portion of the target application's information is encrypted or obfuscated [36], using specific tools such as Proguard, DashO, and DexProtector. These tools rename classes, methods, and variables assigning them meaningless names [37]. Consequently, the parsing activity will be very difficult and sometimes impossible (even with the support of the reflections [30]). In all other cases, if applications release sensitive information, our approach is able to detect these leaks.

3.1. The Xposed Framework. The framework used [32] is identified by four individual components: the *Xposed*, the *XposedBridge*, the *XposedInstaller*, and the *XposedMods* system. Among these, the first two are responsible for preparing the device to accommodate the framework. Let us briefly explain what happens when two generic methods, A and B, are called (Figures 3 and 4).

When the device is switched on,

(1) the boot sequence starts: (a) the Boot ROM code starts executing from a predefined location, loading the bootloader into RAM, (b) the bootloader sets up the necessary resources in two stages—network and memory—needed to run the kernel, (c) the Android kernel sets up a group of resources—cache, protected memory, scheduling, and drivers—and looks for *init* in the system files, (d) *init* is the very first process, which sets the environment for *Zygote* [38] and daemons, and (e) daemons are invoked;

(2) once the daemons are invoked, an extended version of process */system/bin/app_process* [39] is called, which is meant to load the necessary classes designed to perform hooking—*XposedBridge.jar*;

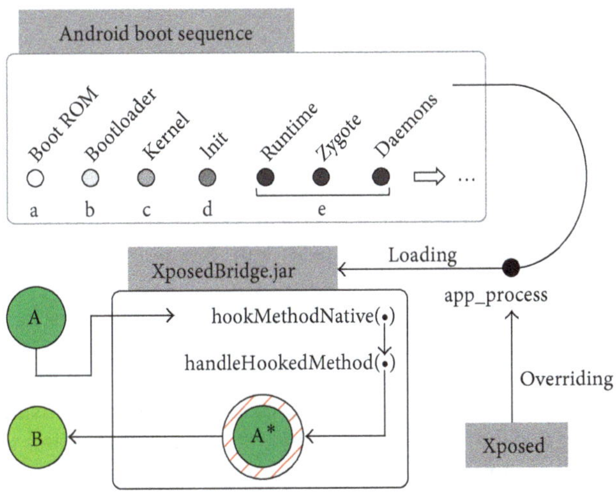

FIGURE 3: A diagram that shows how Xposed works in detail while intercepting a method.

(3) as soon as an application calls a generic method (A), it is intercepted and redirected firstly to *hookMethodNative*, which increases the privilege level of the method received as argument, and secondly to *handleHookedMethod*, which links the method implementation to its own native generic method. In this way, it is possible to read all the arguments;

(4) finally, the flow resumes naturally.

4. Testing Activity

We download and analyze several applications from Google's official Android Market, using two mobile devices—*Wiko Wax* (Android KitKat, rooted with *KingRoot* [40])

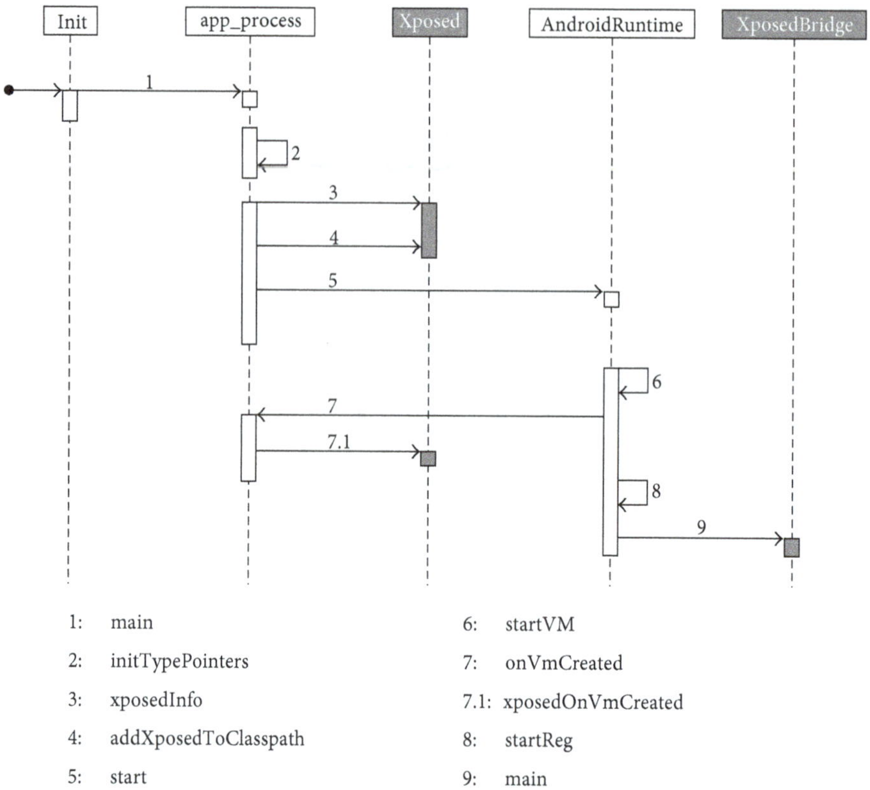

FIGURE 4: The sequence diagram illustrates how the system changes while the framework is active.

and *Samsung Galaxy Nexus* (Android Lollipop, rooted with *Nexus Root Toolkit* [41]) (at time of writing, Android KitKat and Lollipop represent nearly half (about 47%) of the market) [42].

Our analysis follows two main directions. A first approach targets events resulting from data leakage of the method calls. These leaks are usually characterized by an improper use of objects as arguments, for example, using string as passwords, making whole structures visible, and so on. Then, to improve the ability to recognize data leakage, a second approach has been developed with the aim to find leaks on data transmitted over the Internet by phone.

4.1. First Approach. We downloaded 135 Android applications from *Google Play Store*, where 36 applications belong to "TOOLS" category, 54 to "PRODUCTIVITY," 7 to "SOCIAL," 8 to "COMMUNICATION," and 30 to "FINANCE," taking care of the installation count value. Such indicators represent the number of users who installed the chosen application, and it can be found at the information panel of each application [43]. In addition, let us remark that the choice of a particular application was taken relying on the fact that is used for security purposes and deal with data that are particularly sensitive for user side. For each application, we collect and store classes, methods, arguments, and return values.

More precisely, our approach works as follows (Figure 5):

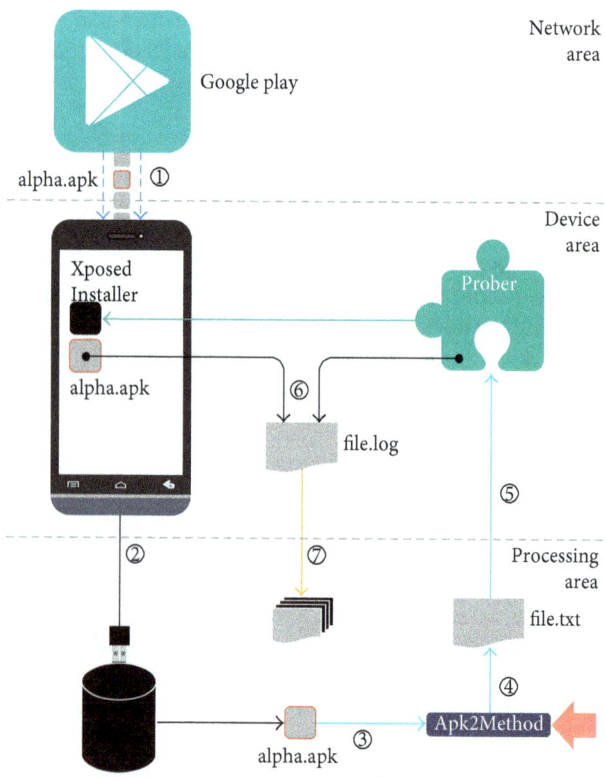

FIGURE 5: The entire project control flow which represents how an Android application is analyzed.

```
luca@epsilon:~$ java -jar ApkToMethod.jar --onlyCrypto alpha.apk
java.security.KeyFactory generatePrivate
java.security.KeyFactory generatePublic
java.security.KeyPair getPrivate
java.security.KeyPair getPublic
...
b.b.d.c.b c
b.b.d.c.c b
b.b.e.e.b.b.e engineGenerateSecret
b.b.e.e.d engineInitSign
...
```

file.txt

```
...
-> Found method called: public final java.security.PrivateKey java.security.K...
• Method Name:
public final native java.security.PrivateKey java.security.KeyPair.getPrivate()
• Class Name:
"java.security.KeyFactory"
• Arguments values:
java.security.spec.PKCS8EncodedKeySpec@41a77130
• Return values:
OpenSSLRSAPrivateKey{modulus=c6ad6617587b87164443b4bf666a20e06d44b25c432ab998...
...
```

file.log

FIGURE 6: A toy example of the outputs obtained by analyzing an application *alpha.apk*.

(1) An application *alpha.apk* is downloaded from *Google Play Store* and installed on the device.

(2) Then *alpha.apk* is transferred to the computer, using the Android Debug Bridge (ADB) [44].

(3) The *Apk2Method* tool inputs *alpha.apk*.

(4) The *Apk2Method* tool outputs classes and methods, storing them in *file.txt* previously mentioned in Section 3. The top of Figure 6 shows a toy example, pointing out that classes and methods of an application might be obfuscated.

(5) Such a file is copied in a specific path of our application *Prober*, and a rebooting of the mobile device is required to apply changes to the system.

(6) When the *alpha* application runs—for example, the user input ID, password, e-mail, personal data, and so on—*Prober* stores methods invoked, arguments, and return values in *file.log*, as shown in the lower part of Figure 6.

(7) Finally, in *file.log*, we are able to identify the presence of data leakage.

All apps analyzed have been cataloged using four levels of granularity: (1) *no leakage*: the application is safe; (2) *abnormal behavior*: the application suddenly freezes or crashes; (3) *privacy concerns*: the application releases unprotected sensitive information, that is, IMEI, phone number, geolocation, OS, and so on; and (4) *account info*: the application reveals account information—login IDs and passwords.

As shown in Tables 2 and 3 and in Figure 7, the testing results suggest that some issues have been identified for the category *tools*, *productivity*, and *finance*. In particular, in such categories, 51.6% of the tested applications suffer from one (at least) of the following issues:

(i) The application does not perceive to be observed.

TABLE 2: The results of the analysis obtained with the Android 5.*x* device.

	No leakage	Abnormal behavior	Privacy concerns	Secret data
Tools	18	0	2	18
Productivity	23	3	1	31
Social	7	5	0	0
Communication	8	3	0	0
Finance	17	16	7	13

(ii) The application does not warn the user about the presence of a jailbroken/rooted device.

(iii) Private keys used during a communication (e.g., the *OpenSSLRSAPrivateCrtKey* or the *RSAPrivateKey* and the associated parameters) are in plaintext.

(iv) Personal data, such as IMEI and geolocation, are not protected.

(v) The master password (of the password manager) or the users account password (login IDs and password) are handled in plaintext.

On the contrary, the applications tested which belong to *social* and *communication* are not affected by the same issues.

4.2. Second Approach. A second issue is related to the leakage of encrypted data transmitted over the Internet and stored in the device itself. To avoid a user being forced to create a new account, a common practice is to exploit a third-party app that handles the authentication phase using a delegation protocol—for example, OAuth 2.0 [45]. In particular, the authentication phase is done through an access token that is stored in the application's internal directory, preventing user from entering the login credentials

TABLE 3: Correlation between the installation count and the 4 levels of granularity.

Installation count	No leakage	Abnormal behavior	Privacy concerns	Secret data
1,000,000,000–5,000,000,000	4	1	0	0
500,000,000–1,000,000,000	3	1	0	0
100,000,000–500,000,000	6	4	0	0
50,000,000–100,000,000	2	2	0	0
10,000,000–50,000,000	2	0	0	1
1,000,000–5,000,000	3	5	2	9
500,000–1,000,000	4	2	0	7
100,000–500,000	19	8	5	11
50,000–100,000	7	3	1	3
10,000–50,000	10	1	2	9
5,000–10,000	3	0	0	4
50–5,000	10	1	0	18

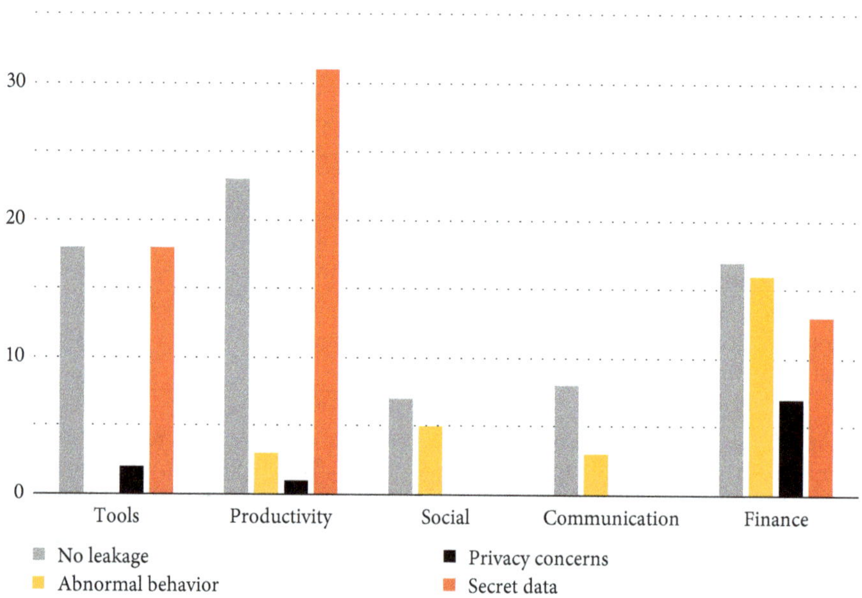

FIGURE 7: The histogram shows the results of all ranges of Table 2.

(see Alice in Figure 8). Since the access token can be seen as a set of user attributes used to prove that a user is authenticated, the client application usually does not use a mechanism to validate the access token, and in rooted devices, this token can be easily found by browsing the application's folder; an attacker may retrieve such a token and inject it during a new authentication phase, stealing the identity of the victim (see Eve in Figure 8). Moreover, for all users who ignore the alerts and unknowingly accept everything, the token may be stolen on the channel through a man-in-the-middle attack.

For this set of users, we also tried to identify different types of possible attacks. Therefore, we downloaded and analyzed 67 Android apps that send data over the Internet and should take care about user sensitive information. As described in Section 4.1, these applications belong to the following categories: 2 apps belong to "TOOLS," 16 to "PRODUCTIVITY," 4 to "SOCIAL," 10 to "COMMUNICATION," and 35 to "FINANCE." The main issue found is that several applications do not perform the SSL/TLS client authentication, thus making them potentially vulnerable to a man-in-the-middle attack. Tables 4 and 5 summarize our testing activities. More precisely, we found leaks on 55.2% of the apps tested, where 50.0% comes from "TOOLS," 75.0% from "PRODUCTIVITY," 25.0% from "SOCIAL," 60.0% from "COMMUNICATION," and 48.6% from "FINANCE."

5. Conclusions

Since mobile devices are widely spread and used for everything, the protection of information, transaction data, and privacy have to be taken into account seriously.

In this paper, we focused on the real case scenario of rooted devices, analyzing the most installed Android applications with the aim to check how safe they are. We

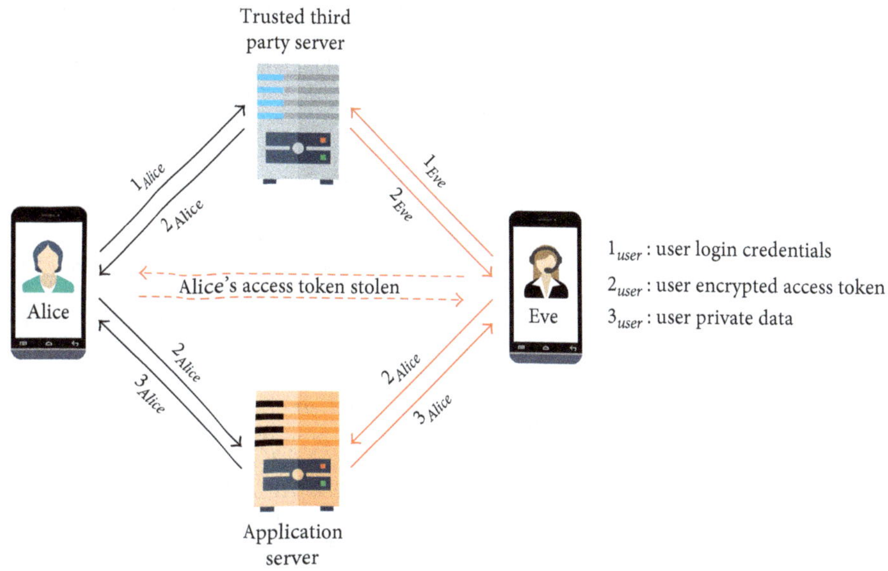

FIGURE 8: A graphical representation of the problem concerning the delegation scheme implemented by some applications.

TABLE 4: Number of apps that are potentially vulnerable to a MITM attack.

	Number of apps	MITM vulnerability
Tools	2	1
Productivity	16	12
Social	4	1
Communication	10	6
Finance	35	17

TABLE 5: Correlation between the installation count and MITM vulnerability.

Installation count	MITM vulnerability
1,000,000,000–5,000,000,000	3
500,000,000–1,000,000,000	2
100,000,000–500,000,000	5
50,000,000–100,000,000	1
10,000,000–50,000,000	3
1,000,000–5,000,000	5
500,000–1,000,000	1
100,000–500,000	11
50,000–100,000	3
10,000–50,000	1
5,000–10,000	0
50–5,000	1

showed that 62 out of 135 apps suffer of data leakage and 37 out of 67 apps, which send sensitive information over the Internet, are potentially vulnerable to man-in-the-middle attacks. The most significant flaws found concern (a) password managers (we assume that password managers store user passwords implementing the minimum requirements for cryptographic applications, for example, adopting a password-based key derivation function [46, 47] and avoiding the well-known issues described in literature [14, 48–50]) that may release ID password of several accounts or the master password of password manager themselves; (b) financial applications that sometimes release secret codes or account credentials; and (c) applications which do not implement a SSL/TLS client authentication, making them potentially vulnerable to a MITM attack. Notice that the issues described in this paper can be easily faced by app developers—for example, exploiting obfuscation/encryption mechanisms, passing sensitive data using objects, or implementing two-step verification techniques—and users—for example, installing a stock ROM instead of a custom one.

Acknowledgments

The authors would like to thank Marco Mauri and Giovanni Intorre who executed part of the testing activity described in Section 4.2.

References

[1] M. Louk, H. Lim, and H. Lee, "An analysis of security system for intrusion in smartphone environment," *The Scientific World Journal*, vol. 2014, Article ID 983901, 12 pages, 2014.

[2] M. Lettner, M. Tschernuth, and R. Mayrhofer, "Mobile platform architecture review: Android, iPhone, QT," in *Proceedings of the International Conference on Computer Aided Systems Theory*, pp. 544–551, Springer, Las Palmas de Gran Canaria, Spain, February 2011.

[3] P. Feng, J. Ma, and C. Sun, "Selecting critical data flows in Android applications for abnormal behavior detection," *Mobile Information Systems*, vol. 2017, Article ID 7397812, 16 pages, 2017.

[4] L. Wu, M. Grace, Y. Zhou, C. Wu, and X. Jiang, "The impact of vendor customizations on Android security," in *Proceedings of the 2013 ACM SIGSAC conference on Computer & communications security*, pp. 623–634, ACM, Berlin, Germany, November 2013.

[5] K. Zhu, X. He, B. Xiang, L. Zhang, and A. Pattavina, "How dangerous are your smartphones? App usage recommendation with privacy preserving," *Mobile Information Systems*, vol. 2016, Article ID 6804379, 10 pages, 2016.

[6] A. P. Felt, E. Ha, S. Egelman, A. Haney, E. Chin, and D. Wagner, "Android permissions: user attention, comprehension, and behavior," in *Proceedings of the Eighth Symposium on Usable Privacy and Security*, p. 3, ACM, New York, NY, USA, July 2012.

[7] J. Alegre-Sanahuja, J. Camacho, J. C. Cortés López, F. J. Santonja, and R. J. Villanueva Micó, "Agent-based model to study and quantify the evolution dynamics of Android malware infection," *Abstract and Applied Analysis*, vol. 2014, Article ID 623436, 10 pages, 2014.

[8] N. Viennot, E. Garcia, and J. Nieh, "A measurement study of Google Play," *ACM SIGMETRICS Performance Evaluation Review*, vol. 42, no. 1, pp. 221–233, 2014.

[9] E. Guzman and W. Maalej, "How do users like this feature? A fine grained sentiment analysis of app reviews," in *Proceedings of the IEEE 22nd International Requirements Engineering Conference (RE)*, pp. 153–162, IEEE, Karlskrona, Sweden, August 2014.

[10] AppBrain, "Number of Android applications," 2017, http://www.appbrain.com/stats/number-of-android-apps.

[11] AppBrain, "Most popular Google Play categories," 2017, http://www.appbrain.com/stats/android-market-app-categories.

[12] Kaspersky Lab, "Rooting your Android: advantages, disadvantages, and snags," 2017, https://www.kaspersky.com/blog/android-root-faq/17135/.

[13] W. Jeon, J. Kim, Y. Lee, and D. Won, "A practical analysis of smartphone security," in *Human Interface and the Management of Information. Interacting with Information*, pp. 311–320, 2011.

[14] S. Bossi and A. Visconti, "What users should know about full disk encryption based on LUKS," in *Proceedings of the 14th International Conference on Cryptology and Network Security*, Marrakesh, Morocco, December 2015.

[15] C. Vorakulpipat, S. Sirapaisan, E. Rattanalerdnusorn, and V. Savangsuk, "A policy-based framework for preserving confidentiality in BYOD environments: a review of information security perspectives," *Security and Communication Networks*, vol. 2017, Article ID 2057260, 11 pages, 2017.

[16] S. T. Sun, A. Cuadros, and K. Beznosov, "Android rooting: methods, detection, and evasion," in *Proceedings of the 5th Annual ACM CCS Workshop on Security and Privacy in Smartphones and Mobile Devices*, pp. 3–14, ACM, Denver, CO, USA, October 2015.

[17] Y. Kim, T. Oh, and J. Kim, "Analyzing user awareness of privacy data leak in mobile applications," *Mobile Information Systems*, vol. 2015, Article ID 369489, 12 pages, 2015.

[18] M. Nauman, S. Khan, X. Zhang, and J. P. Seifert, "Beyond kernel-level integrity measurement: enabling remote attestation for the android platform," in *Proceedings of the Trust and Trustworthy Computing*, pp. 1–15, Berlin, Germany, June 2010.

[19] L. Casati and A. Visconti, "Exploiting a bad user practice to retrieve data leakage on android password managers," in *Proceedings of the International Conference on Innovative Mobile and Internet Services in Ubiquitous Computing*, pp. 952–958, Springer, Torino, Italy, July 2017.

[20] P. Faruki, A. Bharmal, V. Laxmi et al., "Android security: a survey of issues, malware penetration, and defenses," *IEEE Communications Surveys and Tutorials*, vol. 17, no. 2, pp. 998–1022, 2015.

[21] A. Moser, C. Kruegel, and E. Kirda, "Limits of static analysis for malware detection," in *Twenty-Third Annual Computer Security Applications Conference (ACSAC 2007)*, pp. 421–430, IEEE, Miami Beach, FL, USA, December 2007.

[22] I. You and K. Yim, "Malware obfuscation techniques: a brief survey," in *Proceedings of the International Conference on Broadband, Wireless Computing, Communication and Applications (BWCCA 2010)*, pp. 297–300, IEEE, Fukuoka, Japan, 2010.

[23] S. Poeplau, Y. Fratantonio, A. Bianchi, C. Kruegel, and G. Vigna, "Execute this! Analyzing unsafe and malicious dynamic code loading in Android applications," in *Proceedings of the Network and Distributed System Security Symposium (NDSS)*, vol. 14, pp. 23–26, San Diego, CA, USA, February 2014.

[24] R. Lyda and J. Hamrock, "Using entropy analysis to find encrypted and packed malware," *IEEE Security and Privacy*, vol. 5, no. 2, pp. 40–45, 2007.

[25] O. Somarriba, U. Zurutuza, R. Uribeetxeberria, L. Delosières, and S. Nadjm-Tehrani, "Detection and visualization of android malware behavior," *Journal of Electrical and Computer Engineering*, vol. 2016, Article ID 8034967, 17 pages, 2016.

[26] Y. Aafer, W. Du, and H. Yin, "DroidAPIMiner: mining API-level features for robust malware detection in android," in *Proceedings of the International Conference on Security and Privacy in Communication Systems*, pp. 86–103, Springer, Guangzhou, China, October 2013.

[27] T. Azim and I. Neamtiu, "Targeted and depth-first exploration for systematic testing of android apps," in *Proceedings of the 2013 ACM SIGPLAN international conference on Object oriented programming systems languages & applications*, vol. 48, no. 10, pp. 641–660, 2013.

[28] A. Machiry, R. Tahiliani, and M. Naik, "Dynodroid: an input generation system for Android apps," in *Proceedings of the 2013 9th Joint Meeting on Foundations of Software Engineering*, pp. 224–234, ACM, Saint Petersburg, Russia, August 2013.

[29] R. Mahmood, N. Mirzaei, and S. Malek, "Evodroid: segmented evolutionary testing of Android apps," in *Proceedings of the 22nd ACM SIGSOFT International Symposium on Foundations of Software Engineering*, pp. 599–609, ACM, Hong Kong, China, November 2014.

[30] K. Tam, A. Feizollah, N. B. Anuar, R. Salleh, and L. Cavallaro, "The evolution of Android malware and Android analysis techniques," *ACM Computing Surveys*, vol. 49, no. 4, pp. 1–41, 2017.

[31] K. Tam, S. J. Khan, A. Fattori, and L. Cavallaro, "Copperdroid: automatic reconstruction of Android malware behaviors," in *Proceedings of the Network and Distributed System Security Symposium (NDSS)*, San Diego, CA, USA, February 2015.

[32] *Xposed Module Repository*, 2017, http://repo.xposed.info/.

[33] L. Li, D. Li, T. F. Bissyandé et al., "Understanding Android app piggybacking: a systematic study of malicious code grafting," *IEEE Transactions on Information Forensics and Security*, vol. 12, no. 6, pp. 1269–1284, 2017.

[34] H. S. Oh, B. J. Kim, H. K. Choi, and S. M. Moon, "Evaluation of Android Dalvik virtual machine," in *Proceedings of the 10th International Workshop on Java Technologies for Real-time and Embedded Systems*, ACM, New York, NY, USA, October 2012.

[35] D. Octeau, S. Jha, and P. McDaniel, "Retargeting Android applications to java bytecode," in *Proceedings of the ACM SIGSOFT 20th International Symposium on the Foundations of Software Engineering*, p. 6, ACM, Cary, NC, USA, November 2012.

[36] J. Park, H. Kim, Y. Jeong et al., "Effects of code obfuscation on Android app similarity analysis," *Journal of Wireless Mobile Networks, Ubiquitous Computing, and Dependable Applications*, vol. 6, no. 4, pp. 86–98, 2015.

[37] F. Sierra and A. Ramirez, "Defending your Android app," in *Proceedings of the 4th Annual ACM Conference on Research in Information Technology*, ACM, Chicago, IL, USA, October 2015.

[38] J. Andrus and J. Nieh, "Teaching operating systems using Android," in *Proceedings of the 43rd ACM Technical Symposium on Computer Science Education (SIGCSE'12)*, ACM, New York, NY, USA, 2012.

[39] A. Shabtai, Y. Fledel, and Y. Elovici, "Securing Android-powered mobile devices using SELinux," *IEEE Security & Privacy*, vol. 8, no. 3, pp. 36–44, 2010.

[40] KingRoot, 2017, https://kingroot.net/.

[41] Nexus Root Toolkit v.2.1.9, 2016, http://www.wugfresh.com/nrt/.

[42] Dashboards–Platform Versions, 2017, https://developer.android.com/about/dashboards/index.html.

[43] Google Play Console Help Center, 2017, https://support.google.com/googleplay/android-developer/.

[44] Android Debug Bridge, 2017, https://developer.android.com/studio/command-line/adb.html.

[45] D. Hardt, *The OAuth 2.0 Authorization Framework, RFC 6749*, 2012, https://www.rfc-editor.org/info/rfc6749.

[46] Password Hashing Competition, https://password-hashing.net.

[47] K. Moriarty, B. Kaliski, and A. Rusch, *PKCS#5: Password-Based Cryptography Specification Version 2.1. RFC 8018*, 2017, https://www.rfc-editor.org/info/rfc8018.

[48] J. Steube, "Optimising computation of hash-algorithms as an attacker," 2013, https://hashcat.net/events/p13/js-ocohaaaa.pdf.

[49] A. Visconti, S. Bossi, H. Ragab, and A. Calò, "On the weaknesses of PBKDF2," in *Proceedings of the 14th International Conference on Cryptology and Network Security*, Marrakesh, Morocco, December 2015.

[50] A. Ruddick and J. Yan, "Acceleration attacks on PBKDF2: or, what is inside the black-box of oclHashcat?," in *Proceedings of the 10th USENIX Workshop on Offensive Technologies*, Austin, TX, USA, August 2016.

8

MobiSentry: Towards Easy and Effective Detection of Android Malware on Smartphones

Bingfei Ren,[1] **Chuanchang Liu**,[1] **Bo Cheng**,[1] **Jie Guo**,[2] **and Junliang Chen**[1]

[1]*State Key Laboratory of Networking and Switching Technology, Beijing University of Posts and Telecommunications, Beijing, China*
[2]*Beijing Liyun Technology Development Company, Beijing, China*

Correspondence should be addressed to Chuanchang Liu; lcc3265@bupt.edu.cn

Academic Editor: Wenchi Cheng

Android platform is increasingly targeted by attackers due to its popularity and openness. Traditional defenses to malware are largely reliant on expert analysis to design the discriminative features manually, which are easy to bypass with the use of sophisticated detection avoidance techniques. Therefore, more effective and easy-to-use approaches for detection of Android malware are in demand. In this paper, we present MobiSentry, a novel lightweight defense system for malware classification and categorization on smartphones. Besides conventional static features such as permissions and API calls, MobiSentry also employs the *N*-gram features of operation codes (*n*-opcode). We present two comprehensive performance comparisons among several state-of-the-art classification algorithms with multiple evaluation metrics: (1) malware detection on 184,486 benign applications and 21,306 malware samples, and (2) malware categorization on DREBIN, the largest labeled Android malware datasets. We utilize the ensemble of these supervised classifiers to design MobiSentry, which outperforms several related approaches and gives a satisfying performance in the evaluation. Furthermore, we integrate MobiSentry with Android OS that enables smartphones with Android to extract features and to predict whether the application is benign or malicious. Experimental results on real smartphones show that users can easily and effectively protect their devices against malware through this system with a small runtime overhead.

1. Introduction

Smartphones are becoming increasingly popular. Over 334 million smartphones were shipped worldwide in Q1 2017 according to the IDC report [1], and smartphones running Android dominate the mobile market with an 85.0% market share. As there are popularity and openness of Android, numerous applications (apps) are developed on the Android platform to provide new or enhanced services. The success of Android also has a dark side: more and more attackers turn their eyes on Android owing to its openness. As many security companies have recently reported [2–4], the overall volume of Android malware has been increasing significantly in 2016, and 39,000 new malware variants are developed using sophisticated detection avoidance techniques to elude detection.

As smartphones are increasingly used to process sensitive personal information and applications in Android often have the ability to access sensitive resources on mobile devices, it becomes easier for malware to breach privacy data and steal money of users without their consent [5]. For instance, sending SMS to premium-rate numbers secretly and encrypting data and locking devices using ransomware [6, 7]. Android provides permission mechanism to restrict the installation and resources usage of Android malware [5], which forces applications to explicitly request permissions when accessing protected resources (data, system services, peripheral, etc.). However, it can be easily bypassed through collusion attacks, and many users tend to blindly grant permissions due to lack of security knowledge. Therefore, it is a crucial and challenging issue to directly detect malware and novel malware variants on smartphones.

Traditional literature on the detection of Android malware mainly focused on static and dynamic analysis. For example, Apposcopy [8], DroidMat [9], and RiskRanker [10] are methods that use a combination of static features (permissions, deployment of components, and API calls) to characterize applications. However, these methods mainly rely on manually crafted detection patterns and can be rendered largely ineffective due to the production of novel malware variants. TaintDroid [11] and DroidRanger [11] use dynamic features through monitoring the behaviors of the running apps to identify malicious events of them. Unlike static analysis methods, approaches using dynamic features are efficient to detect new malware but usually suffer a significant overhead on smartphones. Machine learning is a buzzword in this area of malware detection. Various methods have been proposed based on the static features (API calls and permissions) analyzing approach that utilizes machine learning methods [12–14]. However, these approaches require specific static features beforehand through expert analysis, and most of them can be bypassed with the use of obfuscation techniques like repackaging [15]. Other works have been proposed for the categorization of malware families [16, 17]. However, most of these methods either use outdated datasets or focus on large families for which training data are available.

To address issues identified above, we present a novel lightweight defense system (MobiSentry) for malware classification and categorization on smartphones based on our preliminary version of this study [18]. Besides conventional static features such as permissions and API calls, MobiSentry also employs the N-gram features of operation codes (n-opcode), which can be used to deal with known obfuscation techniques like rename obfuscation (https://www.preemptive.com/obfuscation). We first investigate n-opcode analysis for both application classification and categorization on real datasets utilizing several state-of-the-art classification algorithms. Then, we conduct the performance comparisons of malware detection and categorization among these classification algorithms with multiple evaluation metrics. We then utilize the ensemble of these supervised classifiers to design MobiSentry and evaluate its performance. In addition, we integrate MobiSentry with Android OS to enable smartphones to extract features from Android Package (APK) files and to predict whether the application is benign or malicious with a small run-time overhead. The main contributions are the following:

(i) We propose a lightweight defense system for malware detection on smartphones. Unlike all related works [8, 11–13, 19–21], we systematically analyze the combination of conventional static features and n-opcode to make full use of them on malware detection and categorization.

(ii) We propose a novel ensemble learning algorithm based on five state-of-the-art supervised classifiers to overcome the drawbacks of them and improve the performance on malware detection.

(iii) We firstly investigate and evaluate several approaches on extracting features from APK files to reduce the run-time overhead on smartphones.

(iv) We thoroughly evaluate our proposed system under a large scale of dataset to present a more real-world performance.

The remaining parts of this paper are organized as follows: we briefly refer to some related studies in Section 2. The dataset is presented in Section 3. Section 4 describes conventional static features and n-gram features of the opcode of applications. Section 5 introduces the methods of designing MobiSentry utilizing the ensemble of five supervised classifiers. Detailed comprehensive performance comparisons of malware detection and categorization are presented in Section 6 and Section 7. In Section 8, we evaluate the effectiveness and performance of MobiSentry integrated with Android OS on real smartphones. Finally, we conclude the paper and discuss the future work in Section 9.

2. Related Work

The area of malware detection on smartphones is an emerging issue. The related work that is mainly being focused and discussed falls into three areas: malware analysis and classification, malware family identification (categorization), and malware detection on smartphones.

2.1. Malware Analysis and Classification. Previous approaches [22–28] have been proposed for malware detection, which mainly use static and dynamic features or the combination of them. AndroidLeaks [29] is a static analysis framework for automatically finding potential leaks of sensitive information in Android applications. Chen [30] focused the work on detecting malicious behavior that can be characterized by the temporal order in which an application uses APIs and permissions. He constructed a permission event graph (PEG) with these static features. 152 malicious and 117 benign applications were used to evaluate the work. AppIntent [31] provides a sequence of GUI manipulations corresponding to the sequence of events that lead to the data transmission. With the help of AppIntent, researchers can determine whether the data transmission is malicious. Zhang et al. [32] introduced graph similarity metrics to detect zero-day malware. Over 2000 malware samples and 13,000 benign applications were used in their experiments. The experimental result shows that their signature-based method can achieve 93% accuracy on malware detection with a low false-negative rate. Wang et al. [16] explored the permission-induced risk in Android apps in a systematic manner. They evaluated the malware detection result with permission ranking on a very large data, and the results show a satisfying performance (the detection accuracy was 94.62% with a false-positive rate being 0.6%). Apposcopy [8] employs a combination of static taint analysis and intercomponent call graph (ICCG) to detect Android applications. Apposcopy is evaluated on a corpus of real-world applications and achieves superior results. There are some related works that used dynamic features to analyze and detect malware. Yuan et al. [33] proposed VetDroid to reconstruct sensitive behaviors in Android apps through dynamic analysis from permission usage perspective. TaintDroid [11] is a dynamic taint tracking and analysis system with the use of the Android's virtualized execution environment. DroidDolphin [34] proposed a dynamic malware analysis framework that extracts useful static and dynamic features dynamically. These methods largely rely on expert analysis to manually extract features and detection patterns through expert analysis. Unlike that, we present a systematical analysis of five state-of-the-art classification methods with the combination of conventional static features and n-opcode, which can automatically learn features from raw data directly.

2.2. Malware Family Identification. Some researchers [35–37] focus their attention on the identification of malware variants of specific malware families. A fact is that 5% to 13% of apps hosted on the studied marketplaces are repackaged according to the experiments performed by DroidMOSS [38], an application similarity measurement system that applies a fuzzy hashing technique to effectively detect the changes from app-repackaging behavior. ViewDroid [15] performs a user interface-based mobile application-repackaging detection from a higher level abstraction. Experimental results show that ViewDroid can effectively and efficiently detect repackaged applications with low false-positive and false-negative rates. Battista et al. [39] presented a model checking-based approach in detecting Android malware families. With the analysis and verification of Java bytecode, a preliminary investigation shows that it can classify OpFake and DroidKungFu families, while in our work, we consider 179 families with 5560 malware family samples. Monet [17] combined runtime behaviors with static structures to detect malware variants based on a fact that the runtime behaviors of malware's core functionalities within a malware family are usually similar in a high degree. Arp [12] is the first one to use comprehensive static features (permissions, API calls, intent filters, etc.) for malware detection. He achieved a TPR of 94% at a low FPR of 1%. Suarez-Tangil [40] proposed a text mining-based approach to measure the similarity between malware samples, which is then used to automatically classify them into families. However, he used the MalGenome dataset [41] to evaluate the method, which is outdated and only has small subset of DREBIN that is used in our work. EC2 [42] performs Android malware families prediction through static and dynamic features with the ensemble of supervised and unsupervised classifiers.

The closest research studies to our work are presented by Kang et al. [19, 20], Shen et al. [21], and Arp et al. [12]. Several efforts [19–21] present n-opcode analysis-based approaches that utilize machine learning to categorize Android malware. Taking the work of Kang et al. as an example, experiments over 2000 samples show that a high accuracy ($F1$-score of 98%) is achievable. Different from Kang et al. [19], Arp et al. [12] only used conventional static features based on expert analysis to manually extract features, and the evaluation result with the SVM classifier presents a malware detection accuracy of 94%. Compared with these approaches, we use the combination of conventional static features and n-opcode, which can specifically describe the characterization of Android applications. Our work also provides systematic analysis of the combination features under a large scale dataset (184,486 benign applications and 21,306 malware samples) with five state-of-the-art classifiers and the ensemble of them and outperforms these approaches (the detection rate was 98.4% with a false-positive rate being 0.5%) and categorization (overall accuracy of 98.17% for all malware families: 99.37% for large families and 48.86% for small families).

2.3. Malware Detection on Smartphones. There are few approaches that aim to detect malware on mobile devices [43–47]. Taking [43] as an example, it presents a method for analysis and detection of malware utilizing static features like permissions and hardware and software features on 1377 malware and 1377 benign samples, which runs natively in the device to predict applications with 94.48% of accuracy and 36-millisecond time overhead.

Our work introduces n-opcode to learn features automatically from raw data to identify novel malware or malware variants. Meanwhile, in order to validate our approach systematically, we conduct several comparison experiments using five supervised classifiers under a large scale of samples. Furthermore, we integrate our work with Android OS to enable smartphones the ability to detect malware. In addition, we investigate and evaluate several methods on extracting features from APK files to further reduce the run-time overhead on smartphones.

3. Dataset

In order to present a realistic performance result that can reflect the real world, we leverage a large scale of app samples to evaluate our approach, consisting of benign applications and malware samples. As the only official app store, the Google Play store contains millions of applications uploaded by developers. However, it does not provide API for analyzers to download a large scale of free apps. Thus, collecting a big dataset of applications is a strict challenge. We collected benign and malicious applications from multiple sources: (i) PlayDrone [48], the first scalable Google Play store crawler that can scan and download applications from the app store. It also shared the whole Android APKs crawled from the Google Play store using PlayDrone online (https://archive.org/details/playdrone-apks). (ii) Anzhi [49], the biggest third-party app market in China. With the help of Wang et al. [16], we got free apps provided by Anzhi. (iii) AndroTotal [50], a publicly available tool (https://bitbucket.org/andrototal/tools) for detecting malware that is used to submit APKs for scanning and retrieve detecting results. It exposes a free-of-charge API to let researchers gain access to all the uploaded samples. (iv) DREBIN, the largest third-party malware dataset with labeled families.

The collected dataset is impure, and further work needs to be done to prepare reliable ground truth data. We utilized VirusTotal [51], a free online service that analyzes suspicious files and URLs to detect types of malware including viruses, worms, and Trojans. It uses different (currently 52) antivirus engines to scan for potential malware: if at least 10 antivirus engines recognize the app as malicious, we then treat it as a benign app to get a high-quality ground truth. Otherwise, the app is considered as malware. In total, we get 205,792 samples consisting of 184,486 benign applications and 21,306 malware samples. As 80/20 is a commonly used ratio for the size of the training set and test set, we split the whole dataset for the subsequent evaluations according to this ratio.

In this work, we utilize the whole dataset or the subset of it for all of our subsequent experiments including malware classification. As DREBIN is the largest labeled dataset of malware families that contains 179 malware families with 5560 samples, we select and analyze it for malware family categorization in our work. In addition, we consider malware families that have more than 15 samples as large families (considering families fewer than 15 samples as small). The number of samples of each malware family in DREBIN is unbalanced: Figure 1 presents the cumulative distribution of the size of malware families. There are 33 large families in total that account for almost 89% of samples in the dataset, which will largely affect the detection result. This suggests us to analyze and evaluate malware family categorization on large families, small families, and the whole dataset separately.

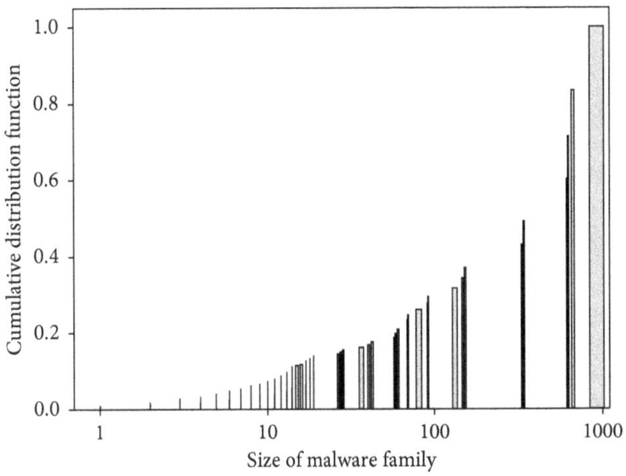

FIGURE 1: Cumulative distribution of the size of malware families in the DREBIN dataset.

4. Feature Extraction

In this section, we introduce the conventional static features and n-opcode that are used in our work. Furthermore, we explain how to extract and select these features from Android application package files to optimize malware classification and categorization in our work. We first present some fundamentals about the Android environment.

4.1. Android Fundamentals. Android is an open-source mobile operating system, and users install applications for their smartphones with Android through the Google Play store, third-party marketplaces, and even unknown sources. Smartphones use APK files to install Android apps. Android applications are archive files (APK files) with an *.apk* suffix that contains all the contents like compiled codes (.dex suffix files) and all the data and resources used in the application. Meanwhile, each Android application must have a manifest file named "AndroidManifest.xml" in its APK file to provide essential information like the following:

(a) *Package name* serves as a unique identifier for the application

(b) *Components* compose the application including *Activities* that are used to interact with users, *Services* that are utilized to execute background tasks, *Broadcast Receivers* mainly for interprocess communications, and *Content Providers* that manage a shared set of apps

(c) *Permissions* required by the application to access correspondingly resources provided by devices

(d) *Hardware or software features* used by the application, like camera and Bluetooth services

Each application runs in its own Linux process in isolation from other apps (usually we call it the sandbox environment), and the Android system assigns a unique user ID to each application to protect data created by the app. Apps with the same user ID can run in the same process and share their data easily. In addition, in order to access device data such as the user's contacts, camera, and Bluetooth, apps need to explicitly request permissions in their AndroidManifest.xml file and have to get explicitly granted these permissions by users.

4.2. Conventional Static Features. Several static features have been proposed in previous approaches [12, 16, 30–32]. We leverage some of them that are proposed in [12] to characterize the applications.

4.2.1. Requested Permissions. One of the most significant security mechanisms in Android is the permission system. With permissions granted at installation time, applications can access specific resources provided by devices. The previous work [16] has shown that malware tends to request certain permissions or the combination of permissions than benign applications. For instance, malicious malware often requests SEND_SMS permission and sends SMS to premium-rate numbers. Thus, we obtain all permissions listed in the manifest file.

4.2.2. Hardware Components. In order to access specific hardware resources (GPS, microphone, etc.), applications have to explicitly declare these features in the manifest file, which may have security or privacy issues. Applications that have access to GPS are able to collect and leak location information of users. In addition, the combination of hardware components and requested permissions even can lead to a heavy hazard. Take the Internet permission and microphone as an example, assuming that an application has access to both of them when in a private business conversation, it can easily record the conversation and expose it through the Internet without the user's consent.

4.2.3. API Calls. Certain API calls found in malicious applications represent that they are allowed to access sensitive resources of devices. Like getDeviceId(), a special API provided by Android can be used to obtain the International Mobile Equipment Identity (IMEI) of devices.

We ignore other static features like app components and filtered intents used in [12]; this is because the name of them can be easily obfuscated and they are not discriminative in malware classification as they introduce more noisy information [42].

4.3. n-Opcode Features. Android applications are generally developed in Java and then compiled and converted to optimized bytecode for the Dalvik virtual machine (Dalvik bytecode) (https://source.android.com/devices/tech/dalvik/dalvik-bytecode). Dalvik bytecode is the common executable format for Android applications and can be disassembled to provide information about process flows and API calls [12].

In order to elude detection, malicious Android applications often utilize sophisticated detection avoidance techniques like repacking and rename obfuscation, which mainly change the name of components, classes, and methods to make them unrecognized. As new malware variants tend to have the same core malicious codes as previous versions, the process flows and logic of the application may not be modified as they raise little suspiciousness in malware identification. Therefore, Dalvik bytecode of compiled applications can be used to distinguish malware from benign applications. We consider the concept of n-gram originated from Natural Language Processing [52] to represent the opcode sequences of applications. The extraction flow of n-gram of opcodes (n-opcode) is shown below in Figure 2.

In order to extract the conventional static features and n-opcode features from the manifest and the source code (actually smali files that are decompiled from the apk file), we need the Android application reverse engineering techniques. There are several optional tools that can be used to extract metadata from apk files and the manifest files. As we integrate our work with Android and run it on smartphones, we firstly investigate and evaluate these approaches on extracting features from APK files to further reduce the run-time overhead on smartphones; detailed description is presented in Section 8. Finally, we choose and leverage Baksmali (https://github.com/JesusFreke/smali) (an Android application assembler/disassembler) and zipfile (https://docs.python.org/2/library/zipfile.html) (a Python library for decompressing apk files) to extract features for further malware classification and categorization.

4.4. Feature Selection. In order to achieve the most optimized malware detection and categorization accuracy, we need to select the best combination of conventional static features and our n-opcode. As the conventional static features are decided, the only variable factor in our features is the value for n in n-opcode. Thus, we perform experiment and analysis with n-opcode of up to 9-opcode for both application classification and categorization on a subset of our real datasets utilizing five state-of-the-art classification algorithms: SVM, KNN, AdaBoost (Ada), random forest (RF), and gradient boosting machine (GBM). We utilize these five supervised classifiers in the subsequent experiments in this paper. There are several open-source libraries that have already implemented these classification algorithms, and we leverage SVM, RF, and AdaBoost classifier provided by scikit-learn (http://scikit-learn.org/stable/) and GBM classifier provided by the LightGBM framework (https://github.com/Microsoft/LightGBM). Then, we adopt the 5-fold cross-validation approach on each combination of conventional static features and n-opcode (n ranges from 1 to 9).

4.4.1. Malware Classification. We first evaluate the combination features for malware classification on a subset of our datasets consisting of 1963 malware and 1963 benign applications. Figure 3 shows the results of the evaluation.

As shown in Figure 3, the performances of these algorithms are similar except for KNN. The accuracy increases to the best as n equals 2, and SVM achieves the best accuracy of 97.06%, but the accuracy decreases when n is greater than 3. This may have a reason that more noisy information is introduced and affect the classification performance when the length of the opcode sequence (n-opcode) is increased. KNN shows the worst performance, but the increasing trend is also similar to others.

4.4.2. Malware Categorization. We then evaluate the combination features for malware family identification on the DREBIN dataset that contains 179 families with 5560 malware family samples. As the purpose in this section is to select features, we evaluate the whole dataset (large families and small families) without considering large and small families separately. The performance result is shown in Figure 4, which shows a similar trend with the result of malware classification in Figure 3. Each algorithm achieves its best performance when n equals 2 or 4.

Tables 1 and 2 show the top ten important n-opcode features for malware classification and malware family categorization, in which we omit the prefix "0x" of the opcode (e.g., using d8 represents 0xd8). And there is a correspondence (http://pallergabor.uw.hu/androidblog/dalvik_opcodes.html) between opcode and the name of opcode in Dalvik. In Table 1 which shows top ten n-opcode features for malware classification, we found one 3-opcode ("390f6e") and two 4-opcode ("6e0c086e" and "08546e0c") which only appeared in malware samples from the study in [19] in our correspondingly selected 3-opcode and 4-opcode as 19999th, 4833th, and 13527th ranked features. Meanwhile, we also found that one 5-opcode

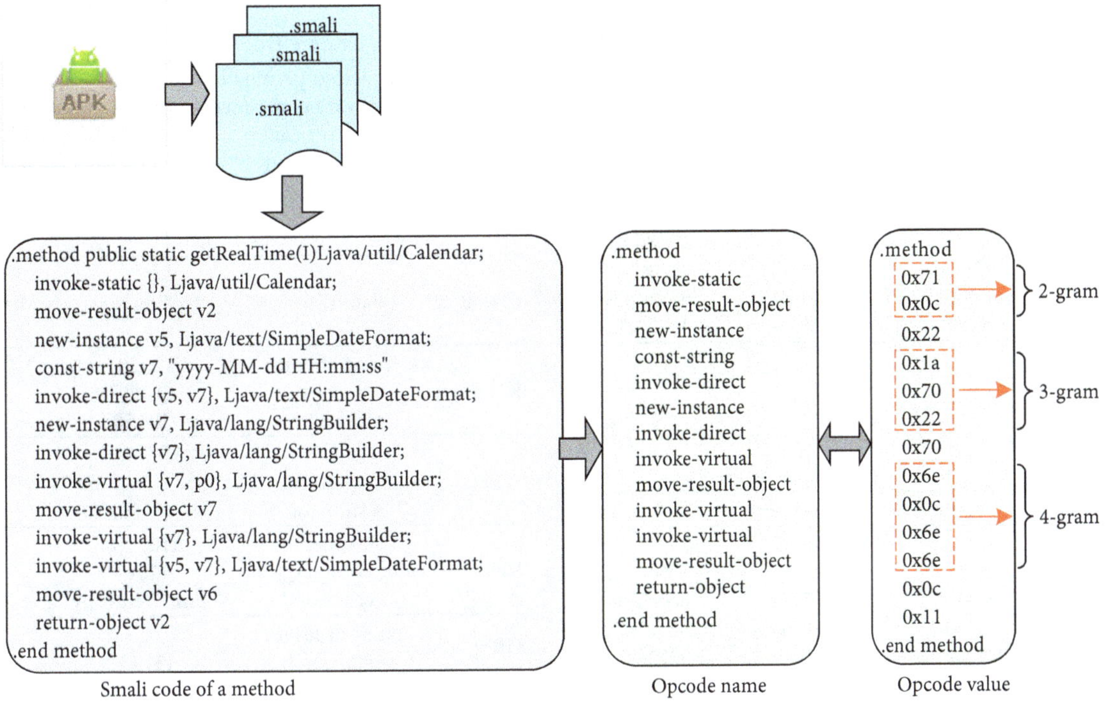

FIGURE 2: Extraction flow of n-opcode.

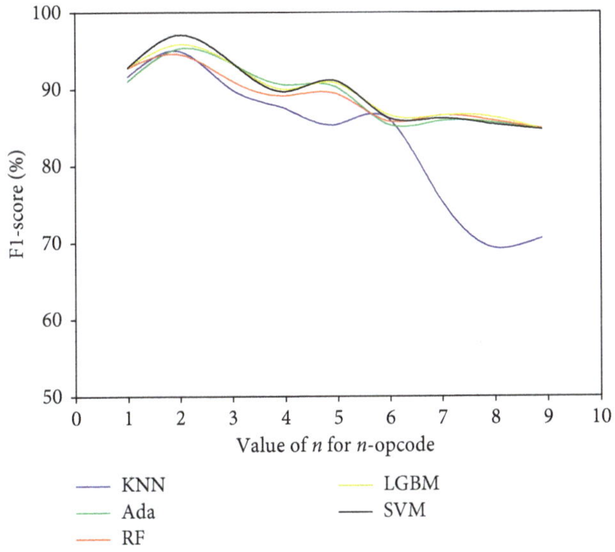

FIGURE 3: Malware classification results with combination features.

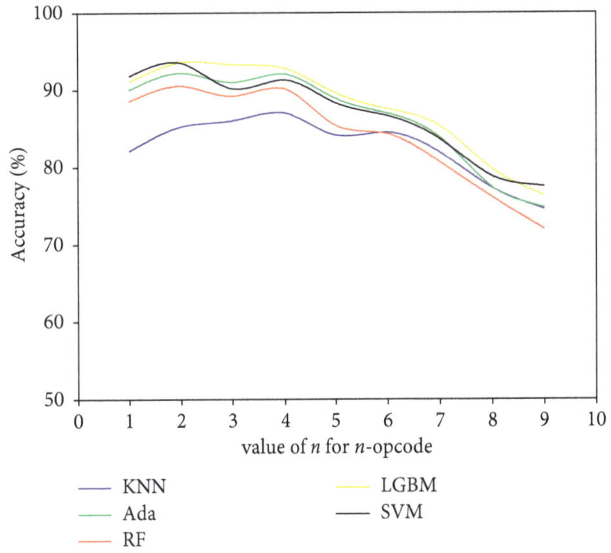

FIGURE 4: Malware categorization results with combination features.

"0c1a6e0a39" from the study in [53] is in our selected 5-opcode as the 19992th ranked feature. Another observation is that one 4-opcode "0d076e12" is extension of the second-ranked 3-opcode "0d076e." In Table 2, which presents top ten n-opcode features for malware categorization that was only previous researched in [19], only one 5-opcode "0c22702271" has been found in our selected 5-opcode as the 15563th ranked feature. Respectively, the similarity and difference in results among the studies in [19, 53] and our work are mainly according to a fact that different benign samples were used between these works, and the malicious datasets used in [19, 53] are just the subset of datasets used in our work.

Besides accuracy, time and memory overhead are other significant factors when choosing features for malware detection. The number of selected n-opcode features increased as n becomes larger [19], which means more time and memory space are needed when extracting n-opcode features. Thus, we select the combination of conventional static features and 2-opcode for our subsequent experiments, although KNN has the highest malware categorization accuracy when n equals 4.

Through feature extraction, more than 2,000,000 features are extracted from each app. As the number of features is too large to be efficiently processed, we employ a linear SVM

TABLE 1: Top ten n-opcode features for malware classification.

2-Opcode	3-Opcode	4-Opcode	5-Opcode
12b1	6e0c11	0d72280d	1f6e1a6ed8
0774	0d076e	d84545bc	6e280d281a
b78e	1f3907	0d076e12	1c70156e6e
1768	221a54	0d6e0728	1a126e0c54
d281	69126e	62707222	0c710c6e07
d81a	5b1a5b	136e0c1a	6e0c6e0c69
461f	0e1211	221a7054	62dd495012
0707	69541a	541a126e	0c28070707
6a1a	720b71	1c6e1c6e	0e52d85952
6174	710a81	0e70386e	12326e0c6e

TABLE 2: Top ten n-opcode features for malware family categorization.

2-Opcode	3-Opcode	4-Opcode	5-Opcode
6e13	6e0c74	6211136e	28541a720a
286e	0e156e	70742913	125c0e6239
520f	28541a	22221222	6e0c213407
34d8	715454	12710b17	d86e54710c
7272	086e0a	0c76086e	591239740c
1215	590e70	6e0e1f6e	3808541a6e
3370	1f391a	7008710b	1215710c6e
1f1a	382270	22705b22	7022080708
0ad8	712854	5b085408	0e1f6e0c11
5b29	700e1f	1f5b541a	0a39391a6e

TABLE 3: Number of feature sets.

	Feature type	Number of features
1	Requested permissions	98
2	Hardware components	30
3	API calls	17,212
4	*N*-gram	35,000

TABLE 4: Optimized parameters for each classifier.

Classifier	Parameters	Range	Parameters that generate best results
KNN	n_neighbors	(2, 20)	8
	n_jobs	(1, 20)	1
	leaf_size	(10, 50)	30
Ada	Criterion	("gini," "entropy")	"gini"
	splitter	("best," "random")	"random"
	estimator	(1, 20)	18
RF	max_depth	(10, 18)	18
	min_samples_leaf	(2, 6)	2
	criterion	("gini," "entropy")	"entropy"
GBM	Num of leaves	(10, 35)	11
	max_depth	(5, 13)	35
SVM	Kernel	("linear," "rbf")	"rbf"
	C	$(2^{-3}, 2^{7})$	2
	gamma	$(2^{-5}, 2^{2})$	2^{-4}

algorithm to sort the importance of each feature and finally use the top-ranked 52,340 features to perform the following evaluations. The feature sets are described in Table 3.

5. Methods

In this section, we introduce our methodology for malware classification and categorization. First, we consider five state-of-the-art supervised classifiers to evaluate the performance comparisons based on the combination of conventional static features and our n-opcode with multiple evaluation metrics. Second, we propose MobiSentry which utilizes the ensemble of these classifiers for classification and categorization separately.

5.1. Supervised Classifiers. We consider five off-the-shelf classifiers mentioned before in Section 4.4 for both malware classification and malware family identification: KNN, Ada, RF, GBM, and SVM. Beforehand, we adopt the k-fold cross-validation approach based on the grid search to find the parameters that generate the best results. In order to perform a systematic evaluation to find the optimized parameter, we vary the range of each significant parameter for each classifier separately. Taking SVM as an example, we vary the kernel function from "rbf" to "linear," set the range of the penalty parameter of the error C varying from 2^{-3} to 2^{7}, and also change the value of gamma from 2^{-5} to 2^{2}. The experimental result is shown in Table 4, and we employ the parameters after the process.

5.2. Ensemble of Classifiers. To further improve the accuracy of malware detection, we employ the ensemble of multiple classifiers to take the advantage of each algorithm. All the prediction results of each classifier will then vote to generate a final decision. In total, we consider the combination of the five algorithms described above with two ensemble rules: *majority voting* and *malware override.*

Definition 1 (majority voting). We consider a sample as the detection result which has the maximum frequency. For instance, if an application is detected as malware by 3 of the 5 classifiers, which means 2 of the 5 classifiers treat it as a benign application, the final decision of it is malware.

Definition 2 (malware override). We treat a sample as a benign application only if the detection results of all the classifiers are benign. Otherwise, we consider this sample as malware.

As described in Algorithm 1, we train these five supervised classifiers on training dataset S1 in the initial step (Step 1), while the other steps are largely dependent on the type of malware detection (Step 2), which can be divided into two categories: malware classification and categorization.

5.3. Malware Classification. For malware classification, we then count the frequency of each unique malware detection result for each test sample (Step 4), and we regard malicious applications as 1 and let 0 represents benign applications in our subsequent experiments, which indicates that the optional value is 0 or 1 in Step 7 and Step 8. If one detection result of a test sample achieves a maximum frequency, it is assigned to

```
Algorithm 1: ensemble of classifiers
Input: classifier set: C, training dataset: S1, and test dataset: S2
Output: labeled detection results of S2

(1) Run C on S1;
(2) TD ⟵ Type of Detection: Classification or Categorization;
(3) ER ⟵ Ensemble rule of multiple classifiers: majority voting or malware override;
(4) for each s ∈ S2 dO
(5)   FMD[s,r] ⟵ Frequency for each unique Malware Detection result of s obtained from C;
(6)   fmd* ⟶ max_∀r FMD[s, r];
(7)   r* ⟶ argmax_r MD[s, r];
(8)   fmd** ⟶ min_∀r FMD[s, r];
(9)   r** ⟶ argmin_r MD[s, r];
(10)  switch TD
(11)  case Classification
(12)    if ER ≡ majority voting then
(13)      LDR[s] ⟶ r*
(14)    else if ER ≡ malware override then
(15)      if r* ≡ 0 AND r** ≡ 0 then
(16)        LDR[s] ⟶ 0
(17)      else
(18)        LDR[s] ⟶ 1
(19)      end if
(20)    end if
(21)  case Categorization :
        LDR[s] ⟶ r*
(22)  end switch
(23) end for
(24) return LDR
```

ALGORITHM 1: MobiSentry algorithm.

the corresponding result (malicious or benign) when using the *majority voting* rule (Step 13). Meanwhile, if all the detection results of a test sample are 0, it is considered to be benign when using the *malware override* rule (Step 16), which means all the classifiers identify this sample as a benign application. Otherwise, we identify this sample as malware (Step 18).

5.4. Malware Categorization. For malware categorization, we measure the frequency of each identified family for each test sample (Step 4), and if the test sample meets a maximum frequency (Step 6), then it can be assigned to the corresponding family (Step 21).

According to the definition of the *malware override* rule, it is clear that this rule mainly focuses on binary classification. Therefore, we use the two rules in the ensemble of classifiers for malware classification and only utilize the *majority voting* rule for malware family categorization.

6. Performance Comparison of Malware Classification

In this section, we present detailed comprehensive performance comparisons of malware classification. First, we evaluate the detection performance of the five selected supervised classifiers with multiple evaluation metrics. Then, we provide the performance of the ensemble of classifiers with the following algorithm:

Malware override and *majority voting* rules are described in Section 5. Meanwhile, we perform an experiment to compare the performance of our MobiSentry with that of some baseline methods.

6.1. Evaluation Metrics. As the evaluation metrics are not standardized and different approaches rely on different metrics [13], we use multiple metrics to evaluate our work compared to previous works that only use few metrics [12, 19, 53]. We consider true-positive rate (TPR), false-positive rate (FPR), and accuracy as the metrics (as in the following formulas) used in our experiments:

$$\begin{aligned} \text{TPR} &= \frac{\text{TP}}{\text{TP}+\text{FN}}, \\ \text{FPR} &= \frac{\text{FP}}{\text{FP}+\text{TN}}, \\ \text{precision} &= \frac{\text{TP}}{\text{TP}+\text{FP}}, \\ \text{recall} &= \frac{\text{TP}}{\text{TP}+\text{FN}}, \end{aligned} \tag{1}$$

where TP, FP, TN, and FN indicate true positive, false positive, true negative, and false negative separately. Meanwhile, we also utilize the receiver-operating characteristic (ROC) curve plot and the precision-recall-curve

(PRC) plot to represent a more generalized performance result clearly.

6.2. Performance Comparisons. We conduct our analysis based on a very large dataset consisting of 184,486 benign applications and 21,306 malware samples for the evaluation of malware detection, and we randomly split the dataset into two subsets for training (80%) and testing (20%). Then, these samples are trained under those five classifiers mentioned before (KNN, Ada, RF, SVM, and GBM), and the GBM classifier achieves the best accuracy of 96.77%. As the result depicted in Figure 5, we can easily find that the value of PRC or ROC of each classifier is closer to 1, which implicates our method based on conventional static features and n-opcode can achieve a high accuracy and stability in malware detection.

Furthermore, we evaluate the performance of the ensemble of these classifiers with *majority voting* and *malware override* rules based on our combinational features. In contrast, we consider the following three methods as baselines (https://github.com/renbingfei/MobiSentry): (i) the best classifier (GBM) of the five basic classifiers in our experiment with the optimized parameters introduced before (Table 4 in Section 4); (ii) the first work that uses comprehensive static features proposed in DREBIN [12]. We employed static features used in DREBIN and utilized a linear support vector machine (SVM) for malware detection. As the parameters of SVM used in DREBIN are not specified, we kept the parameters as default values; and (iii) a highly cited Android malware detection method proposed in recent years [54]. We chose the "max" method with 5 hidden classes for the mixture of naïve Bayes model (MNB) used in [54]. The comparison result is depicted in Table 5.

Table 5 shows the performance comparisons of MobiSentry. Although DREBIN [12] has a better result compared to the method in [54], the accuracy of GBM classifier based on our combinational features provides the best performance among baselines with a higher false-positive rate (3%) than them. The ensemble of basic classifiers gains significant performance (97.4% under the *majority voting* rule and 98.4% under the *malware override* rule) and outperforms other benchmarks, even roughly 10% higher than the method in [54], with a lower false-positive rate.

We also compare MobiSentry with another lightweight research tool named "Kirin" [55] for Android malware detection under the same set of malware samples. Kirin reports only 1057 apps out of 2,000 malicious apps to be malware. This corresponds to a false-negative rate of 47%, which is quite high compared to the less than 1% false-negative rate of MobiSentry. The main reason for the higher accuracy of malware detection of MobiSentry is that Kirin uses manually crafted security rules for malware detection, which can be rendered largely ineffective due to the production of novel malware variants, while MobiSentry employed n-opcode to learn features automatically from raw data to identify novel malware or malware variants.

6.3. False-Positive Analysis. In the experiment, MobiSentry misses nearly 0.6% malware of the whole test samples when employing the ensemble rules. After manually analyzing these positives, we discover that the reason for mislabeling benign applications as malware mainly comes from the following three aspects:

(i) Mislabeled by VirusTotal: we leveraged VirusTotal to prepare the reliable ground truth data before evaluations after collecting the dataset, and then we relabeled the datasets according to the reports returned by VirusTotal. The reports contain false positives, which will make the dataset impure. Meanwhile, in order to have better ground truth data, we considered apps with a higher number of malicious labels in the reports as malware, which miss 12 malware. For instance, a malware with the package name *com.getfreerecharge.mpaisafreerecharge* is labeled as malware by 8 engines.

(ii) Affected by adware: as the adware is in a twilight area between benign application and malware, these kinds of apps are hardly determined according to their behavior. Thus, the existence of adware will affect the detection result [13].

(iii) Ensemble rules: the *majority voting* rule relies on the maximum detection result, while the *malware override* rule considers a sample as benign application only when all the results detected by the five basic classifiers are benign; this will cause a smaller value of true negative (TN) and, correspondingly, will achieve a larger FPR.

7. Performance Comparison of Malware Categorization

This section describes the performance comparisons of malware categorization. As introduced in Section 5, the *malware override* rule mainly focuses on binary classification. Therefore, we only utilize the *majority voting* rule for malware family categorization. The malware family categorization actually is a multiclass/multilabel classification, which is different from binary-class classification. We start by explaining the metrics used to evaluate the performance of the multiclass classification.

7.1. Evaluation Metrics. The balance of malware families in the dataset largely affects the multiclass categorization performance when using default metrics to evaluate it. For example, the previous work frequently leverages a metric $F1$-score calculated using a method provided by the scikit-learn library to measure the accuracy in binary classification and ignores a parameter *average* of the method even in multiclass classification. By default, the method returns the score for each class when the value of *average* is none, and the returned score may be highly inaccurate when the number of samples for each class is highly skewed. When setting the value of this parameter as "weighted," the method will calculate $F1$-score taking label imbalance into account. As the dataset for our malware categorization is largely imbalanced (as introduced in the following section), we consider $F1$-score and precision with the parameter *average*

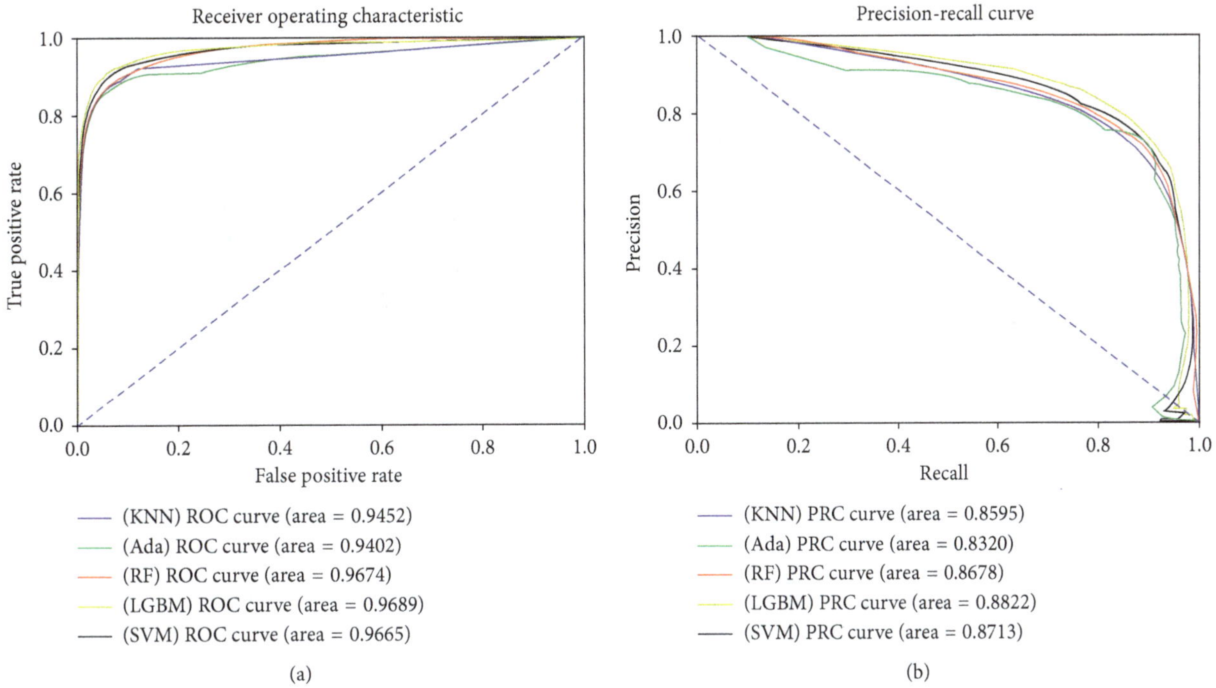

FIGURE 5: Detection performance of MobiSentry. (a) The ROC curve of the performance. (b) The metric of PRC.

TABLE 5: Detection performance of baselines and MobiSentry.

Metrics	Baselines			MobiSentry	
	[54]	[12]	GBM	Majority voting	Malware override
Accuracy (%)	88.2	94	96.76	97.4	98.4
FPR (%)	1.7	1	3	0.55	0.67

assigned to "weighted" to evaluate the performance of malware categorization.

7.2. Performance Comparisons and Analysis. As DREBIN is the largest labeled dataset of malware families that contains 179 malware families with 5560 samples, we consider the dataset of DREBIN to evaluate the performance in the malware categorization study. Like introduced in Section 3, the number of samples of each malware family in DREBIN is imbalanced, and we analyze and evaluate malware family categorization on large families (more than 15 samples), small families (less than 15 samples), and the whole dataset separately.

The result for performance comparisons of our five basic classifiers is shown in Figure 6. As shown in Figures 6(a) and 6(b), the performances of malware categorization using five basic classifiers on both large families and the whole dataset are greatly satisfied, especially on large families, and the GBM classifier achieves the best accuracy (95.26% for the whole dataset and 98.67% for large families). However, the accuracy of small families is relatively low (maximum 47.79% achieved by the Ada classifier), and even the worst accuracy achieved by the KNN classifier is less than 35%. Figure 6(d) shows the detection accuracy on large families, and we observe that a malware family named "*Hamob*" achieves the lowest accuracy, which is less than 70% and roughly 20% lower compared with other families. After analyzing a few samples belonging to *Hamob*, we find that this malware family is more a type of adware than malware. The most prominent characteristics of *Hamob* are pushing and showing advertisements to users. This makes it hard to distinguish it from other malware families.

Furthermore, we evaluate the performance of the ensemble of these classifiers with the majority voting rule based on our combinational features extracted from the DREBIN dataset. In addition, we consider the similar methods used in Section 6 as baselines: (i) the best classifier (GBM) of the five basic classifiers in this experiment of malware categorization on the whole dataset and large families; (ii) the best classifier (Ada) in the experiment of malware categorization on small families; (iii) the first work that uses comprehensive static features proposed in [12]; and (iv) a highly cited Android malware detection method proposed in recent years [54]. The result is shown in Table 6. The accuracy of GBM and Ada classifiers based on our combinational features provides the best performance among baselines, and the reason for outperforming [12] may be based on the fact that the work presented in [12] only uses the SVM classifier under conventional static features, and we leverage five classifiers to detect malware under the combination of n-opcode and conventional static features. The ensemble of basic classifiers gains significant performance and outperforms other benchmarks.

8. Run-Time Performance

In order to enable smartphones to extract features from APK files and to detect malware at the time of installation, we

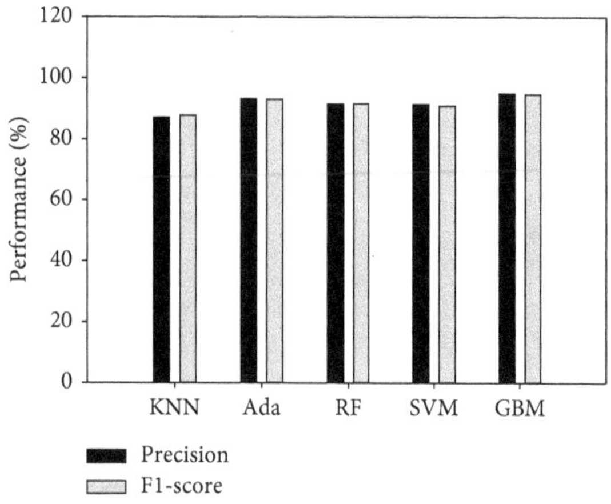

(a)

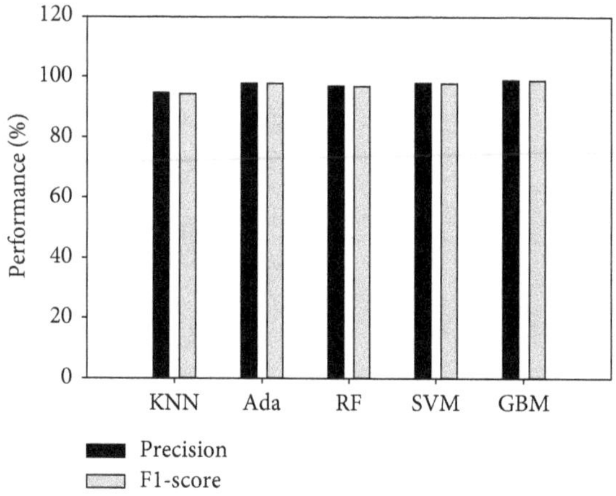

(b)

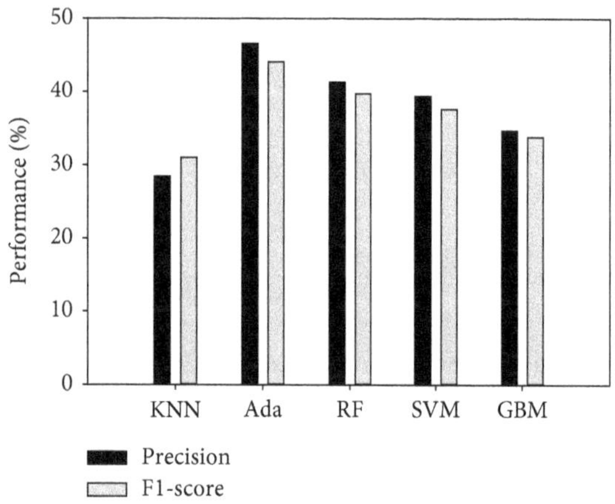

(c)

Malware family	Precision	Malware family	Precision
FakeInstaller	0.99	ExploitLinuxLotoor	0.95
DroidKungFu	0.98	MobileTx	1
Plankton	0.99	Glodream	1
OpFake	1	FakeRun	1
GinMaster	0.97	SendPay	0.89
BaseBridge	0.99	Gappusin	0.93
Iconosys	1	Imlog	1
Kmim	0.97	SMSreg	1
FakeDoc	1	Yzhc	1
Geinimi	1	Jifake	1
Adrd	1	Hamob	0.67
DroidDream	1	Boxer	0.88

(d)

FIGURE 6: Performance of malware categorization using five state-of-the-art classifiers. (a) Five classifiers on the whole dataset. (b) Five classifiers on the large families. (c) Five classifiers on the small families. (d) Precision for large families using the GBM classifier.

TABLE 6: Categorization accuracy of baselines and MobiSentry.

Dataset	Baselines		MobiSentry (%)
	[12] (%)	Basic classifiers (%)	
All	90	95 (GBM)	98
Large families	95	98 (GBM)	99
Small families	42	47 (Ada)	49

integrate MobiSentry with Android OS. The most significant time consumption occurs when extracting features; thus, we firstly investigate and evaluate several methods on extracting features from APK files to further reduce the run-time overhead on smartphones. We mainly consider four frequently used approaches for reverse engineering of Android APKs: (i) Androguard [56], a tool written in python that is mainly used to play with Android files, such as APK, DEX, and ODEX files; (ii) ApkTool [57], a tool for reverse engineering Android APK files; (iii) Baksmali, an assembler/disassembler for the dex format files; and (iv) the combination of zipfile (a python library) and Baksmali, which we mentioned before in Section 5.

We have conducted two experiments to evaluate the time consumption for feature extraction by using different sizes of apk files and dex files in our preliminary work [18], and the result showed that the size of apk files has no direct influence on time consumption for feature extraction because of the existence of resource files in apk files, but time consumption can be directly affected by the size of the dex file. Under this premise, we conduct contrast experiments on these four approaches with 500 applications, and the result is shown in Figure 7.

Experimental results show that extracting features utilizing the combination of zipfile and Baksmali takes the least amount of time.

To analyze the run-time overhead of MobiSentry, we integrate it with Android OS and envelope the learned detection model from our large scale of dataset to the smartphone to enable it the ability of malware detection. Then, we evaluate the detection performance of it on both PC (Intel® Xeon(R) CPU E5620 @ 2.40 GHz × 16 and 132 GB of main memory) and smartphone (Xiaomi 4). In contrast, we utilize the five basic classifiers and the ensemble of them (MobiSentry) as usual. The results are presented in Figure 8 and Table 7. Figure 8 shows the time consumption of malware detection after the process of extracting features from a given application, and we observe that MobiSentry

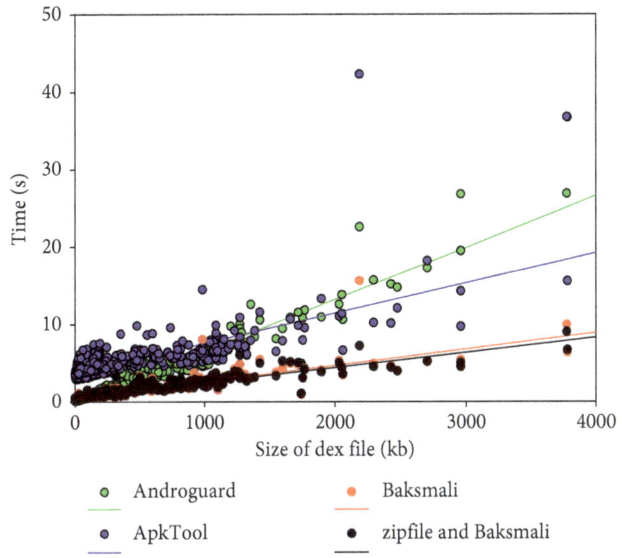

FIGURE 7: Time consumption comparisons for feature extraction among four highly used approaches.

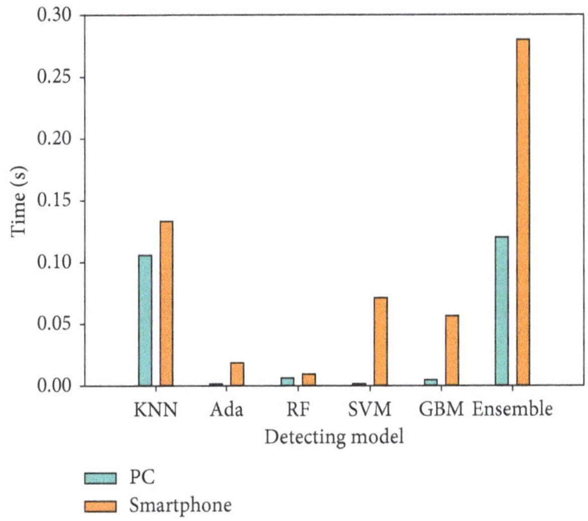

FIGURE 8: Run-time performance comparisons.

TABLE 7: Comparisons of model size and malware detection accuracy.

	KNN	Ada	RF	SVM	GBM	Ensemble
Model size (MB)	73.52	77.67	32.8	26.34	55.88	266.21
Detection accuracy (%)	96.46	96.77	96.06	96.39	96.76	98.45

is able to detect application within less than 1 second. Meanwhile, the ensemble of basic classifiers also achieves the best and satisfying detection accuracy. The limitation of the ensemble approach is that the learned model is larger than five basic classifiers, which is mainly because of the large scale of dataset (more than 200,000 samples) and selected features (more than 180,000). As the storage space on smartphones is largely increased by now, even up to 256 GB for some new Android smartphones like Samsung S8, the model size of the ensemble approach is acceptable. In addition, although there exists false positive using MobiSentry in the real world, the FPR is relatively low (about 0.6% introduced in Section 6), which indicates one false alarm will be informed to users when installing nearly 200 applications.

9. Conclusions

In this paper, we present a lightweight defense system for malware classification and categorization on smartphones.

Besides conventional static features such as permissions and API calls, we systematically analyzed the N-gram features of operation codes (n-opcode), which can be used to deal with known obfuscation techniques like rename obfuscation. Next, we proposed a novel ensemble learning algorithm based on five state-of-the-art supervised classifiers to overcome the drawbacks of them and improve the performance of malware detection. Moreover, we firstly investigated and evaluated several methods on extracting features from APK files to further reduce the run-time overhead on smartphones. Finally, we thoroughly evaluated our proposed system under a large scale of dataset to present a more real-world performance.

Future work might focus on analyzing cluster algorithms to improve the detection accuracy of novel malware variants identification. Moreover, as we observed that some 4-opcode features are just extensions of several 3-opcode features, we may also investigate the feasibility of detecting malware with the combination of different lengths n-opcode features.

Acknowledgments

This work is based on our preliminary version of this study [18] and was supported in part by the National Natural Science Foundation of China under Grant no. U1536112, Special Funds for Public Industry Research Projects of State Administration of Grain of China under Grant no. 201513002, and National High-tech R&D Program of China (863 Program) under Grant no. 2013AA102301.

References

[1] https://www.idc.com/promo/smartphone-market-share/os, 2017.

[2] https://www.symantec.com/security-center/threat-report, 2017.

[3] https://tinyurl.com/yadrfrbx, 2017.

[4] Malwarebytes Labs, "State of malware report," https://www.malwarebytes.com/pdf/white-papers/stateofmalware.pdf.

[5] W. Enck, M. Ongtang, and P. McDaniel, "Understanding android security," *IEEE Security and Privacy*, vol. 7, no. 1, pp. 50–57, 2009.

[6] S. Song, B. Kim, and S. Lee, "The effective ransomware prevention technique using process monitoring on android platform," *Mobile Information Systems*, vol. 2016, Article ID 2946735, 9 pages, 2016.

[7] J. K. Lee, S. Y. Moon, and J. H. Park, "CloudRPS: a cloud analysis based enhanced ransomware prevention system," *Journal of Supercomputing*, vol. 73, no. 7, pp. 3065–3084, 2017.

[8] F. Yu, S. Anand, I. Dillig, and A. Aiken, "Apposcopy: semantics-based detection of Android malware through static analysis," in *Proceedings of 22nd ACM SIGSOFT International Symposium on Foundations of Software Engineering (FSE 2014)*, pp. 576–587, ACM, Hong Kong, China, November 2014.

[9] D. J. Wu, C. H. Mao, T. E. Wei, H. M. Lee, and K. P. Wu, "DroidMat: android malware detection through manifest and API calls tracing," in *Proceedings of 2012 Seventh Asia Joint Conference on Information Security*, pp. 62–69, Tokyo, Japan, August 2012.

[10] M. Grace, Y. Zhou, Q. Zhang, S. Zou, and X. Jiang, "RiskRanker: scalable and accurate zero-day android malware detection," in *Proceedings of the 10th international conference on Mobile systems, applications, and services (MobiSys'12)*, pp. 281–294, ACM, Ambleside, UK, June 2012.

[11] W. Enck, P. Gilbert, S. Han et al., "TaintDroid: an information-flow tracking system for realtime privacy monitoring on smartphones," *ACM Transactions on Computer Systems*, vol. 32, no. 2, pp. 1–29, 2014.

[12] D. Arp, "DREBIN: effective and explainable detection of android malware in your pocket," in *Proceedings of Symposium on Network and Distributed System Security (NDSS)*, San Diego, CA, USA, February 2014.

[13] S. Roy, "Experimental study with real-world data for android app security analysis using machine learning," in *Proceedings of 31st Annual Computer Security Applications Conference (ACSAC 2015)*, pp. 81–90, ACM, Los Angeles, CA, USA, December 2015.

[14] T. Vidas, N. Christin, and L. Cranor, "Curbing Android permission creep," in *Proceedings of Web 2.0 Security and Privacy 2011 workshop (W2SP 2011)*, Oakland, CA, USA, May 2011.

[15] F. Zhang, H. Huang, S. Zhu, D. Wu, and P. Liu, "ViewDroid: towards obfuscation-resilient mobile application repackaging detection," in *Proceedings of 2014 ACM conference on Security and privacy in wireless & mobile networks (WiSec'14)*, pp. 25–36, ACM, New York, NY, USA, July 2014.

[16] W. Wang, X. Wang, D. Feng, J. Liu, Z. Han, and X. Zhang, "Exploring permission-induced risk in android applications for malicious application detection," *IEEE Transactions on Information Forensics and Security*, vol. 9, no. 11, pp. 1869–1882, 2014.

[17] M. Sun, X. Li, J. C. S. Lui, R. T. B. Ma, and Z. Liang, "Monet: a user-oriented behavior-based malware variants detection system for android," *IEEE Transactions on Information Forensics and Security*, vol. 12, no. 5, pp. 1103–1112, 2017.

[18] B. Ren, C. Liu, B. Cheng, Y. Feng, and J. Chen, "Poster: EasyDefense: towards easy and effective protection against malware for smartphones," in *Proceedings of 23rd Annual International Conference on Mobile Computing and Networking(MobiCom'17)*, pp. 570–572, ACM, New York, NY, USA, June 2017.

[19] B. J. Kang, "N-opcode analysis for android malware classification and categorization," in *Proceedings of 2016 International Conference On Cyber Security And Protection Of Digital Services (Cyber Security)*IEEE, London, UK, June 2016.

[20] B. Kang, S. Y. Yerima, S. Sezer, and K. McLaughlin, "N-gram opcode analysis for android malware detection," *International Journal on Cyber Situational Awareness*, vol. 1, no. 1, pp. 231–254, 2016.

[21] F. Shen, J. D. Vecchio, A. Mohaisen, S. Y. Ko, and L. Ziarek, "Android malware detection using complex-flows," in *Proceedings of 2017 IEEE 37th International Conference on Distributed Computing Systems (ICDCS)*, pp. 2430–2437, Atlanta, GA, USA, June 2017.

[22] F. D. Cerbo, A. Girardello, F. Michahelles et al., "Detection of malicious applications on android OS," in *Proceedings of International Workshop on Computational Forensics*, pp. 138–149, Springer, Tokyo, Japan, November 2010.

[23] Y. Zhou, Z. Wang, W. Zhou et al., "Hey, you, get off of my market: detecting malicious apps in official and alternative android markets," in *Proceedings of Annual Network & Distributed System Security Symposium*, San Diego, CA, USA, June 2014.

[24] W. Zhou, Y. Zhou, X. Jiang et al., "Detecting repackaged smartphone applications in third-party android marketplaces," in *ACM Conference on Data and Application Security and Privacy, ACM*, pp. 317–326, San Antonio, TX, USA, Feburary 2012.

[25] I. Burguera, U. Zurutuza, and S. Nadjm-Tehrani, "Crowdroid: behavior-based malware detection system for Android," in *Proceedings of 1st ACM Workshop on Security and Privacy in Smartphones and Mobile Devices (SPSM'11)*, pp. 15–26, ACM, New York, NY, USA, July 2011.

[26] B. Amos, H. Turner, and J. White, "Applying machine learning classifiers to dynamic Android malware detection at scale," in *Proceedings of 2013 9th International Wireless Communications and Mobile Computing Conference (IWCMC)*, pp. 1666–1671, Sardinia, Italy, July 2013.

[27] T. Petsas, G. Voyatzis, A. Elias, M. Polychronakis, and S. Ioannidis, "Rage against the virtual machine: hindering dynamic analysis of Android malware," in *Proceedings of Seventh European Workshop on System Security (EuroSec'14)*, ACM, New York, NY, USA, April 2014.

[28] S. Wang, Q. Yan, Z. Chen, B. Yang, C. Zhao, and M. Conti, "Detecting android malware leveraging text semantics of network flows," *IEEE Transactions on Information Forensics and Security*, vol. 13, no. 5, pp. 1096–1109, 2018.

[29] C. Gibler, J. Crussell, J. Erickson, and H. Chen, "AndroidLeaks: automatically detecting potential privacy leaks in android applications on a large scale," S. Katzenbeisser, E. Weippl, L. J. Camp, M. Volkamer, and M. Reiter, Eds., in *Proceedings of 5th international conference on Trust and Trustworthy Computing (TRUST'12)*, pp. 291–307, Springer-Verlag, Vienna, Austria, June 2012.

[30] K. Z. Chen, "Contextual policy enforcement in android applications with permission event graphs," in *Proceedings of NDSS Symposium*, pp. 1–19, San Diego, CA, USA, Feburary 2013.

[31] Z. Yang, M. Yang, Y. Zhang, G. Gu, P. Ning, and X. Sean Wang, "AppIntent: analyzing sensitive data transmission in android for privacy leakage detection," in *Proceedings of 2013 ACM SIGSAC conference on Computer & communications security(CCS'13)*, pp. 1043–1054, ACM, Chicago, IL, USA, November 2013.

[32] M. Zhang, Y. Duan, H. Yin, and Z. Zhao, "Semantics-aware android malware classification using weighted contextual API dependency graphs," in *Proceedings of 2014 ACM SIGSAC Conference on Computer and Communications Security (CCS'14)*, pp. 1105–1116, ACM, Berlin, Germany, November 2014.

[33] Y. Zhang, M. Yang, B. Xu et al., "Vetting undesirable behaviors in android apps with permission use analysis," in *Proceedings of the 2013 ACM SIGSAC conference on Computer & communications security(CCS'13)*, pp. 611–622, ACM, Berlin, Germany, November 2013.

[34] W.-C. Wu and S.-H. Hung, "DroidDolphin: a dynamic android malware detection framework using big data and machine learning," in *Proceedings of 2014 Conference on Research in Adaptive and Convergent Systems (RACS'14)*, pp. 247–252, ACM, Towson, MA, USA, November 2014.
[35] W. Wang, Y. Li, X. Wang, J. Liu, and X. Zhang, "Detecting android malicious apps and categorizing benign apps with ensemble of classifiers," *Future Generation Computer Systems*, vol. 78, pp. 987–994, 2018.
[36] R. Schlegel, K. Zhang, X. Zhou, M. Intwala, A. Kapadia, and X. Wang, "Soundcomber: a stealthy and context-aware SoundTrojan for smartphones," in *Proceedings of 18th Annual Network & Distributed System Security Symposium (NDSS)*, San Diego, CA, USA, Feburary 2011.
[37] M. Ahmadi, D. Ulyanov, S. Semenov, M. Trofimov, and G. Giacinto, "Novel feature extraction, selection and fusion for effective malware family classification," in *Proceedings of Sixth ACM Conference on Data and Application Security and Privacy (CODASPY'16)*, pp. 183–194, ACM, New Orleans, LA, USA, March 2016.
[38] Z. Wu, Y. Zhou, X. Jiang, and P. Ning, "Detecting repackaged smartphone applications in third-party android marketplaces," in *Proceedings of second ACM conference on Data and Application Security and Privacy (CODASPY'12)*, pp. 317–326, ACM, Antonio, TX, USA, February 2012.
[39] P. Battista, F. Mercaldo, and V. Nardone, "Identification of android malware families with model checking," in *Proceedings of 2nd International Conference on Information Systems Security and Privacy*, pp. 542–547, Rome, Italy, February 2016.
[40] G. Suarez-Tangil, "Dendroid: a text mining approach to analyzing and classifying code structures in android malware families," *Expert Systems with Applications*, vol. 41, no. 4, pp. 1104–1117, 2014.
[41] Android Genome Project: http://www.malgenomeproject.org/, 2017.
[42] T. Chakraborty, F. Pierazzi, and V. S. Subrahmanian, "EC2: ensemble clustering and classification for predicting android malware families," *IEEE Transactions on Dependable and Secure Computing*, vol. 99, p. 1, 2017.
[43] S. Morales-Ortega, P. J. Escamilla-Ambrosio, A. Rodriguez-Mota, and L. D. Coronado-De-Alba, "Native malware detection in smartphones with android OS using static analysis, feature selection and ensemble classifiers," in *Proceedings of 2016 11th International Conference on Malicious and Unwanted Software (MALWARE)*, pp. 1–8, Fajardo, PR, USA, Octomber 2016.
[44] B. Dixon and S. Mishra, "On rootkit and malware detection in smartphones," in *Proceedings of 2010 International Conference on Dependable Systems and Networks Workshops (DSN-W)*, pp. 162-163, Chicago, IL, USA, July 2010.
[45] M. Zhao, "RobotDroid: a lightweight malware detection framework on smartphones," *Journal of Networks*, vol. 7, no. 4, pp. 715–722, 2012.
[46] A. Shabtai, U. Kanonov, Y. Elovici, C. Glezer, and Y. Weiss, ""Andromaly": a behavioral malware detection framework for android devices," *Journal of Intelligent Information Systems*, vol. 38, no. 1, pp. 161–190, 2012.
[47] K. Xu, Y. Li, and R. H. Deng, "ICCDetector: ICC-based malware detection on android," *IEEE Transactions on Information Forensics and Security*, vol. 11, no. 6, pp. 1252–1264, 2016.
[48] N. Viennot, E. Garcia, and J. Nieh, "A measurement study of google play," in *Proceedings of 2014 ACM international conference on Measurement and Modeling of Computer Systems (SIGMETRICS'14)*, pp. 221–233, ACM, Austin, TX, USA, June 2014.
[49] Anzhi: http://www.anzhi.com/.
[50] F. Maggi, A. Valdi, and S. Zanero, "AndroTotal: a flexible, scalable toolbox and service for testing mobile malware detectors," in *Proceedings of Third ACM workshop on Security and privacy in smartphones & mobile devices (SPSM'13)*, pp. 49–54, ACM, New York, NY, USA, November 2013.
[51] Virus Total: https://www.virustotal.com/, 2017.
[52] http://text-analytics101.rxnlp.com/2014/11/what-are-n-grams.html, 2017.
[53] Q. Jerome, "Using opcode-sequences to detect malicious Android applications," in *Proceedings of IEEE International Conference on Communications*, pp. 914–919, IEEE, Sydney, Australia, June 2014.
[54] H. Peng, C. Gates, B. Sarma et al., "Using probabilistic generative models for ranking risks of Android apps," in *Proceedings of the 2012 ACM conference on Computer and communications security(CCS'12)*, pp. 241–252, ACM, Raleigh, NC, USA, October 2012.
[55] W. Enck, M. Ongtang, and P. McDaniel, "On lightweight mobile phone application certification," in *Proceedings of 16th ACM conference on Computer and communications security (CCS'09)*, pp. 235–245, ACM, Chicago, IL, USA, November 2009.
[56] Androguard: https://github.com/androguard/androguard.
[57] Apktool: https://ibotpeaches.github.io/Apktool/.

9

Conjunctive and Disjunctive Keyword Search over Encrypted Mobile Cloud Data in Public Key System

Yu Zhang,[1] **Yin Li**,[1] **and Yifan Wang**[2]

[1]*School of Computer and Information Technology, Xinyang Normal University, Xinyang 464000, China*
[2]*Wayne State University, 42 W. Warren Ave., Detroit, MI 48202, USA*

Correspondence should be addressed to Yu Zhang; willow1223@126.com

Academic Editor: Paolo Bellavista

The searchable encryption scheme can perform keywords search operation directly over encrypted data without decryption, which is crucial to cloud storage, and has attracted a lot of attention in these years. However, it is still an open problem to develop an efficient public key encryption scheme supporting conjunctive and a disjunctive keyword search simultaneously. To achieve this goal, we introduce a keyword conversion method that can transform the query and index keywords into a vector space model. Through applying a vector space model to a predicate encryption scheme supporting inner product, we propose a novel public key encryption scheme with conjunctive and disjunctive keyword search. The experiment result demonstrates that our scheme is more efficient in both time and space as well as more suitable for the mobile cloud compared with the state-of-art schemes.

1. Introduction

With the rapid development of the cloud computation accompanied by the boosting amount of data, more and more enterprises and individuals are willing to share their own data on the cloud platform. Because the data stored in the cloud may be sensitive, such as medical records, the popularity of the cloud storage inevitably brings its users security concern. Specifically, hacker attack and administrator theft can lead to data leakage. In order to protect the data privacy, encrypting data before outsourcing it on the cloud server is a common way. However, users still confront the problem of how to search the encrypted data stored on the cloud efficiently. A straightforward approach is to download all the encrypted data to the clients and then decrypt them all. After obtaining all the unencrypted data, users can search the document by using common information retrieval technical. Nevertheless, this strategy needs tremendous cost of transportation, storage, and computation, which brings a new issue: how to search encrypted data efficiently without decrypting it first.

Many searchable encryption (SE) schemes were proposed to realize keyword search over encrypted data with various search functions. There are two main categories of SE according to its applications: searchable public key encryption and searchable symmetric key encryption. Over the last few years, many searchable symmetric key encryption schemes have been proposed, which achieve complex search conditions such as Boolean keyword search, personal keyword search, and query result ranking [1–4]. However, the development of searchable public key encryption is relatively slow since it is difficult to support advanced search function without sacrificing security. Specifically, a scheme supporting Boolean keyword search in the public key setting is still in urgent demand. To meet this need, one should solve two subproblems: public key encryption with conjunctive keyword search (PECK) and public key encryption with disjunctive keyword search (PEDK). Park et al. presented the formal model and the security definition of PECK followed by two constructions [6]. Then Hwang and Lee [5] introduced a concept of multiuser PECK as well as a more efficient scheme with less search time. But all these schemes require keyword fields. To eliminate the keyword fields, Boneh and Waters [11] presented a hidden vector encryption (HVE) scheme supporting conjunctive search, comparison queries, and subset queries on encrypted data. Recently, Zhang and Zhang proposed a new PECK scheme with less storage space and query time [12].

However, the research process of PEDK is very slow. In order to support disjunction formulae, Katz et al. gave a predicate encryption supporting inner product (IPE) scheme [7]. This scheme enables more complex evaluations on disjunction, conjunction, and polynomial formulae. In an IPE scheme, the secret key corresponding to a predicate vector $\overrightarrow{v}$ can decrypt the ciphertext associated with an attribute vector $\overrightarrow{x}$ if and only if $\overrightarrow{x} \cdot \overrightarrow{v} = 0$. Unfortunately, the IPE scheme mentioned above was proved in a selective security model. To improve its security level, Lewko et al. presented a fully secure IPE scheme [23] by using dual system encryption introduced in [13].

Although we can create an PEDK scheme by making use of an IPE scheme and a trivial method presented in [6], the time complexities of the encryption and the test in this PEDK scheme are both $O(2^n)$, where n is the number of the keywords in a document. In addition, the storage cost of an index also encounters exponential increase. Therefore, this scheme is not practical when n is large. Besides, if a user wants to perform a conjunctive and a disjunctive keyword query simultaneously, a PEDK and a PECK scheme should be constructed and maintained together. To construct a more efficient PEDK scheme supporting conjunctive keyword search simultaneously, Zhang and Lu proposed an approach of converting an IPE scheme into a public key encryption with conjunctive and disjunctive keyword search (PECDK) scheme and gave a concrete scheme [15]. In their scheme, the size of each document's index, the size of a trapdoor, and the time cost of pairing operations in the test process are all $O(N)$. Since N is a large integer and seen as the number of keywords in a dictionary, there is still a great room to improve the efficiency of the previous PECDK scheme.

In this paper, we first propose a new method that can change an IPE scheme into a PECDK scheme and then give an instance. Our contributions are summarized as follows.

(1) We design a new approach, which converts an index keyword set and a query keyword set into an attribute matrix and a predicate vector, respectively. Technically, we first use the index keyword set to construct an equation of degree n with one unknown. Then, we apply coefficients and the roots of the equation to create a predicate vector and an attribute matrix, respectively.

(2) We propose a construction of PECDK based on the method mentioned in (1) and an efficient IPE scheme proposed in [23]. We also prove the security of our PECDK scheme according to the security definition introduced in [15]. The experiment shows that compared with the previous PECDK scheme, the time complexity of keyword search and that of index construction are both reduced from $O(N)$ to $O(n^2)$, where $n \ll \sqrt{N}$, so is the space cost of encrypted index. Since N is much larger than n, for example, n is less than 20 while N is large than 10000, we can argue that the proposal is suitable for the mobile setting.

1.1. Organization. The rest of this paper is organized as follows. Related work is discussed in Section 2, and Section 3 gives the model of PECDK together with its security model. Section 4 firstly introduces our transformation method, then proposes a concrete PECDK scheme based on our approach, and finally presents its security proof. We present the theoretical and experimental analysis in Section 5. Section 6 covers the conclusion.

2. Related Work

Searchable encryption schemes enable the clients to store the encrypted data to the cloud and execute keyword search over ciphertext domain. Thus, our solution belongs to this field. Due to different cryptography primitives, searchable encryption schemes can be classified into public key system and symmetric key system.

Song et al. first introduced the definition of searchable symmetric encryption and proposed a concrete scheme [1]. Then Goh defined the concept of conjunctive keyword search over encrypted data and presented an effective scheme by taking advantage of the Bloom filter [2]. In addition, he also gave a formal security definition of searchable symmetric encryption. According to this scheme, some improved schemes with less computation and communication cost were proposed in [3–5]. However, the search time cost in these schemes is linear with the number of the documents in a dataset since each document needs an encrypted keyword index. To reduce the time cost of search, some works utilized the tree structure, such as R-tree and kd-tree, to obtain a sublinear search efficiency [8, 9]. Considering the issues that returning all related documents will bring network traffic and the previous schemes fail to sort search results, rank search schemes realizing a quick search of top-k relevant documents were proposed in [16–18]. These works only supported single keyword search due to using order-preserving encryption (OPE) [14]. Recently, some schemes achieving multikeywords rank search were presented in [19, 20].

With slower development than searchable symmetric encryption, searchable public key encryption is also difficult to support complex query condition. Boneh et al. brought up the new concept of PEKS and provided several constructions [10] related to the Identity-Based Encryption (IBE) presented in [21]. Based on that, Abdalla et al. specified the computational and statistical consistency of PEKS and proposed a statistically consistent scheme [22]. However, their works only supported a single keyword search. Park et al. raised a concept of PECK in [6]. They defined the security model of this mechanism followed by two constructions. One needs more bilinear pairing operations, while the other needs more private keys. Then Hwang and Lee designed a more efficient scheme as well as introduced a multiuser PECK scheme that enables multiusers keyword search [5]. To avoid the usage of keyword field just like all the constructions mentioned above did, Boneh and Waters presented the hidden vector encryption (HVE), a public key system supporting conjunctive keyword search, comparison queries, and subset queries on encrypted data [11]. For achieving disjunctive keyword search without keyword field,

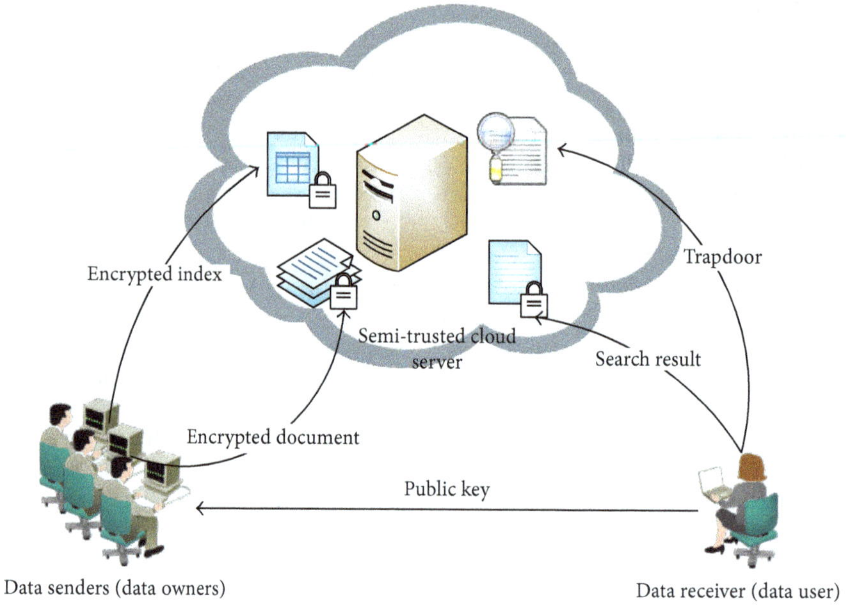

FIGURE 1: Architecture of the search over encrypted cloud data.

Katz et al. proposed an IPE scheme [7]. To improve the security level and the decryption efficiency, fully secure IPE schemes were proposed in [23, 24]. In addition, the previous schemes fail to support conjunctive keyword search and disjunctive keyword search at the same time. Addressing this issue, Zhang and Lu proposed a PECDK scheme [15].

Another class of searchable encryption is called range search over encrypted data. It can be used to test whether a multidimension point is inside in a hyperrectangle. Related works were presented in [25–27].

3. Preliminaries

3.1. The System Model. Consider a data storage service in cloud, where a data owner has a set of documents D to be outsourced to the cloud server in an encrypted form. To enable efficient query over encrypted documents, we consider the keyword-based index structure for storing the outsourced files. Specifically, the data owner builds an encrypted searchable index set C with each document's keywords, and then both the encrypted index set C and the encrypted document set $\mathrm{Enc}(D)$ are outsourced to the cloud server. For each query of an arbitrary keyword set Q, a data user computes a search token TK_Q of the query Q and sends it to the cloud server. Upon receiving TK_Q from the data user, the cloud server queries over the encrypted index set C and returns the candidate encrypted documents. Finally, the data user decrypts the candidate documents and verifies each document by checking the existence of the keyword.

The application scene of the searchable public key encryption involves three roles: data senders, a data receiver, and a cloud server, as illustrated in Figure 1. In this scene, the data receiver performs keyword search, generates a public key (pk) and a secret key (sk), and sends the pk to the public. Anyone who can access the pk is recognized as a data sender. Data senders send their encrypted documents with the related encrypted index to the cloud server. The cloud server stores the encrypted documents and the encrypted searchable indexes for data senders. In addition, we assume that the cloud server is "honest-but-curious" employed by many works on searchable encryption. When the data receiver processes keywords search, one generates a trapdoor of these keywords and sends it to the cloud server. Upon receiving trapdoor from the data receiver, the cloud server queries over each document's encrypted index and returns the candidate encrypted documents.

In this paper, we focus on the searchable public key encryption supporting conjunctive and disjunctive keyword search. Strictly speaking, we present a formal definition PECDK model derived from the model proposed in [15].

Definition 1. There are four polynomial time algorithms in the PECDK scheme: *KeyGen*, *IndexBuild*, *Trapdoor*, and *Test*:

(i) *KeyGen(γ)*: Choosing a security parameter γ, the algorithm outputs the parameter (pk, sk), where pk is the public key and sk is the secret key.

(ii) *IndexBuild(pk, W)*: The algorithm is performed by the sender to produce an encrypted index I_W by using a keyword set $W = \{w_1, w_2, \ldots, w_n\}$ and pk.

(iii) *Trapdoor(sk, Q, sym)*: The receiver runs this algorithm to produce a trapdoor. It takes sk and the keyword query Q, sym as input to generate a trapdoor $T_Q = \{t_Q, \mathrm{sym}\}$.

(iv) *Test(pk, T_Q, I_W)*: It takes the trapdoor $T_Q = \{t_Q, \mathrm{sym}\}$, the secure index I_W, and the public key pk as input. If sym is $\wedge$, it means that the trapdoor is for conjunctive keyword search. In this case, the output is 1 if $Q \subseteq W$ or 0 otherwise. If sym is $\vee$, disjunctive keyword search should be performed. In this case, if $Q \cap W \neq \phi$, it outputs 1 otherwise 0.

3.2. Security Model. Generally speaking, the security of a searchable encryption means that the cloud server can infer as little information as possible from the encrypted data and the search process without sacrificing the search ability. Before introducing the adaptive security definition of our scheme, we first define the privacy leakage, which is revealed to the cloud server inevitably.

3.2.1. Size Pattern. Since the encrypted documents and queries are submitted to the cloud server, the cloud server can obtain the basic size information of these encrypted data easily. This is called the leakage of size pattern.

3.2.2. Access Pattern. For each query, the cloud server can obtain the identifiers of data records that match this query. This is called the leakage of access pattern.

3.2.3. Search Pattern. Given a record set $D = \{D_1, D_2, \ldots, D_n\}$ and a query set $Q = \{Q_1, Q_2, \ldots, Q_t\}$, cloud server can create a $n \times t$ matrix, where the element in the ith row and jth column is 1, if the record D_i matches the query Q_j. The matrix can be seen as the leakage of search pattern.

Actually, Oblivious RAM can be utilized to preserve access and search pattern, but this technique is too inefficient to be used in the real applications. In this paper, we do not consider the problem of how to protect access pattern and search pattern in our scheme.

3.2.4. Query Privacy. The leakage of query privacy means that keywords in the encrypted query will be revealed to the cloud server. It commonly exists in the public key setting since anyone can construct an encrypted index for arbitrary keywords. Because of belonging to public key encryption category, our scheme fails to protect query privacy.

As previous works, we denote the information leakage including size pattern, access pattern, search pattern, and query privacy as leakage function $L(D, I, Q)$, where D is a record set, I is a secure index, and Q is a query set.

3.2.5. Adaptive Security. With the leakage function mentioned above, we introduce an adaptive security definition of the PECDK scheme related to the one proposed in [15] as follows.

Definition 2. A PECDK scheme is adaptively index-hiding against chosen plaintext attacks if for all probabilistic polynomial time adversaries $\mathbb{A}$, the advantage of $\mathbb{A}$ in the following game is negligible.

(1) Setup: The challenger $\mathbb{C}$ runs *KeyGen*(γ) algorithm to generate the public key pk and the secret key sk. Then $\mathbb{C}$ gives pk to the attacker $\mathbb{A}$.

(2) Phase 1: The attacker $\mathbb{A}$ can adaptively ask the challenger $\mathbb{C}$ for the trapdoor T_Q for any query $\{Q, \text{sym}\}$ of his choice.

(3) Challenge: $\mathbb{A}$ selects two keyword sets W^0 and W^1 and sends them to $\mathbb{C}$. Suppose that $\{Q_1, \text{sym}_1\}$, $\{Q_2, \text{sym}_2\}, \ldots, \{Q_t, \text{sym}_t\}$ are the keyword queries in Phase 1, the only restriction is that, for each $i \in [1, t]$, there is $W^0 \cap Q_i = \varnothing$ and $W^1 \cap Q_i = \varnothing$. Then, $\mathbb{C}$ picks a random bit $\beta \in \{0, 1\}$, produces $I_\beta =$ IndexBuild (pk, W^β), and sends $\{I_\beta, W^0, W^1\}$ to $\mathbb{A}$.

(4) Phase 2: $\mathbb{A}$ can continue to ask for trapdoors for any query of his choice, and these trapdoors subject to the same restriction as before.

(5) Response: $\mathbb{A}$ outputs $\beta' \in \{0, 1\}$ and wins the game if $\beta' = \beta$.

We define $\mathbb{A}$'s advantage in the above game as

$$\text{Adv}_{\text{Game}}^{\mathbb{A}} = \left|\Pr[\beta' = \beta] - \frac{1}{2}\right|. \tag{1}$$

Generally speaking, as long as the information leakage of W^0 and that of W^1 are the same under the leakage function $L(D, I, Q)$, we should insure that the encrypted form of W^0 and that of W^1 are computationally indistinguishable to the adversary.

4. Proposed PECDK Scheme

Based on the system and security model descried in the previous section, in this section, we present the method that converting index and query keyword sets into a vector space model. This model can be applied to an IPE scheme easily.

4.1. Conversion Method. We suppose that any keyword w can be expressed as $\{0, 1\}^*$ and define a function H_1: $\{0, 1\}^* \to Z_p^*$. Since p is a large prime and is larger than the number of the all words, H_1 can be collision-resistance. This means that, if $i \neq j$, then $H_1(w_i) \neq H_1(w_j)$, where w_i and w_j are two distinct keywords.

We first construct an equation of degree n with one unknown by using the index and query keyword sets. After that, we use the roots and coefficients of the equation to create a query vector and an index matrix. Let $W = \{w_1, w_2, \ldots, w_n\}$ and $Q = \{q_1, q_2, \ldots, q_m\}$ are two keyword sets, where $m < n$. The approach is described as follows:

(1) For the keyword set $Q = \{q_1, q_2, \ldots, q_m\}$, the following function can be constructed:

$$\begin{aligned} f(x) &= (x - H_1(q_1))(x - H_1(q_2)) \ldots (x - H_1(q_m)) \\ &= a_m x^m + a_{m-1} x^{m-1} + \ldots + a_0 x^0. \end{aligned} \tag{2}$$

According to the coefficient of the $f(x)$, a vector $\overrightarrow{a} = \{a_0, a_1, \ldots, a_m\}$ can be obtained.

(2) For the keyword set $W = \{w_1, w_2, \ldots, w_n\}$, the following function can be constructed:

$$\begin{aligned} g(x) &= (x - H_1(w_1))(x - H_1(w_2)) \ldots (x - H_1(w_n)) \\ &= b_n x^n + b_{n-1} x^{n-1} + \ldots + b_0 x^0. \end{aligned} \tag{3}$$

It is not difficult to find that the roots of the equation $g(x) = 0$ are $H_1(w_1), H_1(w_2), \ldots, H_1(w_n)$. According to the above roots, a matrix can be constructed:

$$M = \begin{pmatrix} M_{11} & M_{12} & \cdots & M_{1n} \\ M_{21} & M_{22} & \cdots & M_{2n} \\ \vdots & \vdots & \ddots & \vdots \\ M_{n+1,1} & M_{n+1,2} & \cdots & M_{n+1,n} \end{pmatrix} = \begin{pmatrix} H_1(w_1)^0 & H_1(w_2)^0 & \cdots & H_1(w_n)^0 \\ H_1(w_1)^1 & H_1(w_2)^1 & \cdots & H_1(w_n)^1 \\ \vdots & \vdots & \ddots & \vdots \\ H_1(w_1)^n & H_1(w_2)^n & \cdots & H_1(w_n)^n \end{pmatrix}. \quad (4)$$

(3) Note that if there is a keyword $q_i \in Q$ such that $q_i \in W$, according to (2–4), it is not difficult to verify that $f(w_i) = \overrightarrow{a} \cdot \overrightarrow{M_i} = 0$ where $\overrightarrow{M_i} = \{M_{1i}, M_{2i}, \ldots, M_{m+1,i}\}$

As a result, if we can make sure there is $\overrightarrow{a} \cdot \overrightarrow{M_i} = 0$ for each i or some i, a conjunctive or disjunctive query can be obtained when applying the method above. Based on this, a concrete PECDK scheme will be proposed in next section.

4.2. Construction Details. According to the definition of IPE [13], suppose that $\text{Setup}_{\text{IPE}}$, $\text{Enc}_{\text{IPE}}(\text{pk}_{\text{IPE}}, M, \overrightarrow{x})$, $\text{KeyGen}_{\text{IPE}}(\text{pk}_{\text{IPE}}, \text{msk}_{\text{IPE}}, \overrightarrow{v})$, and $\text{Dec}_{\text{IPE}}(\text{pk}_{\text{IPE}}, c, \text{sk}_{\overrightarrow{v}})$ be the four algorithms in the IPE scheme, where pk_{IPE} and msk_{IPE} are the public key and the master secret key generated by using $\text{Setup}_{\text{IPE}}$, $\overrightarrow{x}$ is the attribute vector, $\overrightarrow{v}$ is the predicate vector, c is the ciphertext generated by using Enc_{IPE}, and $\text{sk}_{\overrightarrow{v}}$ is the secret key generated by using $\text{KeyGen}_{\text{IPE}}$. Our PECDK scheme works as follows:

(i) *KeyGen*: By using the $\text{Setup}_{\text{IPE}}$ algorithm, pk_{IPE} and msk_{IPE} can be obtained. The algorithm sets $\text{pk} = \text{pk}_{\text{IPE}}$ and $\text{sk} = \text{msk}_{\text{IPE}}$ and outputs pk and sk.

(ii) *IndexBuild*: Given a keyword set $W = \{w_1, w_2, \ldots, w_n\}$, the algorithm generates a matrix $M = \{\overrightarrow{M_1}, \overrightarrow{M_2}, \ldots, \overrightarrow{M_n}\}$ by using (4). Then it generates $I_W = \{c_1, c_2, \ldots, c_n\}$ in which $c_i = \text{Enc}_{\text{IPE}}(\text{pk}, 1, \overrightarrow{M_i})$.

(iii) *Trapdoor*: Given a keyword query $\{Q, \text{sym}\}$, the algorithm creates $\overrightarrow{a}$ by using (3). Then it generates t_Q by making use of $\text{KeyGen}_{\text{IPE}}(\text{sk}, \overrightarrow{a})$. Then it outputs a trapdoor $T_Q = \{t_Q, \text{sym}\}$.

(iv) *Test*: Given a T_Q, a I_W, and the pk, there are two situations.

(1) If the symbol in T_Q is $\vee$, the algorithm works as follows.

(a) Choosing a counter i and setting $i = 1$.

(b) If $i > n$, then go to step (c), otherwise the algorithm computes $R = \text{Dec}(\text{pk}, I_W, T_Q)$. If $R = 1$, then the algorithm outputs 1 and ends. Otherwise, it sets $i = i + 1$ and goes to the step (b).

(c) The algorithm outputs 0 and ends.

(2) If the symbol in T_Q is $\wedge$, the algorithm works as follows.

(a) Choosing two counters i and j and setting $i = 1$ and $j = 1$.

(b) If $i > n$, then go to step (c), otherwise the algorithm computes $R = \text{Dec}(\text{pk}, I_W, T_Q)$. If $R = 1$, then the algorithm sets $j = j + 1$. Otherwise, it does nothing. After that, it sets $i = i + 1$ and goes to the step (b).

(c) If $j = m$, the algorithm outputs 1 and ends. Otherwise, it outputs 0 and ends.

4.3. Security Proof. The proposed PECDK scheme can be constructed by making use of the fully secure IPE scheme. Therefore, we have the following proposition.

Proposition 1. *If the IPE scheme is secure, then our PECDK scheme is secure.*

Proof Sketch: If there is a PPT algorithm $\mathbb{A}$ which can break the PECDK scheme, we can say that $\mathbb{A}$ can break the IPE scheme. To create pk and sk in the PECDK scheme, the challenger $\mathbb{C}$ uses the $\text{Setup}_{\text{IPE}}$ algorithm to generate pk_{IPE}, sk_{IPE}, and sets $\text{pk} = \text{pk}_{\text{IPE}}$ and $\text{sk} = \text{sk}_{\text{IPE}}$. $\mathbb{A}$ can adaptively query trapdoors of keyword set $\{Q_1, Q_2, \ldots, Q_t\}$. It is not difficult to find that these trapdoors can be seen as a group of decryption keys for the IPE scheme. Then $\mathbb{A}$ outputs two challenge keyword sets W^0 and W^1, under a constraint that $Q_i \cap W^0 = \emptyset$ and $Q_i \cap W^1 = \emptyset$, where $i \in [1, t]$. $\mathbb{C}$ flips a coin $\beta \in \{0, 1\}$ and gives an index I_{W^β} to $\mathbb{A}$. This index can be seen as a set of challenge ciphertexts of IPE. After this, $\mathbb{A}$ can continue querying trapdoors which subject to the restriction described above. These trapdoors can still be seen as a group of decryption keys for IPE. Finally, $\mathbb{A}$ gives a guess β'. Note that if $\mathbb{A}$ can break the PECDK scheme, the value of $|\Pr[\beta' = \beta] - (1/2)|$ is not negligible. It means that the two challenge indices can be distinguished. Because the challenge indices in the PECDK scheme is equal to the challenge ciphertext in the IPE scheme, according to the security definition for IPE, it means that $\mathbb{A}$ can break the IPE scheme.

5. Performance Evaluation

5.1. Cost Analysis. We denote the previous PECDK scheme [15] as PECDK-2 and the proposal as PECDK-1. Let $F = \{f_1, f_2, \ldots, f_D\}$ be the file set, and W_i be the keyword set for the corresponding file f_i, where $i \in [1, D]$. For each keyword set W_i, we build an index of file f_i. We assume that n is the number of keywords in W_i, m is the number of keywords in a query Q, and N is a large number such as the number of keywords in a dictionary, where $N \gg n$. The index and query structures for PECDK-1 and PECDK-2 are illustrated in Figures 2(a) and 2(b), respectively. Note that the index

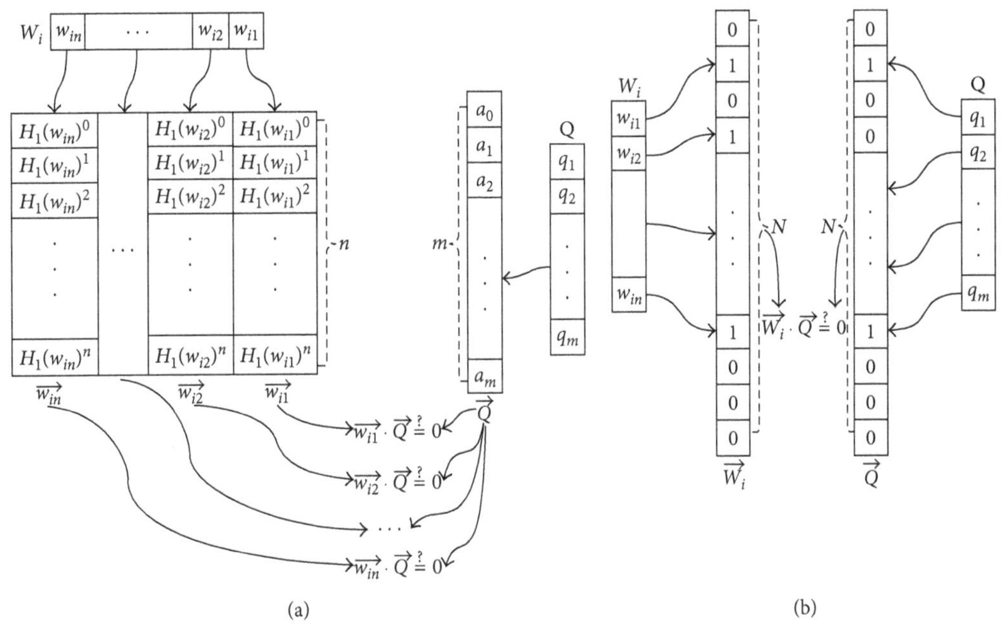

FIGURE 2: Index structure for a single file f_i's keyword set W_i, query structure for a query keyword set Q, and testing process for W_i and Q in PECDK-1 (a) and PECDK-2 (b), where $i \in [1, D]$.

structure in Figure 2 is only for one file f_i, and the index for whole files is a simple combination for each file's index.

From Figure 2(a), we can find that W_i is converted into a matrix in which each column is associated with a keyword w_{ij} in W_i and represented by a vector $\overrightarrow{w_{ij}}$, where $j \in [1, n]$. Correspondingly, the query Q is changed to a vector $\overrightarrow{Q}$ based on the approach introduced in Section 4. The test algorithm is to verify whether there exists at least one j such that $\overrightarrow{w_{ij}} \cdot \overrightarrow{Q}$ equals 0. Therefore, PECDK-1 is independent with the vocabulary length N. According to the above analysis, we know that the time complexity for index building, trapdoor generation, and testing is $O(n^2)$, $O(m)$, and $O(m \cdot n)$, respectively.

From Figure 2(b), we suppose that the vocabulary is defined as $w_1, w_2, \ldots, w_N$. For the keyword set W_i, the index of W_i is initialed by a zero vector with length N which is denoted as $\overrightarrow{W_i}$. After initialization, if $w_{ij} = w_k$, the k-th position of $\overrightarrow{W_i}$ is set to be 1, where $j \in [1, n]$ and $k \in [1, N]$. Moreover, the vector building process for query Q is the same as that for W_i. The test process in PECDK-2 is to check whether $\overrightarrow{W_i} \cdot \overrightarrow{Q}$ equals 0. According to the above description, it is clear that the time complexities of index building, trapdoor generation and testing are all $O(N)$.

Let $|G|$ represents the size of an element in G, $|T_e|$ be the time cost for a pairing operation [28], and $|T_G|$ be the time cost for the power operation on G. The results of the comparison with the PECDK-2 scheme are shown in Table 1.

Since, in the encryption and test phrase, the time cost of the pairing and the power operation are much more than other operations, we do not take account of other operations. According to Table 1, we can argue that compared with the previous PECDK (PECDK-2) scheme, the proposed one (PECDK-1) has better performance on time and space complexity when $N > n^2$.

In the following, we will argue that our statement where $N \gg n$ is true in real world. We investigate data collections used in recent works [31, 32] for information retrieval. The statistical information for these data sets are shown in Table 2. From this table, we found that the vocabulary size (N) is commonly linear

TABLE 1: Comparison with the previous PECDK scheme.

	PECDK-2	PECDK-1
pk size	$O(N^2)\|G\|$	$O(n^2)\|G\|$
sk size	$O(N^2)\|G\|$	$O(n^2)\|G\|$
Trapdoor size	$O(N)\|G\|$	$O(m)\|G\|$
Index size	$O(N)\|G\|$	$O(n^2)\|G\|$
Encryption time	$O(N)\|T_G\|$	$O(n^2)\|T_G\|$
Test time	$O(N)\|T_e\|$	$O(m \cdot n)\|T_e\|$

TABLE 2: Statistics for data collections used in the domain of information retrieval.

Dataset name	Documents	Vocabulary size
AP88-89	164,597	247,350
WSJ87-92	173,252	216,539
DOTGOV	1,247,442	3,051,601
MedTrack	100,866	55,065
Yelp2013	335,018	211,245
Yelp2014	1,125,457	476,191
Yelp2015	1,569,264	612,636
IMDB review	348,415	115,831
Yahoo answer	1,450,000	1,554,607
Amazon review	3,650,000	1,919,336

TABLE 3: Statistics for documents' information in the OHSUMED collection (#w denotes the number of words per document).

Field name	#w	Vocabulary size
Title	5~20	33059
Abstract	50~200	83496

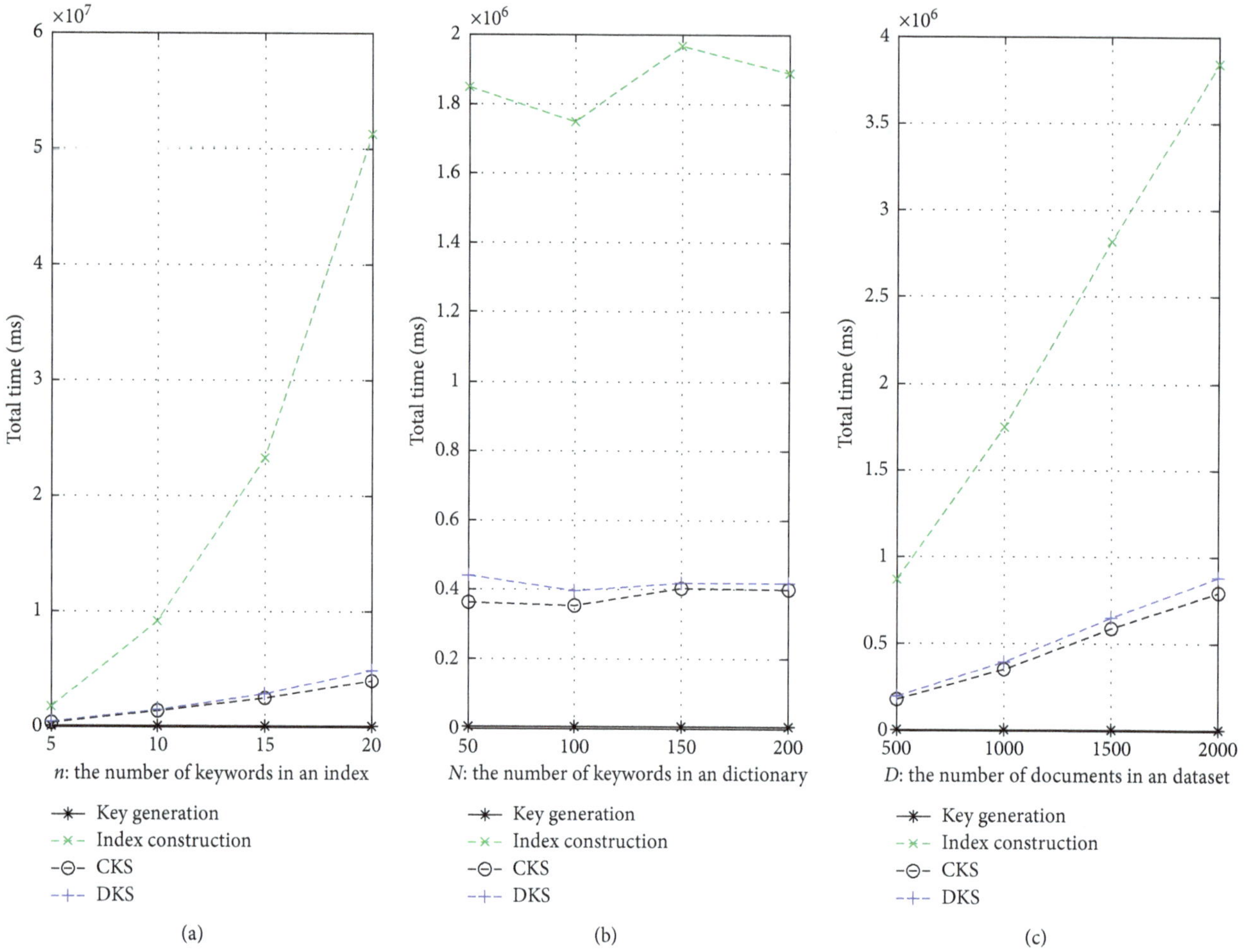

FIGURE 3: Impact of n, N, and D on the time cost of key generation, index construction, conjunctive keyword search (CKS), and disjunctive keyword search (DKS) in PECDK-1 (a) for different sizes of n with $D = 1000$ and $N = 100$, (b) for different numbers of N with $D = 1000$ and $n = 5$, and (c) for different numbers of D with $N = 100$ and $n = 5$.

with $O(10^6)$. However, the number of keywords in a document (n) is usually only 3~5 (e.g., the scientific paper). Even if the abstract field is used as keywords, the number of keywords in a single document is approximately 200. So we can empirically think that N is much bigger than n.

Moreover, we have investigated the OHSUMED collection [30] that was created for information retrieval research. In this collection, each document contains several fields, such as title, abstract, sequential identifier, and so on. In our experiment, we use the field of "title" and "abstract," as well as a file (file name: ohsumed.87) containing more than 54000 documents for statistics. The results of this experiment are described in Table 3. Whether we use the "title" or "abstract" field as keywords, the keywords size n is far less than the vocabulary size N. In conclusion, we think that the statement of $N \gg n$ is true.

In addition, since PECDK-1 needs to test each keyword w_{ij} in the W_i against the query Q one by one, it will leak the information of the keyword set $Q \cap W_i$. Therefore, the proposed scheme is proven to be secure under a weaker security definition (Definition 2) than the previous one [15].

5.2. Experiment Results. For our experiments, we build artificial plaintext index with different number of keywords in a dictionary (i.e., $N = 50, 100, 150, 200$), different number of documents (i.e., $D = 500, 1000, 1500, 2000$), and different number of keywords in a document (i.e., $n = 5, 10, 15, 20$). In this index, we denote each keyword as a unique integer in a range $[0, N]$. We encrypted the indices with PECDK-1 and PECDK-2, respectively, and the encrypted indices were stored on our machine. We then executed random queries over these encrypted indices. We implemented our constructions in JAVA with Java Pairing Based Cryptography library (JPBC) [29]. Our experiments were run on Intel(R) Core(TM) i7-3520M CPU at 2.90 GHz processor and 3537 MB memory size. In our implementation, the bilinear map is instantiated as Type A pairing (base field size is 512-bit), which offers a level of security equivalent to 1024-bit DLOG [29].

5.2.1. Performance Comparison Between PECDK-1 and PECDK-2. For a query with five keywords, Figures 3 and 4 show the following:

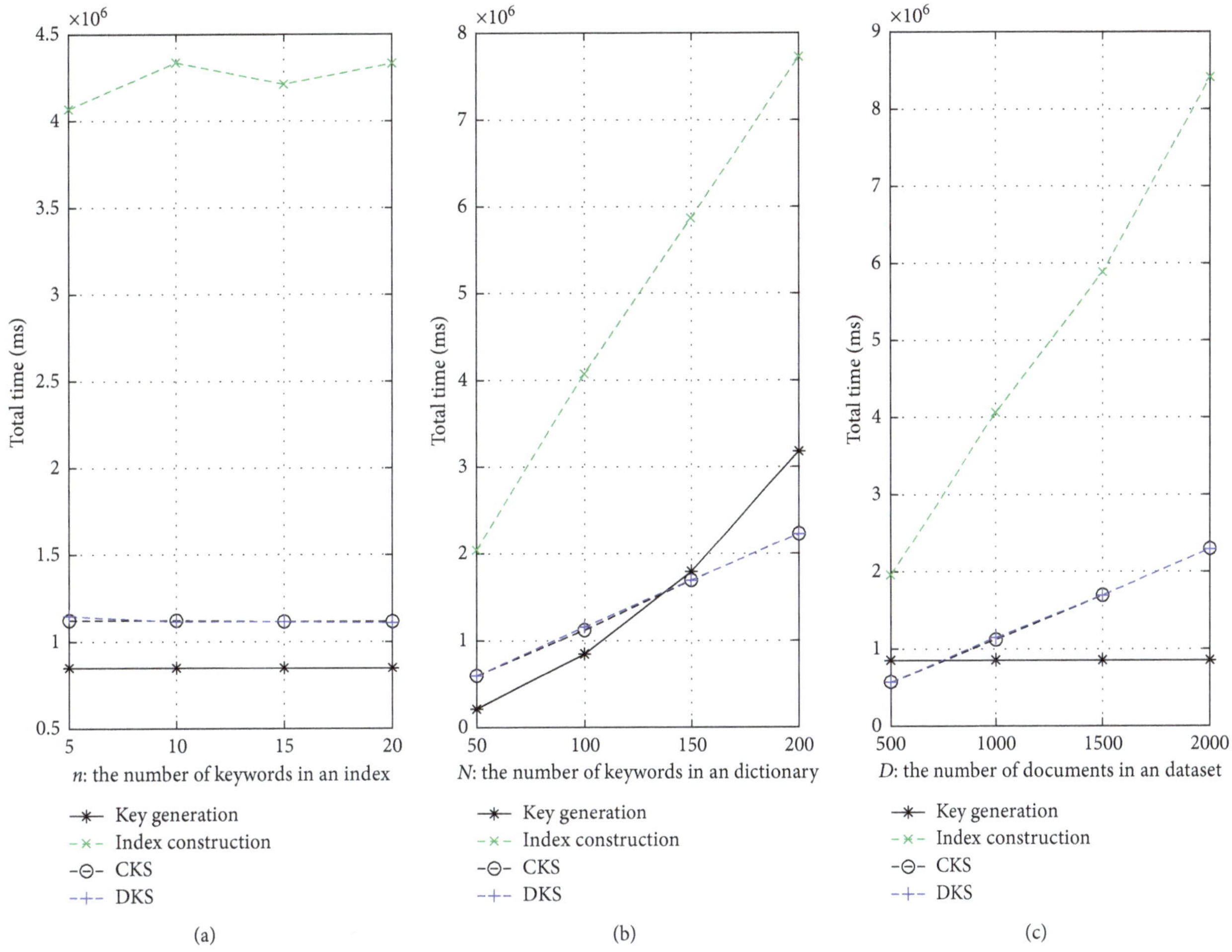

FIGURE 4: Impact of n, N, and D on the time cost of key generation, index construction, conjunctive keyword search (CKS), and disjunctive keyword search (DKS) in PECDK-2 (a) for different sizes of n with $D = 1000$ and $N = 100$, (b) for different numbers of N with $D = 1000$ and $n = 5$, and (c) for different numbers of D with $N = 100$ and $n = 5$.

(1) Figures 3(b) and 3(c) and Figures 4(a) and 4(c) denote that the execution time of key generation in PECDK-1 is irrespective of N and D, while that in PECDK-2 is independent with n and D.

(2) Figures 3(b) and 4(a) indicate that the running time of index construction and keywords search in PECDK-1 is independent with N, and that in PECDK-2 is not connected with n.

(3) Figures 3(c) and 4(c) illustrate that both the search time and the indexing time in PECDK-1 and PECDK-2 are nearly linear with D.

(4) Figures 3(a) and 4(b) shows that PECDK-1 is influenced mainly by n, while PECDK-2 is affected primarily by N.

According to the item (4), since n and N are two different parameters, we need to investigate the influence extent of the time and space complexity from these two parameters.

5.2.2. *Key Generation.* The running time of key generation in PECDK-1 is less than $O(n^2)$ while in PECDK-2 is less than $O(N^2)$. Since n is much less than N, PECDK-1 is more efficient than PECDK-2 in key generation phase.

TABLE 4: Impact of n on time consumption of trapdoor generation ($m = 5,\ N = 100$).

Size of n	5	10	15	20
PECDK-1 (ms)	576	963	1338	1805
PECDK-2 (ms)	5564	5562	5565	5603

TABLE 5: Impact of N time consumption of trapdoor generation ($m = 5,\ n = 5$).

Size of N	50	100	150	200
PECDK-1 (ms)	592	576	589	606
PECDK-2 (ms)	2862	5564	8224	10860

5.2.3. *Index Construction.* Figure 3(b) shows that the time cost of index construction in PECDK-2 is linear with N. Specifically, when $n = 5$, $N = 200$ and $D = 1000$, it takes around 7720.292 seconds to build the index. Therefore, we

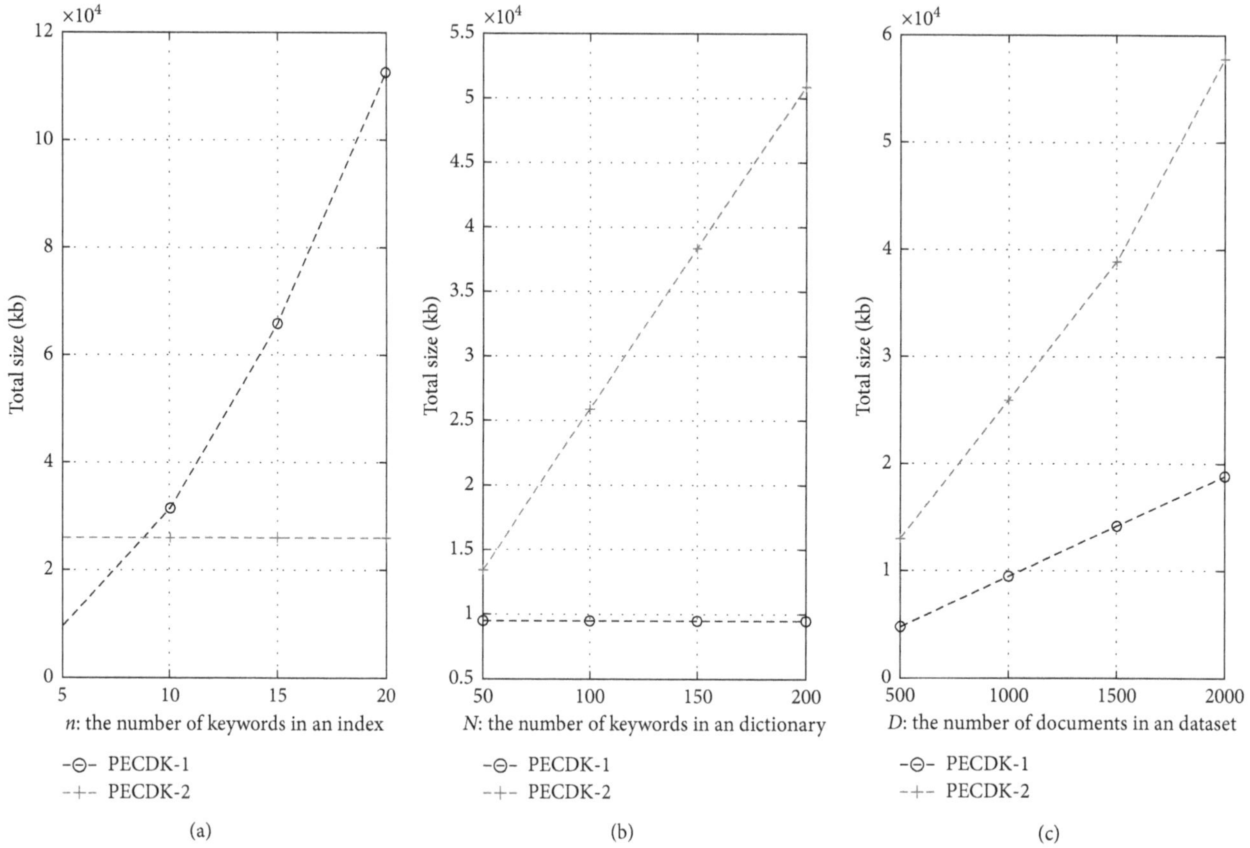

FIGURE 5: Impact of n, N, and D on the storage cost of the index in PECDK-1 and PECDK-2 (a) for different sizes of n with $D = 1000$ and $N = 100$, (b) for different numbers of N with $D = 1000$ and $n = 5$, and (c) for different numbers of D with $N = 100$ and $n = 5$.

can argue that under the condition that $N = 10000$ and $D = 1000$, the time of building index requires 7720.292*500 seconds, which is significantly longer than the time of index construction in PECDK-1 with $n = 20$ and $D = 1000$. According to this, we can conclude that the PECDK-1 scheme is more suitable and practical since n is always less than 20 and N is always larger than 10000.

5.2.4. Trapdoor Generation. Tables 4 and 5 show that the time of trapdoor generation in PECDK-1 is linear with n, while that in PECDK-2 is linear with N. Since n is much less than N, the time cost of building a trapdoor in PECDK-1 is less than that in PECDK-2. The trapdoor generation, unlike the key generation, is not a one-time cost and is performed by the data receiver. Therefore, the PECDK-1 scheme is more suitable on mobile cloud where data users possess little computation capacity.

5.2.5. Search Efficiency. Generally speaking, Figures 3(c) and 4(c) indicate that the time cost of conjunctive keyword search and that of disjunctive keyword search in PECDK-1 are both increasing with n^2, while those in PECDK-2 are both linearly increasing with N. Specifically, when $n = 5$, $N = 200$ and $D = 1000$, it takes around 2220 seconds to make a conjunctive or disjunctive keyword search in PECDK-2. According to the property that search time is linear with N, we can also conclude that the time cost of keyword search needs 2220∗500 seconds when $N = 10000$ and $D = 1000$. But the PECDK-1 scheme only costs approximately 5000 seconds to realize the conjunctive or disjunctive keyword search when $n = 20$ and $D = 1000$. As the same reason mentioned above, the PECDK-1 scheme is more practical than the PECDK-2 one.

5.2.6. Storage Overhead. As shown in Figures 5(a) and 5(b), we put forward the following arguments:

(1) The parameters N and n are independent with the storage cost of the index in PECDK-1 and that in PECDK-2, respectively.

(2) The storage of the index in PECDK-1rises with the square of n while that in PECDK-2 is linearly related to N. Due to the reason that n is less than 20 and N is larger than 10000 in common, we deem that the proposal is more useful in practice. As illustrated in Figure 5(c), we can find that both storage overhead in PECDK-1 and that in PECDK-2 are linear with D. Because n is significantly less than N and the increase rate in PECDK-1 is larger than that in PECDK-2, the storage cost in the proposed scheme is less than the previous one when D is large.

5.3. More Comments. Although the index structure, such as inverted document and R-tree, can raise the search efficiency, it fails to support dynamic operations in the public key system. Because anyone who can access the pk can construct an index in this system as well, it is difficult to combine indices obtained from data senders into a structured index. Our proposal can dynamically support document update in nature since each document is associated with an encrypted index.

In addition, because of the simple index structure, we can easily accelerate the search process by utilizing the technique of parallel computation. Thus, we argue that our scheme is practical in the cloud platform.

6. Conclusion

In this paper, we proposed a new approach to construct an efficient PECDK scheme with better performance in time and space complexity under an adaptive security model. To reveal the efficiency of the proposed scheme, we compared it with the existing PECDK scheme presented in [15] through theoretical analysis and experimental results.

Since n and m are much smaller than N, we think that the proposed scheme is beneficial for mobile applications with computation and memory limitations. In the future, we plan to create a new index structure to reduce the time cost of search and index construction.

Acknowledgments

The authors gratefully acknowledge the National Natural Science Foundation of China under Grant nos. 61402393 and 61601396 and Shanghai Key Laboratory of Integrated Administration Technologies for Information Security (no. AGK201607).

References

[1] D. Song, D. Wagner, and A. Perrig, "Practical techniques for searching on encrypted data," in *Proceedings of the IEEE Symposium on Research in Security and Privacy*, pp. 44–55, Berkeley, CA, USA, 2000.

[2] E. J. Goh, "Secure indexes," *IACR Cryptology ePrint Archive*, vol. 2003, p. 216, 2003.

[3] Z. Fu, X. Wu, C. Guan et al., "Toward efficient multi-keyword fuzzy search over encrypted outsourced data with accuracy improvement," *IEEE Transactions on Information Forensics and Security*, vol. 11, no. 12, pp. 2706–2716, 2017.

[4] Z. Fu, K. Ren, J. Shu et al., "Enabling personalized search over encrypted outsourced data with efficiency improvement," *IEEE Transactions on Parallel and Distributed Systems*, vol. 27, no. 9, pp. 2546–2559, 2016.

[5] Y. H. Hwang and P. J. Lee, "Public key encryption with conjunctive keyword search and its extension to a multi-user system," in *Proceedings of the Pairing*, vol. 4575 of Lecture Notes in Computer Science, pp. 2–22, Springer, Tokyo, Japan, July 2007.

[6] D. J. Park, K. Kim, and P. J. Lee, "Public key encryption with conjunctive field keyword search," in *Proceedings of the WISA*, vol. 3325 of Lecture Notes in Computer Science, pp. 73–86, Springer, Wuhan, China, 2004.

[7] J. Katz, A. Sahai, and B. Waters, "Predicate encryption supporting disjunctions, polynomial equations, and inner products," in *Proceedings of the Advances in Cryptology, EUROCRYPT*, vol. 4965 of Lecture Notes in Computer Science, pp. 146–162, Springer, Istanbul, Turkey, April 2008.

[8] D. Cash, S. Jarecki, C. Jutla, H. Krawczyk, M.-C. Roşu, and M. Steiner, "Highly-scalable searchable symmetric encryption with support for Boolean queries," in *Proceedings of the Advances in Cryptology, CRYPTO*, pp. 353–373, Santa Barbara, CA, USA, August 2013.

[9] D. Cash, J. Jaeger, S. Jarecki et al., "Dynamic searchable encryption in very-large databases: data structures and implementation," in *Proceedings of the Network and Distributed System Security Symposium*, San Diego, CA, USA, February 2014.

[10] D. Boneh, G. Di Crescenzo, R. Ostrovsky, and G. Persiano, "Public key encryption with keyword search," in *Proceedings of the EUROCRYPT*, vol. 3027 of Lecture Notes in Computer Science, pp. 506–522, Springer, Interlaken, Switzerland, May 2004.

[11] D. Boneh and B. Waters, "Conjunctive, subset, and range queries on encrypted data," in *Proceedings of the TCC*, vol. 4392 of Lecture Notes in Computer Science, pp. 535–554, Springer, Amsterdam, Netherlands, February 2007.

[12] B. Zhang and F. Zhang, "An efficient public key encryption with conjunctive-subset keywords search," *Journal of Network and Computer Applications*, vol. 34, no. 1, pp. 262–267, 2011.

[13] B. Waters, "Dual system encryption: realizing fully secure IBE and HIBE under simple assumptions," in *Proceedings of the Advances in Cryptology, CRYPTO*, vol. 5677 of Lecture Notes in Computer Science, pp. 619–636, Springer, Santa Barbara, CA, USA, August 2009.

[14] A. Boldyreva, N. Chenette, Y. Lee, and A. O'Neill, "Order-preserving symmetric encryption," in *Proceedings of the Advances in Cryptology, EUROCRYPT*, pp. 224–241, Cologne, Germany, April 2009.

[15] Y. Zhang and S. Lu, "POSTER: efficient method for disjunctive and conjunctive keyword search over encrypted data," in *Proceedings of the ACM SIGSAC Conference on Computer and Communications Security*, pp. 1535–1537, Hong Kong, China, December 2014.

[16] A. Swaminathan, Y. Mao, G. M. Su et al., "Confidentiality-preserving rank-ordered search," in *Proceedings of the ACM Workshop on Storage Security and Survivability, Storages*, pp. 7–12, Alexandria, VA, USA, October 2007.

[17] S. Zerr, D. Olmedilla, W. Nejdl et al., "Zerber+R:top-k retrieval from a confidential index," in *Proceedings of the International Conference on Extending Database Technology: Advances in Database Technology*, pp. 439–449, Saint Petersburg, Russia, March 2009.

[18] C. Wang, N. Cao, K. Ren, and W. Lou, "Enabling secure and efficient ranked keyword search over outsourced cloud data," *IEEE Transactions on Parallel and Distributed Systems*, vol. 23, no. 8, pp. 1467–1479, 2012.

[19] Z. Fu, X. Sun, Q. Liu et al., "Achieving efficient cloud search services: multi-keyword ranked search over encrypted cloud data supporting parallel computing," *IEICE Transactions on Communications*, vol. 98, no. 1, pp. 190–200, 2015.

[20] Z. Xia, X. Wang, X. Sun et al., "A Secure and dynamic multi-keyword ranked search scheme over encrypted cloud data,"

IEEE Transactions on Parallel and Distributed Systems, vol. 27, no. 2, pp. 340–352, 2016.

[21] B. Dan and M. Franklin, *Identity-Based Encryption from the Weil Pairing*, Society for Industrial and Applied Mathematics, Philadelphia, PA, USA, 2001.

[22] M. Abdalla, M. Bellare, D. Catalano et al., "Searchable encryption revisited: consistency properties, relation to anonymous IBE, and extensions," in *Proceedings of the CRYPTO*, vol. 3621 of Lecture Notes in Computer Science, pp. 205–222, Springer, Barbara, CA, USA, August 2005.

[23] A. Lewko, T. Okamoto, A. Sahai, K. Takashima, and B. Waters, "Fully secure functional encryption: attribute-based encryption and (hierarchical) inner product encryption," in *Proceedings of the Advances in Cryptology, EUROCRYPT*, vol. 6110 of Lecture Notes in Computer Science, pp. 62–91, Springer, French Riviera, France, May-June 2010.

[24] T. Okamoto and K. Takashima, "Achieving short ciphertexts or short secret-keys for adaptively secure general inner-product encryption," *Designs Codes and Cryptography*, vol. 77, pp. 138–159, 2015.

[25] B. Wang, Y. Hou, M. Li, H. Wang, and H. Li, "Maple: scalable multi-dimensional range search over encrypted cloud data with tree-based index," in *Proceedings of the 9th ACM Symposium on Information, Computer and Communications Security*, pp. 111–122, Kyoto, Japan, June 2014.

[26] B. Wang, M. Li, and H. Wang, "Geometric range search on encrypted spatial data," *IEEE Transactions on Information Forensics and Security*, vol. 11, no. 4, pp. 704–719, 2016.

[27] H. Ren, H. Li, H. Chen et al., "Efficient privacy-preserving circular range search on outsourced spatial data," in *Proceedings of the 2016 IEEE International Conference on Communications (ICC)*, pp. 1–7, Kuala Lumpur, Malaysia, May 2016.

[28] A. Joux, "The Weil and Tate pairings as building blocks for public key cryptosystems," in *Proceedings of the International Algorithmic Number Theory Symposium (ANTS)*, vol. 2369, pp. 20–32, Sydney, NSW, Australia, July 2002.

[29] A. D. Caro, "The java pairing based cryptography library (JPBC)," pp. 2–24, 2013, http://gas.dia.unisa.it/projects/jpbc/laatstnagekeken op.

[30] W. Hersh, C. Buckley, T. J. Leone, and D. Hickam, "OHSUMED: an interactive retrieval evaluation and new large test collection for research," in *Proceedings of the International ACM SIGIR Conference on Research and Development in Information Retrieval*, pp. 192–201, Dublin, Ireland, July 1994.

[31] Z. Yang, D. Yang, C. Dyer, X. He, A. Smola, and E. Hovy, "Hierarchical attention networks for document classification," in *Proceedings of the Conference on North American Chapter of the Association for Computational Linguistics: Human Language Technologies*, pp. 1480–1489, San Diego, CA, USA, June 2017.

[32] G. Zuccon, B. Koopman, P. Bruza, and L. Azzopardi, "Integrating and evaluating neural word embeddings in information retrieval," in *Proceedings of the Australasian Document Computing Symposium*, pp. 1–8, Parramatta, NSW, Australia, December 2015.

10

Cryptocurrency Networks: A New P2P Paradigm

Sergi Delgado-Segura, Cristina Pérez-Solà, Jordi Herrera-Joancomartí, Guillermo Navarro-Arribas, and Joan Borrell

Department d'Enginyeria de la Informació i les Comunicacions, Universitat Autònoma de Barcelona, 08193 Bellaterra, Catalonia, Spain

Correspondence should be addressed to Cristina Pérez-Solà; cristina.perez@uab.cat

Academic Editor: L. J. García Villalba

P2P networks are the mechanism used by cryptocurrencies to disseminate system information while keeping the whole system as much decentralized as possible. Cryptocurrency P2P networks have new characteristics that propose new challenges and avoid some problems of existing P2P networks. By characterizing the most relevant cryptocurrency network, Bitcoin, we provide details on different properties of cryptocurrency networks and their similarities and differences with standard P2P network paradigms. Our study allows us to conclude that cryptocurrency networks present a new paradigm of P2P networks due to the mechanisms they use to achieve high resilience and security. With this new paradigm, interesting research lines can be further developed, both in the focused field of P2P cryptocurrency networks and also when such networks are combined with other distributed scenarios.

1. Introduction

Since 2009, when the Bitcoin cryptocurrency [1] was released, a plethora of more than 600 different cryptocurrency proposals have appeared. The market capitalization of cryptocurrencies surpasses 51 billion US dollars (data retrieved from http://coinmarketcap.com/ on May 8, 2017) with Bitcoin leading the market as of today.

Security and robustness are probably the most important properties for a currency, and cryptocurrencies achieve them by using cryptographic techniques and a decentralized approach. Decentralization avoids both a single point of failure and a single trust anchor but potentially introduces discordances between parties. In order to reach consensus among nodes, cryptocurrencies take advantage of a distributed mechanism that allows the system to maintain a single unambiguous view of its state [2], the blockchain.

To support the communication between different entities of a cryptocurrency, a decentralized P2P approach is adopted to deploy the so-called P2P cryptocurrency network, that is, the communication overlay that transports all data needed in the cryptocurrency system. The main goals of such a network are, firstly, to allow members of the network to synchronize their view of the system state and, secondly, to disseminate peer information in order to allow peers to reenter the system after a disconnection.

Although the goals of P2P networks are shared among all blockchain-based cryptocurrencies, there is no standard for P2P cryptocurrency networks. In this paper, we analyze the Bitcoin P2P network to characterize general P2P cryptocurrency networks. Two main reasons made us choose Bitcoin as the subject of analysis. On one hand, far beyond the economic impact, being the largest cryptocurrency also conveys technical implications: both the volume of information flowing through its network and its size and heterogeneity surpass any other deployed cryptocurrency. On the other hand, being Bitcoin the first open-source cryptocurrency proposed, other new cryptocurrencies are developed as a software fork of the Bitcoin reference implementation. Although new cryptocurrencies have tweaked the Bitcoin source code in order to achieve different properties, an in-depth analysis shows that network mechanisms are usually unmodified and, in fact, even multiple cryptocurrencies share exactly the same network behavior as Bitcoin [3].

The first objective of this paper is twofold. On one hand, the paper fully describes the Bitcoin P2P network. On the other hand, it characterizes the network to show how the aforementioned network goals, together with the special format of the information being transmitted through the network, conform to a new paradigm for P2P networks. This characterization will point out how, when considering cryptocurrency P2P networks, some of the well-known problems of P2P networks are not a concern, while other problems pose entirely new challenges.

The second objective of this paper is to analyze to what extent the adoption of cryptocurrencies, and their underlying P2P networks, can be a powerful tool for the development of distributed applications with mobile components. There are three relevant properties of cryptocurrencies that can be used as building blocks for such applications: secure distributed payment mechanisms, distributed storage with integrity by design, and secure transfer and distribution of digital assets. We analyze how these properties can be used to support distributed applications such as mobile crowdsensing or distributed IP/name resolution, to cite just two examples.

The structure of this paper is the following. First of all, in Section 3 and preceded by a basic description of the Bitcoin system, we provide a global description of all the elements in the Bitcoin P2P network, an overview that, to our best knowledge, lacked in the scientific literature (the only reference we are aware of is [4], and it is mainly focused on the economic aspects of the Bitcoin network). Second, in Sections 4 and 5, we perform a deep analysis of the Bitcoin network, which is compared to other existing P2P paradigms through a well-known P2P taxonomy. This characterization allows us to provide enough evidence to show that P2P cryptocurrency networks represent a new paradigm for P2P networks. Finally, in Section 6, we identify different applications in the field of mobile computation where cryptocurrencies may be applied, and we point out some of the opportunities and challenges that such an interaction may entail.

2. A Basic Description of the Bitcoin System

In this section, we point out the main ideas to understand the basic functionality of the Bitcoin cryptocurrency. Such a background is needed to understand the underlying P2P network that supports the communication between Bitcoin entities. However, the complexity of Bitcoin makes it impossible to provide a full description of the system in this review, so interested readers can refer to Narayanan et al.'s book [2] for a detailed and more extended explanation on the Bitcoin system.

Bitcoin is a cryptocurrency based on accounting entries [5]. Therefore, bitcoins should not be seen as digital tokens but as the balance of a Bitcoin account. A *Bitcoin account* is defined by an elliptic curve cryptography key pair. The Bitcoin account is publicly identified by its *Bitcoin address*, obtained from its public key. Using this public information, users can send bitcoins to that address (notice that the terms "public key," "address," or "Bitcoin account" refer to the same concept). Then, the corresponding private key is needed to spend the bitcoins of the account. Special purpose software, commonly referred as *wallets*, has been developed to create and manage those private keys and addresses.

Payments in the Bitcoin system are performed through transactions between Bitcoin accounts. A *Bitcoin transaction* indicates a Bitcoin movement from source addresses to destination addresses. Source addresses are known as *input addresses* in a transaction, and destination addresses are named *output addresses*. As it can be seen in Figure 1, a single transaction can have one or multiple input addresses and one or multiple output addresses.

A transaction (implicitly) details the exact amount of bitcoins to be transferred from each input address. The same applies to the output addresses, indicating the total amount of bitcoins that would be transferred to each account (although in this case, the specification is explicitly made). The Bitcoin protocol forces input addresses to spend the exact amount of a previously received transaction (notice that, in Figure 1, there are two input addresses that are exactly the same, which indicates that bitcoins have arrived to this Bitcoin account in two separate transactions). Therefore, each input must unambiguously indicate the previous transaction identifier (a transaction is identified in the Bitcoin system by its hash value) and the index of the output where the bitcoins were received. As a consequence, at any given moment, an output may be in two states: either already spent or not yet spent. An output that has not been spent is known as unspent transaction output, or *UTXO*.

Finally, the owner of the input addresses should perform a digital signature using his private keys to authorize a Bitcoin transfer, proving that he is the real owner of such accounts (although this is the standard form of Bitcoin verification for regular Bitcoin transfer transactions, the verification of a transaction can be much more complex and is based on the execution of a stack-based scripting language (more details can be found in Chapter 3 of [2])).

Before accepting a payment from a standard transaction, the receiver should

(i) validate that the digital signatures are correct;

(ii) validate that the bitcoins of the input addresses are not previously spent.

The first validation can be performed with the information included in the transaction itself (field *ScriptSig*) together with the information of the transaction identified in the *Previous output (Index)* (field *scriptPubKey*). The second validation prevents double-spending in the Bitcoin system, and it is performed through a ledger, the blockchain, where all previous transactions are annotated.

The *blockchain* is a general append-only ledger containing all Bitcoin transactions performed since the system started to operate (back in 2009), and it is freely replicated and stored in different nodes of the Bitcoin network, making the Bitcoin a completely distributed system.

Transactions are included in the blockchain at time intervals, rather than in a flow fashion, and such an addition is performed by collecting all new transactions of the system, compiling them together in a data structure called block, and including the block at the top of the blockchain. Every time that a block containing a specific transaction is included in the blockchain such a transaction is said to be a *confirmed*

FIGURE 1: Bitcoin transaction example: four input addresses and two output addresses (data from http://blockchain.info).

FIGURE 2: Example of a Bitcoin block (data from http://blockchain.info).

transaction since it has already been included in the blockchain and can be checked for double-spending prevention.

Blocks are data structures that mainly contain a set of transactions that have been performed in the system (Figure 2). To achieve the append-only property, the inclusion of a block in the blockchain is a hard problem, so adding blocks to the blockchain is time- and work-consuming. Furthermore, every block is indexed using its hash value, and every new block contains the hash value of the previous one (see the field *Previous block* in Figure 2). Such a mechanism ensures that the modification of a block from the middle of the chain would imply to modify all remaining blocks of the chain from that point to the top in order to match all hash values.

Adding a block to the blockchain is known as the *mining process*, a process that is also distributed and that can be performed by any user of the Bitcoin network using specific-purpose software (and hardware). The mining process uses a hashcash proof-of-work system, first proposed by Back as an antispam mechanism [6]. The proof of work consists of finding a hash of the new block with a value lower than a predefined target (notice that the value of the target determines the difficulty of the mining process. Bitcoin adjusts the target value depending on the hash power of the miners in order to set the throughput of new blocks to 1 every 10 minutes (in mean)). This process is performed by brute force varying the nonce value of the block. Once the value has been found, the new block becomes the top block of the

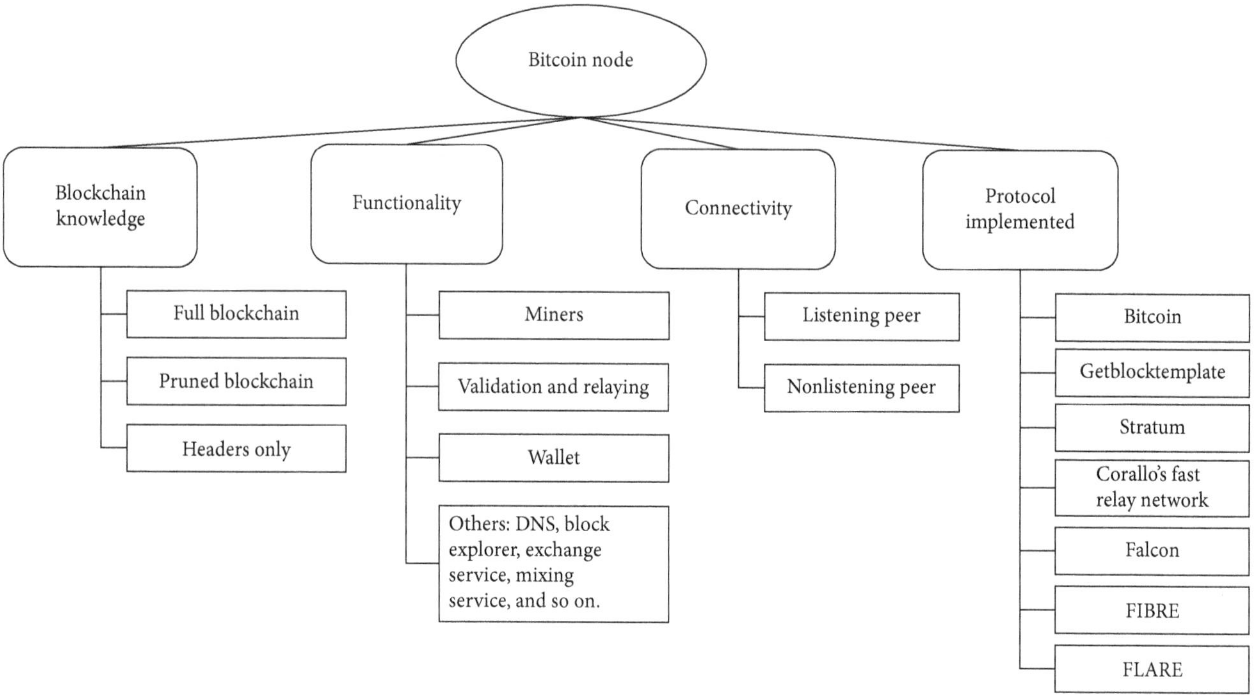

FIGURE 3: Bitcoin node classification.

blockchain, and all miners discard their work on that block and move to the next one.

Mining new blocks is a structural task in the Bitcoin system since it helps to confirm the transactions of the system. For that reason and also assuming that mining implies a hard work, miners have to be properly rewarded. In the Bitcoin system, miners are rewarded with two mechanisms. The first one provides them with newly created bitcoins. Every new block includes a special transaction, called *generation transaction* or coinbase transaction (see the first transaction in Figure 2), in which it does not appear to have any input address and the output address is determined by the miner who creates the block, who obviously indicates one of its own addresses (the amount of a generation transaction is not constant, and it is determined by the Bitcoin system. Such a value, started in 50 bitcoins, is halved every four years, fixing asymptotically to 21 million the total number of bitcoins that will ever be created).

The Bitcoin system needs to disseminate different kinds of information, essentially, transactions and blocks. Since both are generated in a distributed way, the system transmits such information over the Internet through a P2P network, that we describe in detail in the next section.

3. Description of the Bitcoin P2P Network

Bitcoin was first presented to the public in a white paper [1] describing its main concepts. Some months later, an open-source implementation of the Bitcoin client was released, giving birth to the cryptocurrency we now know and the P2P network that supports it. Such P2P network definition and implementation have been cloned in multiple new cryptocurrencies that derive from the Bitcoin implementation. In such new cryptocurrencies, the network configuration has been implemented almost identically. For instance, as described in [3], Litecoin, Dogecoin, Dash, and Peercoin have exactly the same network message types of Bitcoin, being the resulting networks for those cryptocurrencies very similar and in some cases identical to the Bitcoin one.

Since its deployment in 2009, where the only Bitcoin client available was the reference client, the Bitcoin network is now made up of very heterogeneous peers, whose hardware capabilities and software implementations differ largely from each other. Moreover, even new protocols have been created with the goal of optimizing certain tasks the Bitcoin ecosystem needs.

In order to describe the existing Bitcoin network, we first identify some of the properties that characterize Bitcoin peers. After that, we review the most common peer configurations, using the properties described before. Finally, we describe the composition of the current Bitcoin network.

3.1. Properties Describing Bitcoin Nodes. As we have already mentioned, Bitcoin peers are heterogeneous, presenting notable differences in both their hardware and software. In this section, we focus on describing the main properties that define a Bitcoin node: the exact part of the blockchain stored, its main functionalities, its connectivity, and the protocols it uses to communicate with other nodes. Figure 3 summarizes such a classification.

Peers participating in the network store some data about the *blockchain*. However, the exact data they store differ largely, from a few megabytes to dozens of gigabytes. *Full blockchain* peers store a complete and up-to-date version of the blockchain (on September 2016, the total size of block headers and transactions consists of 80 GB of data). *Pruned blockchain peers* store an up-to-date version of the blockchain with complete blockchain data for at least the last 2 days (the number of days for which to store complete blockchain data can be tuned by users. Pruned mode is estimated to reduce disk usage to around 2 GB [7]). Although only storing complete blockchain data for a few days, pruned nodes are able to securely validate transactions because they indeed store the required information from their previous history of the blockchain, that is, metadata about all known blocks and the UTXO set. *Simplified payment verification (SPV) clients* have an up-to-date version of the blockchain headers (a block header is an 80-byte structure. On September 2016, the Bitcoin blockchain has 432,000 blocks, thus needing around 33 MB of the disk space). Additionally, SPV clients may store transaction data from some transactions of interest. SPV clients are usually deployed in mobile devices such as smartphones, where having the full blockchain is generally unaffordable.

Peers can also be classified on the basis of their *functionality*. There are three functionalities needed for the Bitcoin system to work. *Mining* is the computationally expensive task of trying to create blocks. New blocks are appended to the end of the blockchain, thus making the public ledger grow. Peers that perform mining are known as miners. Some peers perform *validation and relaying* of the transactions and blocks they receive, that is, they relay to other peers valid transaction and block data, together with network data. Some peers also have a *wallet* functionality, that is, they store a set of key pairs, they track the amount of bitcoins deposited on addresses associated with those keys, and they are able to create transactions that spend those bitcoins. These functionalities do not necessarily exclude each other, that is, a peer may perform more than one functionality at the same time. Additionally, although not strictly necessary for Bitcoin to work, some peers may provide other functionalities. For instance, they may provide a *DNS* service, that offers information about existing peers; a *block explorer service*, where it is possible to query for transaction and block data through a graphical interface; an *exchange* service, where users can buy or sell bitcoins in exchange for other currencies; and *mixing services*, where users are able to obfuscate the history of their coins.

Depending on their *connectivity*, peers can be classified into *listening* peers or *nonlistening* peers. Listening peers are nodes that accept incoming connections, while nonlistening nodes are those not doing so. Although most Bitcoin full implementations listen for incoming connections, some network configurations do not allow these connections to be created (e.g., peers behind NAT).

Even though the original Satoshi Bitcoin paper implicitly assumed that peers would use only one *protocol*, the Bitcoin economy has grown much bigger than the original specification, giving place for lots of protocols to arise. We will use the term "*Bitcoin* protocol" to refer to the network protocol used by the current standard implementation, the Satoshi client. Other protocols that currently exist on the Bitcoin system are mainly targeted to optimize pooled mining and speed up data propagation. *Getblocktemplate* is the new Bitcoin pooled mining protocol (supersedes the previous mining protocol *getwork*), where the full block data are sent to miners. This allows miners to change the content of the block by themselves, thus gaining autonomy with respect to the pool servers. *Stratum* is a protocol first designed for lightweight clients and later extended to handle pooled mining. With respect to mining, it does not send full blocks to miners, thus better scaling with the number of transactions but providing less autonomy to miners to decide what to include in the blocks. The *Bitcoin relay network* has a protocol for communicating with Corallo's fast relay network backbone, a 6-node network intended to speed up the relaying of Bitcoin data. Similarly, *Falcon* is also a backbone of nodes intended to make Bitcoin data propagation faster. Peers can connect to Falcon using either the Bitcoin protocol or a specially designed network protocol that relays packets as received (instead of waiting for all packets of a full block to be received before starting to relay that block). Again with the purpose of speeding up the block propagation, *FIBRE* (Fast Internet Bitcoin Relay Engine) is a protocol that uses UDP with forward error correction to decrease the delays produced by packet loss. It also introduces the usage of compression to reduce the amount of data sent over the network. The lightning network is arising as one of the solutions to Bitcoin scalability limitations. In this context, *FLARE* is the new proposal for a routing protocol for the lightning network.

3.2. Archetypal Bitcoin Nodes. The term "*full client*" is used to define peers that perform full validation of transactions and blocks. In order to perform this full validation, they need to store either the full blockchain or a pruned version. There currently exist many implementations of full clients.

The reference implementation of Bitcoin is known as the Satoshi client, which is currently used to refer to both the *Bitcoin core* and *bitcoind*. Bitcoin core provides a graphical interface, whereas bitcoind is intended for RPC use and does not have a graphical interface. Currently, the Satoshi client is a thick client that may work either with the full blockchain (this is currently the default option) or with a pruned version. It used to have mining functionalities incorporated, but one of the latest versions [8] removes the internal miner and leaves just a minimal functionality for testing purposes. The Satoshi client performs validation and relaying of blocks and transactions and provides a basic wallet. It serves as a reference for the Bitcoin protocol and also incorporates the Bitcoin mining protocol Getblocktemplate. The software tries to create outgoing connections to the P2P network and also listens for incoming connections from other peers.

In early 2016, there was a vivid debate about a change in the consensus rules to increase the block size limit. From that debate, three different forks of the Satoshi client appeared (which maintain the original properties but change the consensus rules regarding the block size limit). *Bitcoin classic* increased the block size limit to 2 MB. *Bitcoin XT* changed the limit to 8 MB (with subsequent increases over time). *Bitcoin Unlimited* proposed to remove the limit.

Apart from the implementations that appeared from the disagreements on how to handle block size limitations, other

Table 1: Properties of archetypal Bitcoin nodes. The properties described in the table refer to the most common nodes of each type, but due to the vast heterogeneity of Bitcoin nodes, some differences may be found in the real network.

Node	Blockchain	Functionality	Connectivity	Protocol
Full client	F/P	V/R, W	L/NL	B
SPV client	H	W	NL	B
Non-SPV light client	—	W	—	S/SP
Solo miner	F/P	V/R, W, M	L/NL	B
Pool mining server	F/P	V/R, W, M	L/NL	B/S/SP
Pool mining client	—	W, M	—	S/SP

forks from the Satoshi client currently exist, for instance, *Bitcoin Knots*. There also exists some implementation of full clients that are not forks of the Satoshi client. For instance, *Bitcore* (Javascript), *bitcoinj* (Java), or *btcd* (Go).

SPV clients are peers that only have a full copy of the blockchain headers, which allows them to save on space requirements. However, they are not able to perform full validation of transactions and blocks since they lack the needed information to do so. Their main functionality is as wallets. To that end, apart from the blockchain headers, they also store cryptographic keys that allow them to spend bitcoins and the transactions that are related to those keys. SPV clients use the Bitcoin SPV protocol.

There exist many implementations of SPV clients, for instance, *breadwallet*, *Electrum*, *Bither*, *GreenBits*, *Simple Bitcoin*, *Bitcoin Wallet*, or *MultiBit HD*.

SPV clients are said to be lightweight clients because they minimize the resources needed to accomplish their functionality. However, there exist *other lightweight clients* that are not based on SPV. The current alternatives are centralized approaches, where clients connect to a set of predefined servers that relay them the information they need in order to work as wallets. This approach requires to trust the servers. The specific amount of data about the blockchain or cryptographic keys stored by these clients depends on each implementation. Some of them publish their source code for public review, while others do not. The protocol is also specific. Some examples of these kinds of wallets are *Mycelium*, *Coinomi*, *Coin.Space*, or *Copay*.

Solo miners are peers whose main functionality is mining. Initially, they had a full copy of the blockchain (or at least a pruned copy), in order to be able to validate the transactions they include in blocks, and they communicated using the Bitcoin protocol. They also needed a wallet in order to manage their mining rewards. However, as mining has become more and more specialized with the introduction of dedicated hardware, the paradigm has changed, and currently, the mining task is split into two: block structure creation and hashing. The first task is performed by peers that do have a copy of the blockchain and validate the transactions they include in blocks, whereas the second task is performed in specialized hardware, optimized to speed up hashing.

Moreover, lots of miners group together in order to reduce the amount of redundant work and to minimize the variance of the rewards obtained from the mining process. Groups of miners are known as pools and usually operate as client-server architectures, with the pool operator providing a *pool mining server* to which *pool mining clients* connect to in order to retrieve their portion of work. Clients do not need to have any knowledge about the blockchain nor to perform any validation on transactions. Clients communicate with the pool server via specifically designed protocols such as stratum.

Some well-known software implementations for client miners are cgminer or *BFGMiner*. A basic miner server is included in *bitcoind*, and some existing complete mining servers are *CoiniumServ*, *ecoinpool*, or *Eloipool*.

Table 1 summarizes the properties of the aforementioned archetypal Bitcoin nodes. Regarding blockchain knowledge, F stands for full blockchain, P for pruned, and H for headers only. With respect to functionality, W means *wallet*, M mining, and V/R validation and relaying. Concerning connectivity, L means listening, while NL stands for non-listening. Finally, as regards to the protocol, B stands for Bitcoin, S for stratum, and SP for specific protocols.

3.3. Network Description. In order to better characterize the so-called Bitcoin network, let us define three subsets of the overall network, as represented in Figure 4:

(i) The *reachable* Bitcoin network is composed of all listening nodes that talk the Bitcoin protocol. The size of the reachable Bitcoin network is estimated to be in the range of 5,000 to 10,000 nodes [9].

(ii) The *nonreachable* Bitcoin network is made of nodes that talk the Bitcoin protocol, regardless of whether they are listening for incoming connections. The size of the nonreachable Bitcoin network is estimated to be 10 times bigger than that of the reachable Bitcoin network.

(iii) The *extended* network comprises all nodes in the Bitcoin ecosystem, even those not implementing the Bitcoin protocol. This network includes, for instance, pooled miners communicating with the pool server using only the stratum protocol. To our best knowledge, there are no estimations on the number of nodes that belong to the extended network.

Both the *reachable* and *nonreachable* Bitcoin networks are P2P networks: they are distributed systems built without mediation of a centralized server or authority, they can adapt to changes in the network and their participants autonomously, and their nodes contribute to storage, computing power, and bandwidth to the network.

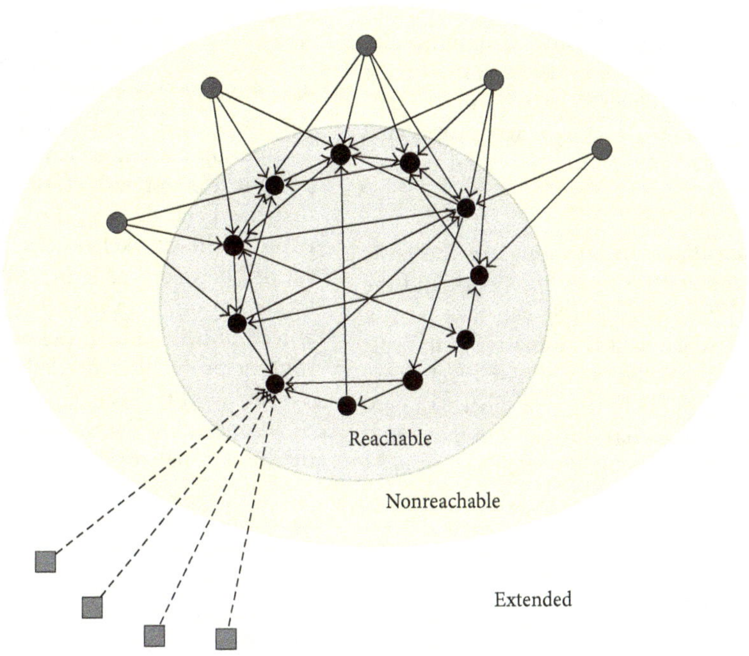

FIGURE 4: Abstraction of the defined network subsets: nodes in the *reachable* network accept incoming connections; nodes in the *nonreachable* network just create outgoing connections; nodes in the *extended* network do not implement the Bitcoin protocol (for instance, miner clients using the stratum protocol).

In this section, we have provided a detailed description of the Bitcoin network by first describing the main properties that define a Bitcoin peer, then identifying the most common Bitcoin peers, and finally providing an overview of the network. Having described the Bitcoin network, the next section provides its characterization as a P2P network.

4. Network Characterization

In order to characterize the new P2P network paradigm that cryptocurrency networks represent, we perform an analysis of the Bitcoin P2P network using the taxonomy defined by Lua et al. [10] for the comparison of different P2P overlay network proposals. Following the same taxonomy, we will be able to stress the differences of such new networks in comparison with the existing ones. The following analysis is performed aiming only at the Bitcoin reachable network, following the classification established in Section 3, since it is the only full P2P part of the Bitcoin network.

4.1. Decentralization. Decentralization assesses to what extend the analyzed network presents a distributed nature or, on the contrary, its configuration shows some centralized characteristics. In that sense, the Bitcoin network is a nonstructured P2P overlay with some similarities with Gnutella. With a flat topology of peers, in the Bitcoin network, every peer is a server or client, and the system does not provide centralized services nor information about the network topology.

4.2. Architecture. The architecture describes the organization of the overlay system with respect to its operation. As we already indicated, the Bitcoin network presents a flat architecture with no layers nor special peers. The network is formed by peers joining the network following some determined basic rules, where randomness is an essential component. Such a random behavior in the network creation intends to generate an unpredictable and uniform network topology, unknown to its users. As we will see in Section 5, such lack of knowledge about topology is needed for security reasons.

4.3. Lookup Protocol. One of the main problems in P2P networks, specially those used for content distribution, is the lookup query protocol adopted by the overlay to find the desired content. However, although the Bitcoin network can be regarded as a content distribution network (where content is transactions and blocks), the information flowing in the network is completely replicated in every node. Hence, there is no need for such a lookup protocol since information is always available at one hop peer at most. However, information propagation has to be performed in order to synchronize all peers of the network with the same data. Such a propagation is performed through the controlled flooding protocol.

Mainly, the controlled flooding protocol works on a push paradigm, propagating the data as they are generated. Instead of being directly sent, data availability is announced to the selected peers, and in case a peer lacks some of the announced information, he requests it back to the announcer. Two types of data structures are propagated through the network in that way: *transactions* and *blocks*.

4.3.1. Transaction Propagation. Transactions are the basic data structure flowing though the Bitcoin network and the one most usually seen. Every single node can take part in a transaction by simply using a wallet, no matter of its type. Transactions flow though the network aiming to reach every single node to, eventually, be included in a block. The average number of transactions per day during September 2017 was 222,458 moving around 1,808,009 bitcoins between different accounts (raw data obtained from https://blockchain.info/charts).

4.3.2. Block Propagation. Blocks are the data structure the blockchain is built from and include some of the transactions that have been created during the block mining process. Unlike transactions, blocks require a tremendous hashrate to be generated, which virtually limits their creation to mining pools. Moreover, the block generation throughput is set by design to 6 blocks per hour, periodically adjusting the block mining difficulty according to the total network hashrate.

Nonetheless, a pull data synchronization mechanism is also performed in the network, and while having a quite specific use, it is fundamental for its proper operation. Its main purpose is to synchronize the blockchain of outdated nodes, that have been off-line when data have been propagated. Outdated nodes request an on-demand synchronization to their peers during the bootstrapping phase, obtaining all the missing blocks in their local blockchain. Such a request does not refer to specific block values but to all blocks above the last block the enquirer is aware of. Besides blocks, on-demand propagation of other types of data, such as transactions, is not set by default. Only nodes that have built a full index of transactions along the blockchain, like *block explorer services*, can provide this type of data since normal nodes only track transactions bounded to their addresses.

4.4. System Parameters. Different P2P network overlays require a set of system parameters for the overlay system to operate. For instance, structured P2P networks require to store information on the distribution of peers in the network in order to improve routing performance. However, the Bitcoin P2P network, in line with other unstructured P2P overlays, does not require any special system parameters for the normal behavior of the network. Every single node could join the network with no prior knowledge of it. Apart from that, some default parameters are used by nodes, such as the maximum connection limit set to 125, although such a value is not a restriction and each node could select the number of connections it wants to maintain.

4.5. Routing Performance. Differing from traditional P2P networks (such as Gnutella [11]), Bitcoin does not follow a multihop routing scheme. As already pointed out before, peers in the network store a replica of all the information that has been flowing through the system up to the date, namely, the blockchain. In that way, no queries are forwarded between peers since all information is supposed to be located at one hop peer at most. Therefore, data are guaranteed to be located if the network is synchronized, and no routing protocol is needed nor used, apart from the synchronization protocol.

Propagation delay is therefore a fundamental factor for the Bitcoin network in order to achieve synchronization at any given time. In that way, Decker and Wattenhofer [12] analyzed the block propagation time for 10,000 blocks and discovered that it followed an exponential curve, being the median block propagation time 6.5 seconds while the mean was located at 12.6 seconds. However, the distribution showed a long tail, implying that a short percentage of the nodes (5%) required more than 40 seconds to receive the blocks. Furthermore, an analysis of how block size influences on the propagation delay was also performed. They reach the conclusion that, for small amounts of data, that is, less than 1 kB, there is a huge overhead during the propagation since the protocol involves multiple messages while negotiating the information that has to be forwarded. This applies basically to transactions, actually to a huge amount of them (96%), while not to blocks. For data size larger than 20 kB, the added overhead is negligible (around 80 ms).

4.6. Routing State. Despite being a content distribution network, the routing state of Bitcoin cannot be directly defined due to the randomness and dynamism of its topology and to the fact that it is not known. Moreover, as we have pointed out before, no multihop routing is performed since data could be found at one hop peer at most.

4.7. Peers Join and Leave. How to build the network is a classic problem P2P networks have to deal with. From building the network from its roots to how nodes deal with peer disconnection, P2P networks need to be highly adaptable to avoid partitioning. In order to deal with this problem and also provide a fair and secure way to choose the peers a node is going to be connected to, the Bitcoin network performs a particular network discovery mechanism.

By default, all peers maintain up to 125 connections with other peers. Each node will start 8 of those connections with other peers (namely, outgoing connections) and will accept up to 117 from potential peers (namely, incoming connections). Despite the name, all connections are bidirectional. In order to pick the outgoing connections, every single node will look for a subset of nodes it stores in a local database. This database is formed by two different tables: *tried* and *new*. *Tried* table contains addresses from peers the node has already connected to, and *new* table contains addresses the node has only heard about. Additionally, when the node tries to establish a connection to the network for the first time, it queries a well-known list of DNS nodes, that will provide a set of online potential peers (further information about how peers are stored and selected can be found in [13]).

Nodes try to always maintain their 8 outgoing connections, selecting new peers from the database if any of the established connections is dropped. Peers are stored and selected from the database following a pseudorandom procedure that gives the network high dynamism and keeps its structure unknown. Peer information can be obtained by a node following two ways. First of all, a node could request such data to its neighbors, in order to fill up its database, through sending a *getaddr* message, or could receive such information spontaneously from one of its peers without any kind of request. In both cases, the information is sent using a set of *addr* messages, containing up to 2,500 peer addresses both from the neighbor's *tried* and *new* tables. Such addresses are stored in the local node's *new* table. On the other hand, an *addr* message containing a single address could be sent to a node when a node wants to start a connection with a potential peer. By sending its address, the node notifies the receiver that it has been picked as a peer, and if the latter has room for more incoming connections, the communication is established. Peer addresses received in that way are stored in *tried* table. All addresses are stored in the database together with a timestamp that helps the node to evaluate the freshness of such an address when selecting a peer.

4.8. Security. Security in P2P networks has always been a broad topic since multiple security threats can be identified in different P2P implementations. The interested reader can refer to Wallach's survey [14] for an introduction to the topic of security in general P2P networks, to Bellovin's paper [15] for a description focused on the security issues affecting specific P2P protocols such as Napster and Gnutella, and to [16] for an introduction to security problems in P2P SIP communications.

However, in P2P cryptocurrency networks, security takes a different twist. At first sight, one could believe that the threats P2P cryptocurrency networks face are a subset of the threats found in standard P2P networks. However, as we will see in detail in the next section, most of the threats encountered in general P2P networks do not apply directly to P2P cryptocurrency networks due to the cryptographical mechanisms used by the currencies and the level of security offered by their protocols.

Additionally, one can also believe that multiple new threats will also arise in cryptocurrencies due to the sensitivity they have as money transfer networks. However, as we will see in the next section, this is not also the case.

In the next section (Section 5), we provide a detailed review of the most common security threats identified for typical P2P networks and discuss to what extent they affect the Bitcoin network.

4.9. Reliability and Fault Resiliency. Reliability and fault resiliency analyze how robust the overlay system is when subjected to faults. Typically, such robustness measurements are related to nonintentional failures, for instance, by a massive disconnection of peers of the network or an increasing volume of information being transferred through the network, but do not include intentional attacks that would be categorized inside the security properties of the network.

Bitcoin implements a distributed consensus protocol resilient to Byzantine faults. That is, the protocol is resistant to arbitrary faults produced in the participating peers, from software errors to adversary attacks. The main idea behind this protocol is to use a proof-of-work system to build the public ledger where transactions are stored. Appending new information to the public ledger requires a huge amount of computer power, thus preventing attackers to monopolize ledger expansion and censuring transactions. In a similar way, changing the content of the blockchain is also computationally expensive, up to the point that transactions are considered secure when they have 6 confirmations (i.e., five blocks have been created on the top of the block that included the transaction). Additionally, the blockchain is replicated on all full blockchain nodes, contributing to the fault resiliency of the system and providing high availability of the ledger data.

Assuming that categorization, the Bitcoin P2P network has been designed with a high level of reliability, thanks to the redundancy that implies the storage of all the relevant information of the network in every peer of the network. With this approach, the high inefficiency level in terms of storage space is translated into a high resilience of the network since the availability of a single node in the network contains the information to keep the system alive. Moreover, the proof-of-work system allows peers to (eventually) reach a consensus state, even in the presence of attackers trying to subvert the system. As a drawback, the consensus protocol is somehow slow, with transactions needing 9 minutes (median confirmation time as of October 13, 2016 [17]) to confirm, and expensive, requiring the consumption of lots of energy for each mined block.

5. Security Concerns in P2P Networks

Security in P2P networks has been extensively studied in the literature. In this section, we provide a broad overview of the main security problems that arise in P2P networks, we review how each of the security problems may affect the Bitcoin network, and if it is the case, we explain the specific countermeasures Bitcoin provides in order to defend from each attack.

The list of reviewed attacks goes over the most typical types of attacks and security flaws found in common P2P networks. It is clear that specific networks and applications might present specialized attacks, but in most cases, they can be seen as a specification of the attacks presented here.

So as to provide a clear picture of how common P2P attacks affect Bitcoin, we first review the three attacks that have been shown to be clearly applicable to Bitcoin. After that, we include a list of attacks identified for common P2P networks, but this does not have such a high impact on Bitcoin, reviewing why the attacks do not apply to the specific

Bitcoin network and detailing the particular cases where those attacks (or some variation) may somehow relate to Bitcoin.

In favor of a clear and concise presentation, we have not explicitly covered some recent attacks such as [18], which do not directly affect or involve the Bitcoin network, or network-related attacks such as [19], which rely on BGP hijacking and are thus out of the scope of our study.

5.1. DoS Flooding. Denial-of-service (DoS) attacks are possible in most P2P scenarios and are especially relevant, for example, in P2P streaming applications [20–22]. Given their dynamic nature, P2P networks are usually more resilient against generic DoS attacks than more static networks. Targeted DoS attacks to specific parts of the P2P network (a given node) or services are usually more important.

There exist several potential DoS flooding attacks in Bitcoin, but the system has countermeasures in place. *Transaction flooding* is prevented by not relaying invalid transactions and imposing fees to valid transactions. On one hand, transactions are signed by the senders in order to demonstrate that they are authorized to transfer those bitcoins. If the signatures of a transaction are not correct, the transaction is considered invalid and is not relayed to the network. On the other hand, the default protocol does not relay transactions without fees (except for a few very specific cases that would also result in very expensive attacks). Moreover, transaction's fees increase for lower input ages (i.e., for bitcoins that have been moved recently), so an attacker trying to generate a huge amount of transactions that move the same bitcoins would have to pay increasing fees. *Block flooding* is prevented by only relaying valid blocks, which must contain a valid proof of work. In order for a block to contain a valid proof of work, its hash must be lower than a given target. Obtaining a block with such a hash is a computationally expensive task, thus performing DoS attacks with block data unfeasible. *Network data flooding* is easier than the previous two cases because it is indeed possible to create valid network messages without paying fees nor spending computation cycles. However, Bitcoin has a banning protocol: peers may ban other peers for one entire day if their misbehavior score crosses a certain threshold. The misbehavior score is increased for sending duplicate version messages, sending large messages, and sending invalid blocks. Given the nature of Bitcoin, *cpu usage DoS* is possible by trying to make peers spend lots of time validating a transaction or a block. In order to prevent this kind of attacks, Bitcoin tries to catch errors before starting to validate a transaction, limits the number of signature operations per transaction and per block, and limits the size of the script. Finally, previous versions of the Bitcoin client were also susceptible to *continuous hard disk read attacks*, where an attacker repeatedly sent double-spend transactions that passed the initial checks and required to retrieve data from disk in order to be fully validated. This attack is now prevented by checking that the inputs of the transaction that is being validated are in the UTXO set (i.e., checking whether the transaction is a double spend) before retrieving any information from disk.

5.2. Eclipse Attacks. An eclipse attack occurs when an attacker creates (or has control of) a large number of distinct nodes that populate the whole neighborhood of the victim node [23]. The attacker can then *eclipse* the view of the network that has the victim. Common solutions for sybil attacks are usually insufficient to defend against eclipse attacks [24].

In a cryptocurrency network, isolating a node from the rest of the network may enable two other attacks to the eclipsed peer. First, an eclipsed peer may undergo a censorship attack because the victim's transactions must pass through the attackers' nodes in order to reach the network. Therefore, the attacker may decide not to forward these transactions, thus censoring the victim's transactions. Second, if the eclipsed victim is a miner, the attacker can drop or delay the propagation of the new blocks found by the rest of the network. As a consequence, the victim wastes computation time trying to mine on the top of old blocks.

Bitcoin has many defense mechanisms to prevent eclipse attacks, some of which were added recently, after a study pinpointed some of the flaws of the then vigent implementation [13]: the client restricts the amount of outgoing connections to addresses in the same network, randomizes the address selection procedure, and maintains a big list of peers, among others.

5.3. User Profiling. In some P2P networks, it is easy to record all the activities of a giving node allowing attackers to easily create identifying profiles of users and their activities. This is relevant in anonymous systems, or systems that want to guarantee a certain degree of anonymity [16, 25].

Bitcoin provides pseudonymity by allowing users to receive payments to their addresses, which are not initially linkable to their identities. The usage of new addresses for each transaction in the system is intended to provide unlinkability between the different actions a single user performs through Bitcoin. Therefore, user profiling in Bitcoin usually consists in attacking the unlinkability between different addresses a single user has. Three different approaches have been taken to perform address clustering: using network layer data [26], performing analysis over the transaction graph [27–29], and analyzing Bloom filters [30]. The idea of using network layer data to cluster addresses is straightforward: if an attacker is able to connect to all the peers of the network, the first node that sends him a given transaction should be the creator of that transaction. Therefore, if the attacker first receives two different transactions from the same peer, he can infer that the source addresses of both transactions belong to the same user. However, as simple the attack may seem conceptually, it is not that easy to perform in practice. It is not trivial to connect to all nodes of the network since most of them do not accept incoming connections. Moreover, some peers anonymize their connections using Tor. Finally, collected data are very noisy, and therefore, it is not easy to make strong claims when analyzing it. Regarding transaction graph analysis, there exist mixing services that are able to effectively break the relationship between an address and its past. Additionally, the use of a secure wallet that tries to minimize the leaked information about addresses clusters helps mitigate the consequences of this kind of analysis. Finally, concerning the usage of bloom filters, users must be very careful when choosing the parameters of the filter and when generating different filters that match the same set of addresses and public keys. Additionally, new protocols are being designed to allow lightweight clients to retrieve their transactions of interest while maintaining privacy.

Bitcoin's scalability problems have triggered the search for new solutions that would allow to increase the transaction throughput of the network. Several proposals provide mechanisms to create off-chain payment channels, such that secure transactions between Bitcoin users may be performed without needing to include all the transactions into the blockchain. In turn, these solutions may also entail privacy problems that are yet to be carefully studied [5].

5.4. Other Attacks. After analyzing the three main attacks that have threatened the Bitcoin network over the last years, we summarize other common P2P attacks that have a lesser impact on Bitcoin. We will show how some of those attacks could be used as a preliminary phase to achieve one of the three previously introduced ones, while others are not harmful for the Bitcoin network due to its design.

5.4.1. ID Attacks. Two different subattacks can be identified in this category:

ID mapping attack: when a node changes its own identifier with malicious purposes. As an example, in DHT-based P2P networks, a node can gain control over given resources by changing its own identifier [31]. These kinds of attacks are more difficult in networks where the identifier is derived from a public key [32].

ID collision attack: similar to previous attacks, here the attack is considered to happen when there are duplicated identifiers. The problem is usually prevented by ensuring the uniqueness of identifiers [16].

There is no clear concept of a peer identifier in Bitcoin. Two different properties could be considered identifiers in Bitcoin, depending on the exact entity one wants to identify: IPs and Bitcoin addresses. IPs allow to identify peers, whereas addresses are linked to users. A malicious peer may benefit from a change of IP if it is banned for misbehavior. Each peer maintains a banscore for each of its neighbors. This banscore is increased whenever the peer misbehaves. If the banscore surpasses a certain threshold, the neighbor is banned for 24 hours. Therefore, being able to change the IP allows a peer to effectively reset its banscore. Regarding the second kind of identifiers, Bitcoin addresses, the recommended behavior for users is indeed to change them frequently. In fact, the suggestion is to not reuse addresses, that is, to create a new address for each transaction made in the system. This allows to protect user privacy.

5.4.2. Sybil Attack. A sybil attack is a well-known attack in P2P networks, where a malicious user creates multiple identities in order to control the system or parts of the system [33]. This has been very extensively studied in the literature in the context of several P2P technologies [34, 35].

Sybil attacks may be a problem in Bitcoin if they are able to eclipse all the connections from a peer (see Section 5.2 for details of eclipse attacks). However, besides its extension to an eclipse attack, a peer with multiple identities cannot harm the system regarding the main content of the network: transactions and blocks. Blocks cannot be counterfeit without the corresponding proof of work, and transaction generation entails an associated fee (in a similar way that was described in *flooding attacks* in Section 5.1). Nevertheless, if lots of sybil nodes start performing a huge amount of connections to the existing network, they may monopolize all available incoming connection slots, and the system decentralization could be reduced.

5.4.3. Fake Bootstrapping. Network access in P2P environments starts by connecting to one or multiple nodes of the network. This first contacted node is known as the bootstrap node. A malicious bootstrap node can influence the view of the network for the new user [23]. Several solutions already exist for this problem such as not relaying in a single bootstrap node, use of cached peers for subsequent connections, random address probing, using external mechanisms, using specific bootstrapping services, or using network layer solutions (e.g., use of a special multicast group for bootstrapping) [16, 36, 37].

Bitcoin deals with bootstraping issues by defining a local peer database on every single node, that is queried following a pseudorandom protocol to obtain a subset of potential peers (see Section 4.7 for details). In that way, Bitcoin applies most of the solutions for the fake boostrapping protocol, such as *not relaying in a simple bootstrap node*, by establishing 8 outgoing connections on every bootstrap, *use of cached peers for subsequent connections*, by using peers stored in *tried* table, *random address probing*, by using a pseudorandom protocol to store and retrieve peer addresses from the database, and *using external mechanisms* by quering a list of well-known DNS nodes or even using a list of hardcoded nodes, if the DNS cannot be reached.

5.4.4. Unauthorized Resource Access. P2P networks often use some sort of private data that have to be protected from unauthorized access. Common solutions are those typically employed for distributed access control [38, 39].

Bitcoin is based on public key cryptography, where private keys are needed to authorize payments. Therefore, private keys must be kept secret, and two methods are usually employed: encryption and off-line storage. By using encryption, private keys remain secure even if an attacker is able to retrieve the key file as long as the encryption key remains secret. As for off-line storage, different approaches can be taken with different technical sophistication levels, from the usage of dedicated hardware devices to paper wallets. Notice that unlike other uses of public key cryptography where private keys need to be online (for instance, in the handshake process in TLS), Bitcoin network operation does not involve private information since validations are performed using public information. For that reason, offline storage of public keys does not impact the network performance.

5.4.5. Malicious Resource Management. A malicious node can deny the existence of a given resource under its responsibility, or claim to have a resource it does not have. This is specially relevant in content distribution applications, and common solutions are replication of resources [40], or use of error-correcting codes to reconstruct missing parts of the resource [41].

Bitcoin network is protected against malicious resource management by, on one hand, the high amount of data redundancy information of the network and, on the other hand, the multiple neighbors a node of the network is connected to. Thanks to the fact that peers establish connections (by default) to 8 other peers, if a given neighbor denies the existence of a certain resource, the peer can learn it from his other neighbors. Moreover, if a neighbor says he has some resource he actually does not have, peers will notice when they try to retrieve it (since transactions and blocks are identified by their hash).

5.4.6. Free Riding. A free-rider user or node in a P2P network is a node that attempts to benefit from the resources of the network (provided by other users) without offering their own resources in exchange [42, 43]. Depending on the application, this might not be an issue or even might not be considered a security problem. It is usually described in content distribution applications, and the main solutions proposed rely on incentive- or penalty-based mechanisms [44].

Bitcoin is sustained by an equilibrium of economic incentives. Miners are remunerated for their work by obtaining a reward for each block they successfully mine. Additionally, transaction senders (and, although indirectly, also transaction recipients) may include a fee to their transactions, which is also collected by the miner of the block that contains the transaction. As a consequence, miners are encouraged not only to create blocks but also to include transactions on those blocks. There is, however, a set of nodes whose role is important in ensuring the decentralization of the network and that do not directly receive economic incentives for their work: full clients. While these clients store the blockchain and perform validation and relaying of transaction and blocks, they do not get a direct economic reward in return for their work.

5.4.7. Man-in-the Middle (MITM). In the context of P2P networks, a MITM attack is usually considered a routing attack, similar to classical network MITM attacks. P2P networks, which require multihop routing, will need to include measures similar to onion routing in order to secure connections between all nodes along the path [45, 46].

MITM attacks in Bitcoin are not a problem for transaction and block integrity because transactions are cryptographically signed and blocks must contain a valid proof of work. Transaction malleability may be a problem (refer to Section 5.4.10 for a detailed explanation) in very specific scenarios, but a solution is currently being deployed. Censorship is neither a problem because a single peer maintains different connections. An attacker must be in the middle of all of them to hide information to the peer (thus resorting in *eclipse attacks*).

5.4.8. Replay Attack. A replay attack is produced when a legitimated transmission is delayed or lately replayed with malicious purposes. This is a very common network attack that can affect P2P networks in several ways but is usually solved at a protocol level.

Replaying transactions or blocks that have been sent to the network does not have any effect on the Bitcoin network. Nevertheless, delaying block propagation may be a beneficial strategy for miners [47]. By not immediately propagating a block the miner has just found, the miner can start working on top of this newly found block while making other miners lose time working on the previous block. This strategy is known as selfish mining and reduces the bound on the percentage of hashing power an attacker must have in order to successfully control the information appended to the ledger.

5.4.9. Routing Dysfunction. Routing dysfunction can be presented in different aspects. On one hand, *incorrect routing* involves attacks where a node routes messages incorrectly (or drops them) [23]. These attacks might not be relevant in P2P networks that do not provide multihop routing. Due to the flooding mechanism used to propagate information through the network, the consequences of a single node dropping messages are negligible.

On the other hand, in a *fake routing update*, the attacker tries to corrupt a given route (equivalent to corrupting a routing table for a given node) [23, 32, 48]. As we have mentioned previously in Section 4, there are no routing tables in the Bitcoin network. The most similar information a peer stores is addresses from other peers. Note that no information about where is this peer in the network nor its connections are stored by the Bitcoin client, just the address and a timestamp. Therefore, the attack that better resembles fake routing updates in Bitcoin is to send fake addresses. These kinds of attacks are usually performed as a first step in eclipse attacks, attacks already described in Section 5.2.

5.4.10. Tampering with Message Bodies. When using multihop routing, intermediate nodes can modify the content of the relaying packets. End-to-end integrity has to be provided in order to, at least, detect these types of attacks [16].

Tampering with the content of a block changes its hash and, with very high probability, invalidates its proof of work. Therefore, tampering with block data is not a feasible attack on Bitcoin. On the other hand, transactions are a signed data structure, with the signature cryptographically protecting its integrity. Therefore, an attacker can not tamper with a transaction to its will, for instance, by changing the destination address of the bitcoins transferred on the transaction. There is, however, a very specific situation where this kind of attack would be possible (although the countermeasures that prevent this attack are already implemented and ready to be deployed). Because Bitcoin transactions are malleable, it is indeed possible for an attacker to change some part of the transaction while keeping the signature valid. This happens mainly because not all parts of the transaction are signed (e.g., the signatures themselves are not signed). The aforementioned situation where malleability is a problem for Bitcoin happens when a user is dealing with 0-confirmation transactions, that is, transactions that have been sent to the network but have not yet been included in a block. Because transactions are not yet in a block, an attacker may change some of the unsigned part of the transaction, creating another valid transaction that spends the same inputs but has a different identifier (recall that transactions are identified by their hash). Then, if this transaction is part of a protocol where transactions are identified by their hash, the attacker may be able to use it at his advantage.

6. Cryptocurrencies as Building Blocks for Decentralized Applications

Cryptocurrencies are indeed a powerful tool for the development of new decentralized applications (currently, the best well-known application for P2P networks is as a content distribution technology [49]), thanks to the distributed trust mechanism in which they are based on. Three relevant properties of cryptocurrencies can be used as a building block for such applications:

(i) Secure distributed payment mechanism
(ii) Distributed storage with integrity by design
(iii) Secure transfer and distribution of digital assets

The obvious use case of cryptocurrencies is, of course, to adopt them as the payment layer in any system where there is the need to transfer money from a payer to a payee in a totally distributed (and uncensored) fashion. Multiple applications could benefit from a flexible payment system, from P2P distributed storage schemes, where users could hire local disk space for an economical incentive, to more sophisticated scenarios, like mobile crowdsensing [50]. Mobile crowdsensing (MCS) is a distributed application where the power of the crowd, jointly with the sensing capabilities of smartphones they wear, provides a powerful tool for data sensing, especially in those scenarios involving user behavior or those that rely on user mobility, where standard sensor networks may not be suitable. However, including human participation in sensing tasks carries, at least, three critical challenges [51]: user participation, data sensing quality, and user anonymity. User participation is extremely important in MCS since the performance and usefulness of such sensor networks heavily depend on the crowd sensor's willingness to participate in the data collection process. Therefore, incentive mechanisms are of utmost importance in MCS scenarios to engage as many crowd sensors and provide the data collection center with a considerable wealth of data. User participation can be promoted by providing a pay-per-sense mechanism. However, standard payment schemes have multiple drawbacks in a pay-per-sense application. First of all, user enrollment in the payment system entails a burden step for user participation. Second, collateral costs of standard payment systems (mainly in the form of fees) prevent their use in a pay-per-sense scenario. Finally, standard payment mechanisms do not provide privacy-preserving properties, specially relevant when such payments could identify sensed data from a particular individual whose identity should not be disclosed. Cryptocurrencies can be successfully used in mobile crowdsensing scenarios as a rewarding mechanism since they allow an affordable pay-per-sense scheme with relevant privacy-preserving properties, as it has been proposed in the Paysense system [52].

Distributed storage is also a very interesting property offered by blockchain-based cryptocurrencies, but despite other distributed proposals, its main advantage is the integrity-by-design property that makes it so attractive for multiple applications. For instance, multiple P2P networks need a distributed IP/name resolution mechanism, and special purpose cryptocurrencies can solve this problem. An example of such an idea is the cryptocurrency Namecoin [53]. Namecoin is a blockchain-based cryptocurrency whose purpose is to provide network address resolution for network identifiers, normally human readable. In such a cryptocurrency, transactions can store data for tying the network address with other identifiers, and such transactions are stored in the blockchain inhering its integrity properties. Keys used for creating the transaction provide an authentication token for the owner who registered the tie. Namecoins could be used as a DNS replacement in a P2P network or even for node authentication when such authentication needs a tie between identity and keys (for instance, using standard public key infrastructure, PKI).

Cryptocurrencies were designed to transfer money, but its use can be extended to transfer other types of digital assets. By using a cryptocurrency as a transport layer, digital assets can be associated with cryptographic keys and can be traded, using the secure information included in the blockchain to determine the legitimate owner of every asset at each specific time. Multiple examples of such digital assets can be found, from shares of a company to DRM where the property of the media object can be determined [54]. Furthermore, extending a bit the concept of asset, cryptocurrencies can also be used to store reputation, conceptualized as an asset that users can store and transfer. Revisiting again the example of mobile crowdsensing, we recall the fact that data-sensing quality was one of the important challenges of such a scenario. In MCS systems, there is no control over the crowd sensors, and it cannot be assumed that all individuals will behave in the exact same manner or will be equally honest. Therefore, the overall quality of the sensor readings can see itself deteriorated if counterfeit data are received from malicious users. Hence, data validation techniques should be properly deployed, and a commonly used approach is to validate the data depending on the trust level of the crowd sensor that reports it. In this particular scenario, cryptocurrencies can be used as an annotation mechanism [52], by which users earn or lose reputation depending on the correctness of previous actions, accounted by the amount of rewards that they previously obtained.

Notice that, in this section, we have pointed out some possible uses of cryptocurrencies in distributed applications to show the broad intersection between both fields. However, an in-depth study on how interaction could be optimally performed between cryptocurrencies and particular scenarios is left for future work.

7. Conclusions and Further Research

In this paper, we have characterized P2P cryptocurrency networks by providing a deep analysis of the most relevant cryptocurrency nowadays: Bitcoin. By characterizing P2P cryptocurrency networks using well-known taxonomy in the field of P2P networks, we can conclude that such networks present a new paradigm due to the main properties that a cryptocurrency has to provide: reliability and security.

P2P cryptocurrency network reliability stands on top of a strong redundant mechanism regarding system information. As a result, every peer of the network stores all the relevant information of the system. With this approach, the availability of a single node in the network contains the information to keep all the systems alive. Notice that this approach turns out to a high inefficiency level regarding storage space, so this strategy is not followed by any other P2P network paradigm. Furthermore, such an approach also demands new synchronization mechanisms to provide all nodes with the same correct information.

Information redundancy is also used in the security plane for network topology protection. As we have seen, the main attacks to cryptocurrencies are eclipse attacks, where a victim or part of the network can be isolated. Such attacks can be performed when an attacker takes advantage of his position on the network topology. To avoid such possibility, the network topology has to be protected, and cryptocurrency networks use two different measures for such protection. On one hand, routing information should not be disclosed, so cryptocurrency networks are not multihop networks, and network nodes only are aware of one-hop neighbors. Using this approach, no routing information has to be provided to network nodes, and there is no restriction regarding information availability since, as we have pointed out above, information is replicated in every network node. On the other hand, network topology disclosure also has to be protected when nodes access the network. In such a phase, cryptocurrency P2P networks use a pseudorandom approach to determine each node connection to hinder the topology structure of the network. Notice that this topological secrecy property of P2P cryptocurrency networks is not so relevant in other P2P network paradigms, and for that reason, the mechanisms to achieve it are also particular of such environments.

Furthermore, some mechanisms specifically designed for other P2P network paradigms are not needed in cryptocurrency networks due to the characteristics of the information flowing in such networks. This is the case of multiple secure protections that try to prevent different attacks. For instance, intrinsic cryptographic properties of blocks and transactions can directly prevent DoS attacks, replay attacks, or tampering with message bodies.

As a new paradigm, P2P cryptocurrency networks open new research opportunities both as a direct field of study and also as a tool for other applications. For instance, a more formal analysis should be performed towards the pseudorandom mechanisms used in this kind of networks for selecting the nodes to connect to verify that network topology is both unknown and uniform. Furthermore, the development of a global P2P cryptocurrency network that could provide service to multiple cryptocurrencies, taking into account different particularities of each cryptocurrency, could also be another interesting research line. On the other hand, analyzing how such a new network paradigm could be efficiently combined and integrated with other distributed applications could also be a relevant topic for future work.

Acknowledgments

This work is partially supported by the Spanish Ministry under Grant no. TIN2014-55243-P and the Catalan AGAUR grant 2014SGR-691.

References

[1] S. Nakamoto, *Bitcoin: A Peer-to-Peer Electronic Cash System*, 2008, https://bitcoin.org/bitcoin.pdf.

[2] A. Narayanan, J. Bonneau, E. Felten, A. Miller, and S. Goldfeder, *Bitcoin and Cryptocurrency Technologies*, Princeton University Press, Princeton, NJ, USA, 2016.

[3] J. A. D. Donet and J. Herrera-Joancomartí, "Cryptocurrency P2P networks: a comparison analysis," in *Actas de la XIV Reunión Espanõla de Criptología y Seguridad de la Información (RECSI 2016)*, J. L. Ferrer and M. Payeras, Eds., pp. 423–428, Menorca, Illes Balears, Spain, 2016.

[4] M. Lischke and B. Fabian, "Analyzing the bitcoin network: the first four years," *Future Internet*, vol. 8, no. 1, p. 7, 2016.

[5] J. Herrera-Joancomartí and C. Pérez-Solà, "Privacy in bitcoin transactions: new challenges from blockchain scalability solutions," in *Modeling Decisions for Artificial Intelligence*, pp. 26–44, Springer, Berlin, Germany, 2016.

[6] A. Back, *Hashcash-A Denial of Service Counter-Measure*, 2002.

[7] Bitcoin Core, *Bitcoin Core Version 0.12.0 Released*, 2016, https://bitcoin.org/en/release/v0.12.0.

[8] Bitcoin Core, *Bitcoin Core Version 0.13.0 Released*, 2016, https://bitcoin.org/en/release/v0.13.0.

[9] Bitnodes: 2016, https://bitnodes.21.co/.

[10] E. K. Lua, J. Crowcroft, M. Pias, R. Sharma, and S. Lim, "A survey and comparison of peer-to-peer overlay network schemes," *IEEE Communications Surveys and Tutorials*, vol. 7, no. 2, pp. 72–93, 2005.

[11] P. Kirk, *Gnutella Protocol Development*, 2011.

[12] C. Decker and R. Wattenhofer, "Information propagation in the bitcoin network," in *Proceedings of the IEEE Internation Conference on Peer-to-Peer Computing (P2P)*, Trento, Italy, 2013.

[13] E. Heilman, A. Kendler, A. Zohar, and S. Goldberg, "Eclipse attacks on bitcoin's peer-to-peer network," in *Proceedings of the 24th USENIX Conference on Secrity Symposium (SEC'15)*, pp. 129–144, USENIX Association, Berkeley, CA, USA, 2015, http://dl.acm.org/citation.cfm?id=2831143.2831152.

[14] D. S. Wallach, "A survey of peer-to-peer security issues," in *Software Security—Theories and Systems*, pp. 42–57, Springer, Berlin, Germany, 2003.

[15] S. Bellovin, "Security aspects of Napster and Gnutella," in *Proceedings of the 2001 Usenix Annual Technical Conference*, Boston, MA, USA, June 2001.

[16] D. S. Touceda, J. M. Sierra, A. Izquierdo, and H. Schulzrinne, "Survey of attacks and defenses on P2PSIP communications," *IEEE Communications Surveys and Tutorials*, vol. 14, no. 3, pp. 750–783, 2012.

[17] Bitcoin Core, *Blockchain Info Website*, https://blockchain.info/.

[18] M. Carlsten, H. Kalodner, S. M. Weinberg, and A. Narayanan, "On the instability of bitcoin without the block reward," in *Proceedings of the 2016 ACM SIGSAC Conference on Computer and Communications Security (CCS'16)*, pp. 154–167, ACM, Vienna, Austria, 2016.

[19] M. Apostolaki, A. Zohar, and L. Vanbever, "Hijacking bitcoin: routing attacks on cryptocurrencies," in *Proceedings of the 2017 IEEE Symposium on Security and Privacy (SP)*, pp. 375–392, San Jose, CA, USA, 2017.

[20] K. Suto, H. Nishiyama, N. Kato, T. Nakachi, T. Fujii, and A. Takahara, "THUP: a P2P network robust to churn and DoS attack based on bimodal degree distribution," *IEEE Journal on Selected Areas in Communications*, vol. 31, no. 9, pp. 247–256, 2013.

[21] W. Wang, Y. Xiong, Q. Zhang, and S. Jamin, "Ripple-stream: safeguarding P2P streaming against DoS attacks," in *Proceedings of the 2006 IEEE International Conference on Multimedia and Expo*, pp. 1417–1420, Toronto, ON, Canada, July 2006.

[22] M. Brinkmeier, G. Schäfer, and T. Strufe, "Optimally DoS resistant P2P topologies for live multimedia streaming," *IEEE Transactions on Parallel and Distributed Systems*, vol. 20, no. 6, pp. 831–844, 2009.

[23] M. Castro, P. Druschel, A. Ganesh, A. Rowstron, and D. S. Wallach, "Secure routing for structured peer-to-peer overlay networks," *SIGOPS Operating Systems Review*, vol. 36, pp. 299–314, 2002.

[24] A. Singh, M. Castro, P. Druschel, and A. Rowstron, "Defending against eclipse attacks on overlay networks," in *Proceedings of the 11th Workshop on ACM SIGOPS European Workshop (EW'11)*, ACM, Leuven, Belgium, 2004.

[25] N. Borisov and J. Waddle, "Anonymity in structured peer-to-peer networks," Tech. Rep. UCB/CSD-05-1390, Computer Science Division (EECS), University of California, Oakland, CA, USA, May 2005.

[26] P. Koshy, D. Koshy, and P. McDaniel, "An analysis of anonymity in bitcoin using P2P network traffic," in *Financial Cryptography and Data Security*, N. Christin and R. Safavi-Naini, Eds., vol. 8437 of Lecture Notes in Computer Science, pp. 469–485, Springer, Berlin, Germany, 2014.

[27] D. Ron and A. Shamir, "Quantitative analysis of the full bitcoin transaction graph," in *Financial Cryptography and Data Security*, A.-R. Sadeghi, Ed., vol. 7859 of Lecture Notes in Computer Science, pp. 6–24, Springer, Berlin, Germany, 2013.

[28] E. Androulaki, G. Karame, M. Roeschlin, T. Scherer, and S. Capkun, "Evaluating user privacy in bitcoin," in *Financial Cryptography and Data Security*, A.-R. Sadeghi, Ed., vol. 7859 of Lecture Notes in Computer Science, pp. 34–51, Springer, Berlin, Germany, 2013.

[29] M. Ober, S. Katzenbeisser, and K. Hamacher, "Structure and anonymity of the bitcoin transaction graph," *Future Internet*, vol. 5, no. 2, pp. 237–250, 2013.

[30] A. Gervais, S. Capkun, G. O. Karame, and D. Gruber, "On the privacy provisions of Bloom filters in lightweight bitcoin clients," in *Proceedings of the 30th Annual Computer Security Applications Conference (ACSAC'14)*, pp. 326–335, ACM, New Orleans, LA, USA, December 2014.

[31] D. Cerri, A. Ghioni, S. Paraboschi, and S. Tiraboschi, "ID mapping attacks in P2P networks," in *Proceedings of the IEEE Global Telecommunications Conference (GLOBECOM'05)*, vol. 3, pp. 1785–1790, St. Louis, MO, USA, 2005.

[32] E. Sit and R. Morris, "Security considerations for peer-to-peer distributed hash tables," in *First International Workshop on Peer-to-Peer Systems (IPTPS 2002)*, Cambridge, MA, USA, vol. 2429 of Lecture Notes in Computer Science, pp. 261–269, Springer, Berlin, Germany, 2002.

[33] J. J. Douceur, "The sybil attack," in *Proceedings of 1st International Workshop on Peer-to-Peer Systems (IPTPS)*, Cambridge, MA, USA, March 2002.

[34] J. Newsome, E. Shi, D. Song, and A. Perrig, "The sybil attack in sensor networks: analysis and defenses," in *Proceedings of the 3rd International Symposium on Information Processing in Sensor Networks (IPSN'04)*, pp. 259–268, Berkeley, CA, USA, April 2004.

[35] B. N. Levine, C. Shields, and N. B. Margolin, "A survey of solutions to the sybil attack," Tech. Rep. 2006-052, University of Massachusetts Amherst, Amherst, MA, USA, October 2006.

[36] J. Dinger and O. P. Waldhorst, "Decentralized bootstrapping of P2P systems: a practical view," in *Proceedings of the 8th International IFIP-TC 6 Networking Conference on NETWORKING 2009*, pp. 703–715, Springer, Aachen, Germany, May 2009.

[37] C. Cramer, K. Kutzner, and T. Fuhrmann, "Bootstrapping locality-aware P2P networks," in *Proceedings of the 12th IEEE International Conference on Networks, 2004 (ICON'04)*, vol. 1, pp. 357–361, Singapore, 2004.

[38] H. Tran, M. Hitchens, V. Varadharajan, and P. Watters, "A trust based access control framework for P2P file-sharing systems," in *Proceedings of the 38th Annual Hawaii International Conference on System Sciences*, p. 302c, Washington, DC, USA, 2005.

[39] Y. Zhang, X. Li, J. Huai, and Y. Liu, "Access control in peer-to-peer collaborative systems," in *Proceedings of the 25th IEEE International Conference on Distributed Computing Systems Workshops (ICDCS'05)*, pp. 835–840, Columbus, OH, USA, June 2005.

[40] A. Ghodsi, L. O. Alima, and S. Haridi, "Symmetric replication for structured peer-to-peer systems," in *Proceedings of the International Workshops on Databases, Information Systems, and Peer-to-Peer Computing (DBISP2P 2005/2006)*, pp. 74–85, Springer, Trondheim, Norway, August 2006.

[41] H. Weatherspoon and J. D. Kubiatowicz, "Erasure coding vs. replication: a quantitative comparison," in *Proceedings of the First International Workshop on Peer-to-Peer Systems (IPTPS 2002)*, pp. 328–337, Springer, Cambridge, MA, USA, March 2002.

[42] D. Hughes, G. Coulson, and J. Walkerdine, "Free riding on Gnutella revisited: the bell tolls?," *IEEE Distributed Systems Online*, vol. 6, no. 6, 2005.

[43] S. J. Nielson, S. A. Crosby, and D. S. Wallach, "A taxonomy of rational attacks," in *Proceedings of the 4th International Workshop on Peer-to-Peer Systems IV (IPTPS 2005)*, pp. 36–46, Springer, Ithaca, NY, USA, February 2005.

[44] M. Feldman and J. Chuang, "Overcoming free-riding behavior in peer-to-peer systems," *SIGecom Exchanges*, vol. 5, no. 4, pp. 41–50, 2005.

[45] C. Jennings, B. Lowekamp, E. Rescorla, S. Baset, and H. Schulzrinne, *Resource Location and Discovery (Reload) Base Protocol*, IETF, RFC 6940, January 2014.

[46] D. Goldschlag, M. Reed, and P. Syverson, "Onion routing for anonymous and private internet connections," *Communications of the ACM*, vol. 42, no. 2, pp. 39–41, 1999.

[47] I. Eyal and E. G. Sirer, "Majority is not enough: Bitcoin mining is vulnerable," in *Financial Cryptography and Data Security, Lecture Notes in Computer Science*, N. Christin and R. Safavi-Naini, Eds., pp. 436–454, Springer, Berlin, Germany, 2014.

[48] S. Marti, P. Ganesan, and H. Garcia-Molina, "SPROUT: P2P routing with social networks," in *Proceedings of the Current Trends in Database Technology EDBT 2004 Workshops: EDBT 2004 Workshops PhD, DataX, PIM, P2P&DB, and ClustWeb*, pp. 425–435, Springer, Heraklion, Crete, Greece, March 2004.

[49] S. Androutsellis-Theotokis and D. Spinellis, "A survey of peer-to-peer content distribution technologies," *ACM Computing Surveys*, vol. 36, no. 4, pp. 335–371, 2004.

[50] W. Zamora, C. T. Calafate, J.-C. Cano, and P. Manzoni, "A survey on smartphone-based crowdsensing solutions," *Mobile Information Systems*, vol. 2016, Article ID 9681842, 26 pages, 2016.

[51] B. Guo, Z. Wang, Z. Yu et al., "Mobile crowd sensing and computing: the review of an emerging human-powered sensing paradigm," *ACM Computing Surveys*, vol. 48, no. 1, pp. 1–31, 2015.

[52] S. Delgado-Segura, C. Tanas, and J. Herrera-Joancomartí, "Reputation and reward: two sides of the same bitcoin," *Sensors*, vol. 16, no. 6, p. 776, 2016.

[53] H. Kalodner, M. Carlsten, P. Ellenbogen, J. Bonneau, and A. Narayanan, "An empirical study of Namecoin and lessons for decentralized namespace design," in *Proceedings of the 14th Workshop on the Economics of Information Security (WEIS'15)*, Delft, Netherlands, June 2015, http://randomwalker.info/publications/namespaces.pdf.

[54] M. Kitahara, J. Kawamoto, and K. Sakurai, "A method of digital rights management based on bitcoin protocol," in *Proceedings of the 8th International Conference on Ubiquitous Information Management and Communication (ICUIMC'14)*, pp. 84:1–84:6, ACM, New York, NY, USA, 2014.

11

An Efficient V2I Authentication Scheme for VANETs

Yousheng Zhou,[1,2,3] **Siling Liu**,[1] **Min Xiao**,[1,3] **Shaojiang Deng**,[2] **and Xiaojun Wang**[4]

[1]*College of Computer Science and Technology, Chongqing University of Posts and Telecommunications, Chongqing 400065, China*
[2]*College of Computer Science, Chongqing University, Chongqing 400044, China*
[3]*School of Cyber Security and Information Law, Chongqing University of Posts and Telecommunications, Chongqing 400065, China*
[4]*School of Electronic Engineering, Dublin City University, Dublin, Ireland*

Correspondence should be addressed to Shaojiang Deng; sj_deng@cqu.edu.cn

Academic Editor: María Calderon

The advent of intelligent transportation system has a crucial impact on the traffic safety and efficiency. To cope with security issues such as spoofing attack and forgery attack, many authentication schemes for vehicular ad hoc networks (VANETs) have been developed, which are based on the hypothesis that secret keys are kept perfectly secure. However, key exposure is inevitable on account of the openness of VANET environment. To address this problem, key insulation is introduced in our proposed scheme. With a helper device, vehicles could periodically update their own secret keys. In this way, the forward and backward secrecy has been achieved. In addition, the elliptic curve operations have been integrated to improve the performance. The random oracle model is adopted to prove the security of the proposed scheme, and the experiment has been conducted to demonstrate the comparison between our scheme and the existing similar schemes.

1. Introduction

Due to the growing demands for a safer and more efficient intelligent transportation system, the development of vehicular ad hoc networks (VANETs) has captured a large amount of attentions from research institutions and industries in recent years. VANETs are deemed to be a variant of the mobile ad hoc network, which is a type of continuously self-configuring, wirelessly connected, and infrastructure-less network of mobile devices [1].

There are two indispensable infrastructure elements in VANETs: on-board units (OBUs), which are mounted in each vehicle, and roadside units (RSUs), which are used to communicate and to assist authentication [2]. In addition, a third trusted party (TA) should also be deployed in VANETs, which mainly provides services of registration and authentication.

A common model of VANETs is exhibited in Figure 1. Communication modes in VANETs could be sorted into two categories: vehicle-to-vehicle (V2V) communication and vehicle-to-infrastructure (V2I) communication. By employing the dedicated short range communication (DSRC) protocol [3], these dynamic nodes (vehicles) could broadcast and exchange traffic information via RSU and other nearby moving vehicles. Upon receiving those messages including location, speed, and traffic conditions, vehicles would take reasonable actions immediately such as rerouting and braking to avoid possible traffic emergency.

As we all know, the communication channels in VANETs are open, so an attacker could capture, modify, replay, and delete messages transmitted in VANETs easily, leading to a large number of security problems, which will have a strong impact on the whole system. Assume that an original message is actually a warning that there is a serious traffic jam ahead, if it is tampered to a different message which tells vehicles that the road is unblocked, a completely opposite result would be caused. Therefore, authentication between vehicle and infrastructure should be employed to guarantee the authenticity of the transmitted messages in this situation.

Moreover, many devices of infrastructure such as RSU are unmanned, and the units installed on the vehicles which are used to run cryptographic algorithms are resource limited; thus, the risk of key exposure is unavoidable and

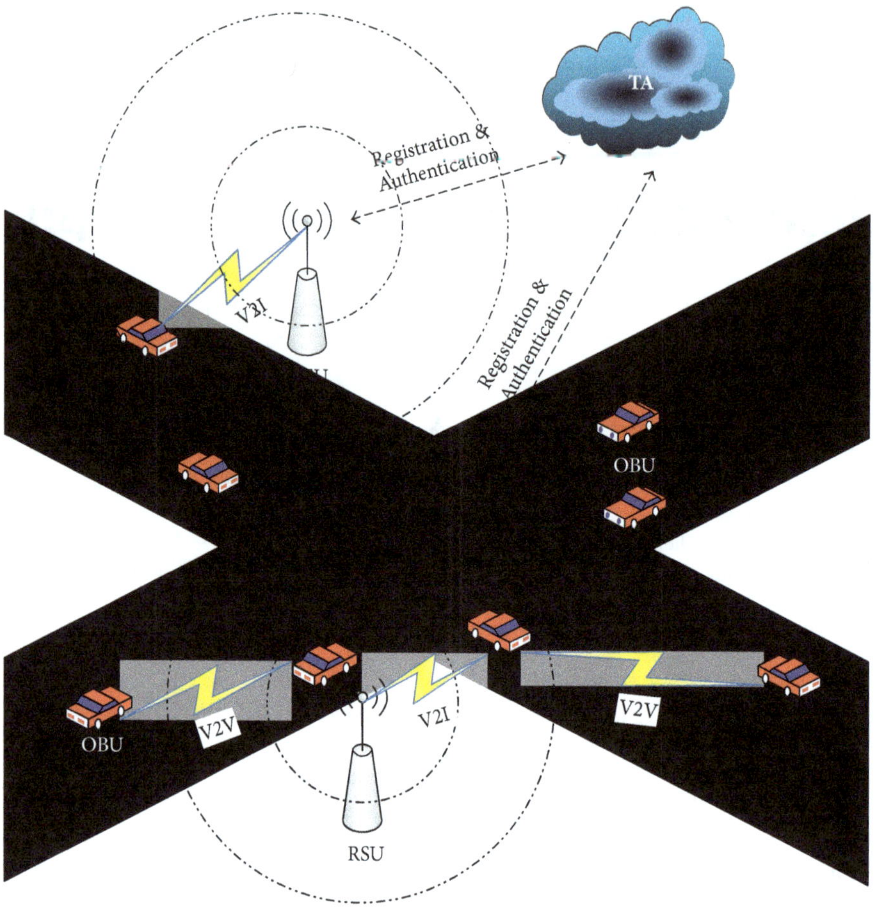

Figure 1: A common model of VANETs.

should not be overlooked. In addition, in the majority of cases, it is much easier for an attacker to obtain a secret key from an insecure device than to get it by breaking cryptogrammic hypothesis which system's security relies on. Once key exposure occurs, it means that security of the whole system loses. Taken into account efficiency, difficulties of construction, and security, key insulation is a desirable method to deal with the key exposure issue.

However, the vast majority of security protocols for VANETs are established on bilinear pairing, which would inevitably cause heavy computation costs. In order to enhance the security and performance of VANET-oriented authentication schemes, a novel practical V2I authentication scheme for VANETs is proposed, which attempts to lower computation complexity and the risk of losing secret keys.

To be specific, the main contributions of this paper represent as follows:

(1) Firstly, the key-insulated method is applied into V2I authentication for VANETs. In our proposed scheme, the user's private key is divided into two portions: one is managed by a secure device called helper or assistant and the other is held by the user, and both of them are updated periodically.

(2) Secondly, ECC, instead of bilinear pairing, is utilized to construct the proposed scheme. As most of devices in VANETs are resource limited, the computation consumption of the adopted schemes should be minimized as much as possible. Operations based on ECC in our construction can save far less time and computation burden than bilinear pairing operations, which is expected to gain higher efficiency.

(3) Finally, the forward and backward secrecy is achieved. The secret key of OBU consists of two fractions in which private information is involved. The secret key must be generated with the helper's participation, and it updates periodically so that malicious attackers cannot obtain the user's private key in the previous periods or in the subsequent periods.

The remainder of this paper proceeds as follows. Section 2 reviews the related work about V2I authentication scheme for VANETs. Section 3 presents the related essential knowledge. Section 4 describes the construction of the proposed scheme in detail. The security analysis and performance evaluation are given in Sections 5 and 6, respectively. Finally, this paper is concluded.

2. Related Works

In recent years, efforts on authentication have been made to address the problems of verification and efficiency. The privacy-preserving scheme [4] introduced by Wang et al. employs membership validity to replace the certificate revocation list and batch verification to improve efficiency, which achieved nonreputation, anonymity, traceability, and forward and backward secrecy. Wang [5] developed a privacy-preserving and accountable authentication protocol for IoT end-devices by adopting short group signature and secret sharing scheme. Shen et al. [6] proposed a multilayer authentication protocol with session key generation for wireless body area networks which is used for one-to-many group authentication scenario. The scheme with group testing towards a secure batch verification was introduced by Lee and Lai [7]. Unfortunately, this scheme is vulnerable to the impersonation attack since a malicious user could generate a fake signature on behalf of other vehicle. Based on this defect, another secure authentication scheme was introduced by Bayat et al. [8] to improve it. Wang and Yao [9] proposed the LIAP scheme, in which the vehicle and RSU are assigned with a long-term certification from the certificate authority (CA). If the vehicle is compromised, CA could easily revoke the vehicle's long-term certificate to terminate its behavior in the network. Jiang et al. [10] proposed an efficient anonymous batch authentication scheme to replace the CRL checking process by calculating the hash message authentication code, which divides the whole area into several domains. However, every vehicle has stored enough pseudonyms; if any of them is revoked, the rest pseudonyms are wasted. Azees et al. [11] proposed another anonymous scheme to avoid malicious nodes attending the activities in VANETs, which provides conditional tracking mechanism, low-cost certificate, and signature verification. Many secure schemes have achieved authentication by various means; however, most of them adopt bilinear pairing to realize their security characteristics. Actually, the bilinear pairing is not efficient for limited VANET devices on account of its vast computation costs. In view of this, pairing-free schemes have been put forward over the past years. For instance, Cui et al. [12] proposed a privacy-preserving scheme, using cuckoo filter and the binary search methods instead of map-to-point hash function and bilinear pairing operations to achieve high efficiency. Xie et al. [13] proposed an ECC-based authentication scheme to realize reliability and integrity of message. Lo and Tsai [14] proposed an efficient authentication scheme for V2I in vehicular sensor networks without bilinear pairing to improve performance, which achieves message integrity, traceability, and unlinkability. He et al. [15] proposed a new ID-based and elliptic curve-based authentication scheme, which withstands diverse types of attacks and yields better performance.

To address the problem of key exposure, Dodis et al. presented the idea of key insulation and came up with the first key-insulated public key cryptosystem [16] and the first strong key-insulated signature scheme [17]. Following the pioneering works, great efforts have been devoted to the key-insulated signature (KIS) schemes [18–20]. The scheme proposed by Gonzlez-Deleito et al. [18] uses numerous power operations, which adopts multiple private keys and master keys to achieve security. Le et al. [19] utilized multiple certification authorities to shorten verification path and mitigated damage. Hanaoka et al. [20] used two helpers to update the secret key and to enhance the system security. Later, quantities of identity-based or attribution-based key insulation schemes have been proposed [21–27], which are all based on bilinear pairing with random key updating. Additionally, key insulation has been applied into various other research fields. Zhou et al. [28] proposed a certificateless key-insulated generalized signcryption scheme without bilinear pairing in the context of cloud, which is proved to be secure under the computational Diffie–Hellman (CDH) assumption and the elliptic curve discrete logarithm (EC-DL) assumption. Hong et al. [29] proposed a key-insulated attribute-based signature without pairings for wireless communications, which attempts to minimize the potential threat and to relieve the computational burden. Kun et al. [30] and Shi et al. [31] put key insulation into peer-to-peer (P2P) networks and electronic commerce environment, respectively. Moreover, key insulation was introduced into mobile ad hoc networks (MANETs) by V. Kumar and R. Kumar [32]. Park et al. [33] proposed the EA2P scheme and first used key insulation in VANET environment. Although EA2P provides anonymity, identity extraction, and traceability, it only isolates the public key certification, not the private key, which actually fails to achieve a practical sense of key insulation.

3. Preliminaries

In this part, some necessary knowledge including system model, KIS framework, the random oracle model, and discrete logarithm (DL) problem is introduced.

3.1. System Model. As shown in Figure 2, the whole system model in this paper consists of three kinds of entities as follows:

PKG: Private key generator, which is deemed to be fully trusted, is responsible for producing keys including secrete keys as well as public keys.

RSU: Roadside unit, the infrastructure of VANET. It is a kind of computing device located on the roadside, which uses DSRC protocol and provides connectivity support for passing vehicles [34].

Vehicle: Each vehicle is equipped with an on-board unit (OBU) and a tamper-proof device (TPD). OBU is used to help vehicle communicate wirelessly with RSU. TPD acts as a helper which is physically secure but computationally limited, and its stored information can never be disclosed.

There are mainly four procedures in our proposed schemes as shown in Figure 2. Firstly, PKG preloads the related keys into TPD, produces, and publishes system parameters in initialization phase (Phase 1). Secondly, TPD helps the OBU to generate the vehicle's temporary secret key in Phase 2. Then OBU generates the signature and sends it to RSU in Phase 3. Finally, RSU validates the signature in Phase 4.

As illustrated in [15], the formalized definition of key-insulated signature (KIS) displays as follows:

Definition 1 (key-insulated signature (KIS)). A 5-tuple of polynomial time algorithm $(Kgen, UpdD, UpdU, Sig, Ver)$ makes up a key-insulated signature scheme as listed below:

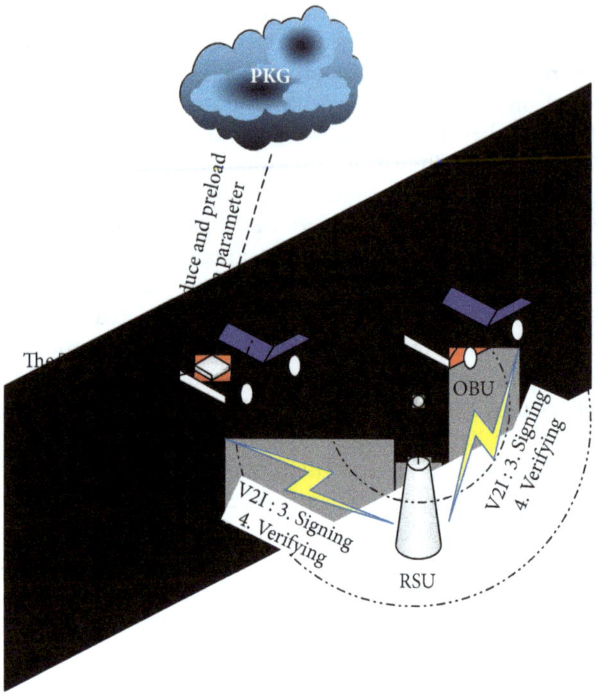

FIGURE 2: Model of VANET in this paper.

Kgen: the key generation algorithm, which falls into the initialized stage, takes a security parameter 1^k and the total number of time periods N as input to return a public key PK, a master key SK^*, and an initial key SK_0.

UpdD: the key update algorithm for the device, which takes indices i, j for time periods (throughout, $1 \le i, j \le N$) and the master key SK^* as input to return a partial secret $SK'_{i,j}\varepsilon$.

UpdU: the key update algorithm for the user, which takes indices i, j, a secret key SK_i, and a partial secret key $SK'_{i,j}$ as input to return the secret key SK^* for the time period j.

Sign: the signing algorithm, which takes an index i of a time period, a message M, and a secret key SK_i as input. Then $Sign_{SK_i}(i, M)$ returns a signature $< i, s >$ constituting the time period i and a signature S.

Verf: the verification algorithm, which takes the public key PK, a message M, and a pair $< i, s >$ as input. Then $Verf_{PK}(M, < i, s >)$ returns a bit b, where $b = 1$ means that the signature is accepted.

If $Verf_{PK}(M, < i, s >) = 1$, we say that $< i, s >$ is a valid signature of M for the time period i.

3.2. Security Model. The random oracle model was first proposed by Bellare and Rogaway [35] to prove the security of cryptographic protocols, and it is quoted in our proof. Oracle is an external device (it is usually being treated as a theoretical black box) that could provide true outputs for any inputs. In the case of inputting x, running a random oracle could be thought to pick a hash function $h(\cdot)$ at random and outputs $h(x)$. Besides, the relationship between the oracle's output and input satisfies properties of function; namely, the same input corresponds to the same output. As a matter of fact, each output is selected from its output domain, and acquired inputs/outputs are completely independent of current inputs/outputs on account of randomness.

Definition 2 (unforgeability). To prove the unforgeability, a game played between a challenger $\mathscr{C}$ and an adversary $\mathscr{A}$ is defined. Our scheme is unforgeable against the malicious OBU if the following condition is satisfied: for any probabilistic polynomial time (PPT) adversary $\mathscr{A}$, the probability that $\mathscr{A}$ wins the following game is negligible. The adversary can adaptively issue a series of undermentioned queries in the game.

Configuration-oracle: this can be considered as an initialization stage. The challenger $\mathscr{C}$ generates the system secret key and public parameters. $\mathscr{C}$ conveys these public parameters to the adversary $\mathscr{A}$.

H_1*-oracle*: upon receiving the query issued by the adversary $\mathscr{A}$ with the message m, the challenger $\mathscr{C}$ picks a stochastic number $e \in Z_q$, puts the tuple (m, e) into the list L_{h_1}, and returns e to the adversary $\mathscr{A}$.

H_2*-oracle*: upon receiving the query issued by the adversary $\mathscr{A}$ with the message m, the challenger $\mathscr{C}$ picks a stochastic number $e \in Z_q$, puts the tuple (m, e) into the list L_{h_2}, and returns e to the adversary $\mathscr{A}$.

TSK-oracle: upon receiving the query issued by the adversary $\mathscr{A}$ on the secret key SK_i, the challenger $\mathscr{C}$ computes it, puts the tuple (m, SK_i) into the list L_{key}, and returns SK_i to $\mathscr{A}$.

H_3*-oracle*: upon receiving the query issued by the adversary $\mathscr{A}$ with the message m, the challenger $\mathscr{C}$ picks a stochastic number $e \in Z_q$, puts the tuple (m, e) into the list L_{h_3}, and returns e to the adversary $\mathscr{A}$.

H_4*-oracle*: upon receiving the query issued by the adversary $\mathscr{A}$ with the message m, the challenger $\mathscr{C}$ picks a stochastic number $e \in Z_q$, puts the tuple (m, e) into the list L_{h_4}, and returns e to the adversary $\mathscr{A}$.

Signing-oracle: the challenger $\mathscr{C}$ generates the message required by the adversary $\mathscr{A}$ and sends it to $\mathscr{A}$.

Verification: the adversary $\mathscr{A}$ inputs the signature $\{U_i, \sigma_i, \lambda_i, M_i, T_i\}$ given by the challenger $\mathscr{C}$, and the verification algorithm returns a bit b, where $b = 1$ means that the signature is valid.

After a polynomial number of queries, if the adversary $\mathscr{A}$ can violate the unforgeability of the proposed scheme by generating a tuple $\{U_i^*, \sigma_i^*, \lambda_i^*, M_i^*, T_i^*\}$ on the condition that the verification phase outputs 1, then we say the adversary $\mathscr{A}$ wins the game.

3.3. Discrete Logarithm (DL) Problem. Provided with two stochastic points P, Q over an elliptic curve E, the DL problem is to compute a number x to meet the equation $Q = x \cdot P$.

4. The Proposed V2I Authentication Scheme

The proposed scheme consists of four phases: system initialization, key generation, signing stage, and verification stage. In the first place, notations used are defined in Table 1.

4.1. System Initialization. In this phase, every appliance in VANETs performs initialization.

(1) PKG generates fundamental system parameters including a group over the chosen elliptic curve $E_p(a,b)$, a random number $SK_{msk} \in Z_q^*$ as the system master key, and the system public key computed as follows:

$$PK_{pub} = SK_{msk} \cdot P. \tag{1}$$

(2) PKG selects a random number $SK_{TPD} \in Z_q^*$ as the private key of the helper (TPD) and calculates its corresponding public key

$$PK_{TPD} = SK_{TPD} \cdot P. \tag{2}$$

All these four parameters should be preloaded into TPD.

(3) RSU selects a random number $SK_{rsu} \in Z_q^*$ as its private key and computes the corresponding public key

$$PK_{rsu} = SK_{rsu} \cdot P. \tag{3}$$

(4) PKG publishes the public parameter set:

$$\text{param} = \{a, b, p, q, P, PK_{pub}, PK_{TPD}, PK_{rsu}, H_1, H_2, H_3, H_4\}. \tag{4}$$

In this paper, we assume that the OBU's public key and its TPD public key have been preloaded.

4.2. Key Generation

4.2.1. Initial Key Generation. Set the parameter α_i corresponding to the time period i as $\alpha_i = H_2(H_1(ID_{obu})SK_{TPD}\cdot PK_{rsu}T_i)$. Note that it is default for TPD to keep its OBU's identity. TPD computes $\alpha_0 = H_2(H_1(ID_{obu})||SK_{TPD}\cdot PK_{rsu}|| T_0)$ and the initial private key of the OBU as

$$SK_{obu}^0 = SK_{msk}\cdot H_1(ID_{obu}) + SK_{TPD}\cdot \alpha_0, \tag{5}$$

which is preloaded into the OBU.

4.2.2. Partial Key Generation. TPD calculates

$$K_{part}^i = SK_{TPD}\cdot(\alpha_i - \alpha_{i-1}), \tag{6}$$

as the partial key corresponding to the time period i, and sends it to the OBU to assist in generating the temporary secret key.

4.2.3. Temporary Secret Key Generation. OBU calculates its own temporary secret key in the time period i

$$SK_{obu}^i = SK_{obu}^{i-1} + K_{part}^i, \tag{7}$$

as soon as it receives K_{part}^i from the TPD.

The temporary public key in the time period i of OBU is set as

$$PK_{obu}^i = SK_{obu}^i \cdot P, \tag{8}$$

and it is published by the OBU, while the partial key K_{part}^i and the initial key SK_{obu}^{i-1} are removed after key updating.

4.3. Signing Stage. An OBU can generate the signature on message M_i in the time period i as follows.

Step 1. Selects the random number $u_i \in Z_q^*$ to compute.

$$U_i = u_i \cdot P. \tag{9}$$

Step 2. Uses the identity ID_{obu}, the temporary secret key in the time period i SK_{obu}^i, the public key of RSU PK_{rsu}, the corresponding time stamp T_i, and hash functions to compute

$$\omega_i = H_1(ID_{obu})\oplus H_3(SK_{obu}^i\cdot PK_{rsu}), \tag{10}$$

$$\eta_i = H_2\cdot(H_1(ID_{obu})SK_{obu}^i\cdot PK_{rsu}T_i). \tag{11}$$

Step 3. Selects another random number $\lambda_i \in Z_q^*$ and uses the identity and η_i to compute

$$\theta_i = \eta_i + \lambda_i \cdot H_1(ID_{obu}). \tag{12}$$

Step 4. Concatenates the hash value of identity $H_1(ID_{obu})$, λ_i, U_i, the message about traffic status M_i and current time stamp T_i to compute

$$\beta_i = H_4(H_1(ID_{obu})||\lambda_i||U_i||M_i||T_i). \tag{13}$$

Step 5. Uses the two random numbers u_i and λ_i, β_i, and the temporary secret key SK_{obu}^i to compute

$$\sigma_i = \lambda_i\cdot SK_{obu}^i + \beta_i\cdot u_i \bmod p. \tag{14}$$

Step 6. Sends the message $\{U_i, \sigma_i, \omega_i, \theta_i, M_i, T_i\}$ to the regional RSU.

4.4. Verification Stage. Upon receiving the signature, RSU proceeds the following steps to verify it.

Step 1. Examines the freshness of T_i. If it is fresh, goes to step 2; otherwise, the signature is rejected.

Step 2. Uses own secret key SK_{rsu}, the private key in the time period i of the vehicle PK_{obu}^i and ω_i to count the hash value of identity of the vehicle:

$$H_1(ID_{obu}) = \omega_i\oplus H_3(SK_{rsu}\cdot PK_{obu}^i). \tag{15}$$

Table 1: Definition of notations.

Notations	Definition
p, q	Two large prime numbers
$E_p(a,b)$	An elliptic curve defined by the equation $y^2 = x^3 + ax + b \bmod p (a, b \in F_p)$
G	An additive group with the order q, where G is constitutive of all points on E and the point at infinity O
P	A generator of the group G
H_1, H_2, H_3, H_4	Four security functions, where $H_1 : \{0,1\}^* \rightarrow Z_q$, $H_2 : \{0,1\}^* \rightarrow Z_q$, $H_3 : \{0,1\}^* \rightarrow Z_q$ and $H_4 : \{0,1\}^* \rightarrow Z_q$
SK_{msk}	The master secret key of system
PK_{pub}	The public key of system
SK_{TPD}	The secret key of assistant device (some papers also called it a base or a helper)
PK_{TPD}	The public key of assistant device
SK_{rsu}	The secret key of RSU
PK_{rsu}	The public key of RSU
T_i	The time of the time period i
ID_{obu}	The identity of OBU
K^i_{part}	The partial key in time period i which is generated by TPD offering assistance to key update of secret key of OBU
SK^i_{obu}	The temporary secret key in time period i of OBU
PK^i_{obu}	The temporary public key in time period i of OBU
σ_i	The signature of OBU
$\oplus$	The exclusive disjunction operation
$\|$	The message concatenation operation

Step 3. Uses the hash value of ID_{obu}, its own secret key SK_{rsu}, the private key of the vehicle, the private key of TPD, and current time stamp T_i to evaluate

$$\alpha_i = H_2(H_1(ID_{\text{obu}})\|SK_{\text{rsu}} \cdot PK_{\text{TPD}}\|T_i), \tag{16}$$

$$\eta_i = H_2\left(H_1(ID_{\text{obu}})\|SK_{\text{rsu}} \cdot PK^i_{\text{obu}}\|T_i\right). \tag{17}$$

Step 4. Uses the hash value of ID_{obu}, θ_i, and η_i to evaluate

$$\lambda_i = \frac{(\theta_i - \eta_i)}{H_1(ID_{\text{obu}})}. \tag{18}$$

Step 5. Concatenates the hash value of ID_{obu}, λ_i, U_i, the message about traffic status M_i, and current time stamp T_i to evaluate

$$\beta_i = H_4(H_1(ID_{\text{obu}})\|\lambda_i\|U_i M_i T_i). \tag{19}$$

Step 6. Checks whether the equation

$$\sigma_i \cdot P = \left(H_1(ID_{\text{obu}}) \cdot PK_{\text{pub}} + PK_{\text{TPD}} \cdot \alpha_i\right) \cdot \lambda_i + U_i \cdot \beta_i \tag{20}$$

holds. If it holds, the signature is valid.

5. Security Analysis

In this part, the correctness and the security analysis under the random oracle model of our proposed scheme are illustrated.

5.1. Correctness Proof

Theorem 1. *A signature from the OBU could pass the verification of the RSU.*

Proof. Actually, given a signature $\{U_i, \sigma_i, \omega_i, \theta_i, M_i, T_i\}$ from an OBU, the RSU could compute

$$\begin{aligned}
\sigma_i &= \lambda_i \cdot SK^i_{\text{obu}} + \beta_i \cdot u_i \\
&= \lambda_i \cdot \left(SK^{i-1}_{\text{obu}} + K^i_{\text{part}}\right) + \beta_i \cdot u_i \\
&= \lambda_i \cdot \left(SK^{i-2}_{\text{obu}} + K^{i-1}_{\text{part}} + K^i_{\text{part}}\right) + \beta_i \cdot u_i \\
&= \lambda_i \cdot \left(SK^{i-3}_{\text{obu}} + K^{i-2}_{\text{part}} + K^{i-1}_{\text{part}} + K^i_{\text{part}}\right) + \beta_i \cdot u_i \\
&= \lambda_i \cdot \left(SK^0_{\text{obu}} + K^1_{\text{part}} + K^2_{\text{part}} + \ldots K^i_{\text{part}}\right) + \beta_i \cdot u_i \\
&= \lambda_i \cdot [SK^0_{\text{obu}} + SK_{\text{TPD}} \cdot (\alpha_1 - \alpha_0 + \alpha_2 - \alpha_1 + \ldots \alpha_{i-1} \\
&\quad - \alpha_{i-2} + \alpha_i - \alpha_{i-1})] + \beta_i \cdot u_i \\
&= \lambda_i \cdot \left[SK^0_{\text{obu}} + SK_{\text{TPD}} \cdot (\alpha_i - \alpha_0)\right] + \beta_i \cdot u_i \\
&= \lambda_i \cdot [SK_{\text{msk}} \cdot H_1(ID_{\text{obu}}) + SK_{\text{TPD}} \cdot \alpha_0 \\
&\quad + SK_{\text{TPD}} \cdot (\alpha_i - \alpha_0)] + \beta_i \cdot u_i \\
&= \lambda_i \cdot [SK_{\text{msk}} \cdot H_1(ID_{\text{obu}}) + SK_{\text{TPD}} \cdot \alpha_i] + \beta_i \cdot u_i,
\end{aligned} \tag{21}$$

$$\begin{aligned}
\sigma_i \cdot P &= \lambda_i \cdot [SK_{\text{msk}} \cdot H_1(ID_{\text{obu}}) \cdot P + SK_{\text{TPD}} \cdot \alpha_i \cdot P] + \beta_i \cdot u_i \cdot P \\
&= \left(H_1(ID_{\text{obu}}) \cdot PK_{\text{pub}} + PK_{\text{TPD}} \cdot \alpha_i\right) \cdot \lambda_i + U_i \cdot \beta_i.
\end{aligned} \tag{22}$$

Therefore, the signature is verified to be valid.

5.2. Security Proof

Theorem 2. *Our proposed V2I authentication scheme is secure under the random oracle model.*

Proof. Assume that there is a PPT adversary $\mathcal{A}$ who could forge a signature to pass the verification successfully. The challenger $\mathcal{C}$ is constructed to tackle the DL problem with a nonnegligible probability by interacting with $\mathcal{A}$. Given a DL instance $(P, PK_{TPD} = x \cdot P = X)$, the game between $\mathcal{A}$ and $\mathcal{C}$ is played as follows.

5.2.1. Query Phase.

$Configuration\text{-}oracle$: The challenger $\mathcal{C}$ allocates $PK_{pub} = s \cdot P$ and $PK_{TPD} = x \cdot P$, generates the public parameter $param = \{a, b, p, q, P, PK_{pub}, PK_{TPD}, PK_{rsu}\}$, and conveys these parameters to the adversary $\mathcal{A}$.

$H_1 - oracle$: A list H_1^{list} is set up and retained by the challenger $\mathcal{C}$, which is initialized to empty. Upon receiving the query about (ID_{obu}, T_i) from the adversary $\mathcal{A}$, $\mathcal{C}$ first examines whether the tuple (ID_{obu}, T_i, R_1) is in H_1^{list}. If so, $\mathcal{C}$ returns $R_1 = H_1(ID_{obu})$ to $\mathcal{A}$; otherwise, $\mathcal{C}$ picks a random number $R_1 \in Z_q^*$, puts the tuple (ID_{obu}, T_i, R_1) into H_1^{list}, and returns $R_1 = H_1(ID_{obu})$ to $\mathcal{A}$.

$H_2 - oracle$: A list H_2^{list} is set up and retained by the challenger $\mathcal{C}$, which is initialized to empty. Upon receiving the query about (ID_{obu}, T_i, D_2) from the adversary $\mathcal{A}$, $\mathcal{C}$ first extracts the tuple (ID_{obu}, T_i, R_1) from the list H_1^{list}. Then, $\mathcal{C}$ examines whether the tuple $(ID_{obu}, T_i, D_2, R_2)$ is in H_2^{list}. If so, $\mathcal{C}$ returns $R_2 = H_2(R_1||D_2||T_i)$ to $\mathcal{A}$; otherwise, $\mathcal{C}$ picks two random numbers $\omega, R_2 \in Z_q^*$ and allocates $D_2 = \omega \cdot PK_{rsu}$. Finally, $\mathcal{C}$ puts the tuple $(ID_{obu}, T_i, D_2, R_2)$ into H_2^{list} and returns $R_2 = H_2(R_1||D_2||T_i)$ to $\mathcal{A}$.

$Temporarysecretkey - oracle$: The adversary $\mathcal{A}$ asks for the temporary private key SK_{obu}^i in the current time period i. The challenger $\mathcal{C}$ first extracts the tuples (ID_{obu}, T_i, R_1) and $(ID_{obu}, T_i, D_2, R_2)$ from lists H_1^{list} and H_2^{list}, respectively. Then, $\mathcal{C}$ computes

$$\begin{aligned} SK_{obu}^i &= SK_{msk} \cdot H_1(ID_{obu}) + SK_{TPD} \cdot \alpha_i \\ &= s \cdot R_1 + SK_{TPD} \cdot R_2. \end{aligned} \tag{23}$$

Finally, $\mathcal{C}$ adds the tuple $(ID_{obu}, T_i, SK_{obu}^i)$ into the list Key^{list} and returns SK_{obu}^i to $\mathcal{A}$.

$H_3 - oracle$: A list H_3^{list} is set up and retained by the challenger $\mathcal{C}$, which is initialized to empty. Upon receiving the query about (ID_{obu}, T_i) from the adversary $\mathcal{A}$, $\mathcal{C}$ first examines whether the tuple (ID_{obu}, T_i, R_3) is in H_3^{list}. If so, $\mathcal{C}$ returns R_3 to $\mathcal{A}$; otherwise, $\mathcal{C}$ picks a random number $R_3 \in Z_q^*$, puts the tuple (ID_{obu}, T_i, R_3) into H_3^{list}, and returns R_3 to $\mathcal{A}$.

$H_4 - oracle$: A list H_4^{list} is set up and retained by the challenger $\mathcal{C}$, which is initialized to empty. Upon receiving the query about $(ID_{obu}, T_i, U_i, \lambda_i, M_i)$ from the adversary $\mathcal{A}$, $\mathcal{C}$ first extracts the tuple (ID_{obu}, T_i, R_1) from the list H_1^{list}. Then, $\mathcal{C}$ examines whether the tuple $(ID_{obu}, T_i, U_i, \lambda_i, M_i, R_4)$ is in the list H_4^{list}. If so, $\mathcal{C}$ returns $R_4 = H_4(R_1||U_i||\lambda_i||M_i||T_i)$ to $\mathcal{A}$; otherwise, $\mathcal{C}$ picks a random number $R_4 \in Z_q^*$, puts the tuple $(ID_{obu}, T_i, U_i, \lambda_i, M_i, R_4)$ into H_4^{list}, and returns $R_4 = H_4(R_1 U_i \lambda_i||M_i||T_i)$ to $\mathcal{A}$.

$Signing - oracle$: Upon receiving the query about the signature on message M_i from the adversary $\mathcal{A}$, the challenger $\mathcal{C}$ picks random numbers $u_i, \sigma_i \in Z_q^*$ and $r_1, r_2, r_4, d_2 \in Z_q^*$. $\mathcal{C}$ sets $U_i = u_i \cdot P$ and $\lambda_i = (r_1 \cdot P_{pub} + r_2 \cdot X)^{-1} \cdot (\sigma_i \cdot P - r_4 \cdot U_i)$ and adds the tuples (ID_{obu}, T_i, r_1), $(ID_{obu}, T_i, d_2, r_2)$, and $(ID_{obu}, T_i, U_i, \lambda_i, M_i, r_4)$ into the lists H_1^{list}, H_2^{list}, and H_4^{list} separately. At last, $\mathcal{C}$ returns $\{U_i, \sigma_i, \lambda_i, M_i, T_i\}$ to $\mathcal{A}$.

$Verification$: The verifier checks if the equation $\sigma_i \cdot P = \lambda_i \cdot (R_1 \cdot PK_{pub} + PK_{TPD} \cdot R_2) + U_i \cdot R_4$ holds. If it does not hold, the verification algorithm outputs 0 and the process is aborted. Otherwise, the verification algorithm outputs 1 and the signature is accepted.

5.2.2. Forgery Phase. If an adversary $\mathcal{A}$ could successfully output a signature which can pass the verification with nonneglectable probability according to the forking lemma [36], then $\mathcal{A}$ could output a second signature in an attack by using a different random oracle with nonneglectable probability. Consequently, $\mathcal{A}$ could output another signature $\{U_i^*, \sigma_i^*, \lambda_i^*, M_i^*, T_i^*\}$ by repeating the process with a different choice of R_2, which leads to a distinguishing σ_i, while the value of λ_i remains unchanged. Such that, the following equation is obtained:

$$\begin{aligned} \sigma_i \cdot P &= \lambda_i \cdot \left(R_1 \cdot PK_{pub} + X \cdot R_2\right) + U_i \cdot R_4 \\ \sigma_i' \cdot P &= \lambda_i \cdot \left(R_1 \cdot PK_{pub} + X \cdot R_2'\right) + U_i \cdot R_4, \end{aligned} \tag{24}$$

$$\begin{aligned} (\sigma_i - \sigma_i') \cdot P &= \lambda_i \cdot X \cdot (R_2 - R_2{'}) \\ x &= (R_2 - R_2')^{-1} \cdot \lambda_i^{-1} \cdot (\sigma_i - \sigma_i'). \end{aligned} \tag{25}$$

At last, the challenger $\mathcal{C}$ outputs $(R_2 - R_2')^{-1} \cdot \lambda_i^{-1} \cdot (\sigma_i - \sigma_i')$ as the answer of the DL problem instance $(P, PK_{TPD} = x \cdot P = X)$. This contradicts with the difficulty of the DL problem. Hence, our proposed V2I authentication scheme is secure against forgery under the random oracle model on the condition of adaptive chosen message attack.

6. Performance Evaluation

To examine the performance of our proposed scheme in reality, an experiment has been conducted on a Windows 10-installed laptop with Intel(R) Core(TM) i7, and the cryptographic operations have been implemented by using the TEPLA library [37], which requires GMP library and OpenSSL [38].

TEPLA Elliptic Curve and Pairing Library is a free C library which provides functions such as finite field arithmetic with 254-bit prime number, elliptic curve arithmetic over Barreto–Neahrig curve, and pairing arithmetic using optimal ate pairing over BN curve. To set up the system

TABLE 2: Definition of symbols.

Symbols	Definition
T_{exp}	The running time of an exponentiation operation
$T_{bili-pairing}$	The running time of a bilinear pairing operation
$T_{bilipair-mul}$	The running time of multiplication operation with two bilinear pairings
$T_{poi-add}$	The running time of a point addition operation correlated with the ECC
$T_{poi-mul}$	The running time of a scale multiplication operation correlated with the ECC
T_h	The running time of a common hash function operation

TABLE 3: Computational costs comparison of the signature and verification phases in theory (ms).

	Signing	Verification
Wang and Yao [9]	$T_{poi-add} + 3T_h + 4T_{poi-mul}$	$2T_h + 3T_{bili-pairing} + T_{bilipair-mul}$
Weng et al. [22]	$T_{poi-mul} + T_h$	$3T_h + 4T_{bili-pairing} + 2T_{bilipair-mul}$
Zhou et al. [23]	$2T_{exp} + T_h$	$3T_h + 4T_{bili-pairing} + 2T_{bilipair-mul}$
Wan et al. [24]	$6T_{exp}$	$5T_{bili-pairing} + 3T_{bilipair-mul}$
Zhao et al. [25]	$2T_{poi-mul} + T_h$	$T_h + 3T_{bili-pairing} + T_{bilipair-mul} + T_{poi-mul} + T_{poi-add}$
Weng et al. [26]	$6T_{exp}$	$6T_{bili-pairing} + 4T_{bilipair-mul}$
Chen et al. [27]	$7T_{exp}$	$T_{exp} + 4T_{bili-pairing} + 2T_{bilipair-mul}$
Our scheme	$2T_{poi-mul} + 4T_h$	$7T_{poi-mul} + 2T_{poi-add} + 4T_h$

TABLE 4: Computational costs comparison of the signature and verification phases in simulation (ms).

	Signing	Verification
Wang and Yao [9]	2.104	4.724
Weng et al. [22]	5.24×10^{-1}	6.305
Zhou et al. [23]	6.3×10^{-2}	6.305
Wan et al. [24]	1.86×10^{-1}	7.882
Zhao et al. [25]	1.047	5.255
Weng et al. [26]	1.86×10^{-1}	9.462
Chen et al. [27]	2.17×10^{-1}	6.333
Our scheme	1.050	3.683

environment as the library required, MinGW64 and MSYS are needed to simulate Linux environment to compile cryptography libraries. To install TEPLA, GNU MP library and OpenSSL are required. GNU MP is a free library for big number related operations, and OpenSSL is used to realize cryptographic operations. After finishing compiling, environment variables for Visual Studio are configured and header files of these cryptography libraries are included to conduct the experiment for our scheme.

The experimental results show that the proposed scheme costs 1.050ms and 3.683ms in terms of signing and verification, respectively. We compare our proposed scheme with seven existing similar authentication schemes [9, 22–27], including a pairing-free authentication scheme [9] for VANETs and six key-insulated authentication schemes [22–27]. Note that in our comparison, only consumption of signing phase and verification phase is put into consideration. For convenience, the description of the symbols used in the comparison is listed in Table 2, the results of the comparison on computational costs of various KIS schemes in theory are listed in Table 3, the real running time is displayed in Table 4, and the intuitive comparison about signing and verification is shown in Figure 3 and Figure 4, respectively.

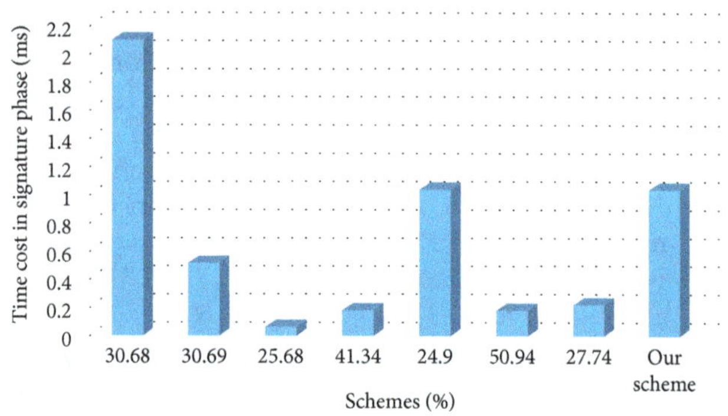

FIGURE 3: Computational costs of signature phase in different schemes.

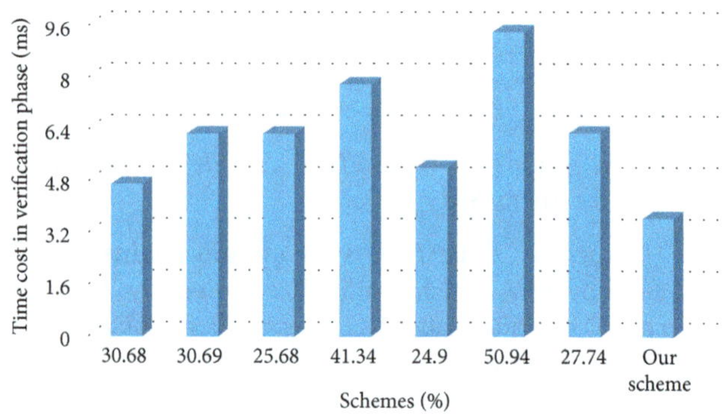

FIGURE 4: Computational costs of verification phase in different schemes.

Except the operations listed in Table 2, other operations have not been considered since their running time is ignorable. In terms of the verification phase, according to Table 3, in the scheme of Wang and Yao [9], RSU needs to run two hash operations, three bilinear pairing operations, and one multiplication operation. In the scheme of Weng [22] and Zhou [23], RSU needs to run three hash operations, four bilinear pairing operations, and two multiplication operations. In the scheme of Wan [24], RSU needs to run five bilinear pairing operations, and three multiplication operations. In the scheme of Zhao [25], RSU needs to run one hash operations, three bilinear pairing operations, one multiplication operation, one point addition operation, and one multiplication operation related to ECC. In the scheme of Weng [26], RSU needs to run six bilinear pairing operations and four multiplication operations. In the scheme of Chen [27], RSU needs to run one exponentiation operation, four bilinear pairing operations, and two multiplication operations. As we all know, the bilinear pairing operation is the most time-consuming, while hash function is the least time-consuming. Furthermore, time consumption for the operations listed could be ranked as follows: $T_{\text{bili-pairing}} > T_{\text{poi-mul}} > T_{\text{exp}} > T_{\text{bilipair-mul}} \approx T_{\text{poi-add}} > T_h$. It could be seen from Table 4 that our scheme possesses comparatively high efficiency in verification since our scheme is constructed by using comparatively light-weighted operations. To show the advantage clearly, the improved ratio of our scheme against other seven schemes is defined as $(T_{[\text{num}]} - T_{[\text{ours}]})/T_{[\text{num}]}$, where $T_{[\text{num}]}$ refers to time costs of the scheme with the reference number "num" and $T_{[\text{ours}]}$ refers to time costs of our scheme. In terms of the verification stage, the improved ratios of our scheme against the schemes [9, 22–27] are $(4.724 - 3.683)/4.724 = 22.04\%$, $(6.305 - 3.683)/6.305 = 41.59\%$, $(6.305 - 3.683)/6.305 = 41.59\%$, $(7.882 - 3.683)/7.882 = 53.27\%$, $(5.255 - 3.683)/5.255 = 29.91\%$, $(9.462 - 3.683)/9.462 = 61.08\%$, and $(6.333 - 3.683)/6.333 = 41.84\%$, respectively. When it comes to the total costs of signing and verifying stages, the improved ratios of our scheme against other seven schemes [9, 22–27] are 30.68%, 30.69%, 25.68%, 41.34%, 24.9%, 50.94%, and 27.74% individually. From Figures 3 and 4, the computation costs of singing phase of our proposed scheme are lightly higher than some schemes, because indispensable operations in this phase are needed to achieve key insulation and to provide better security. Even that our promotion comes at a little price of efficiency of signing, the gain of key-insulated secrecy deserves it, and on the whole, the proposed scheme achieves a better trade-off between security and efficiency than the compared schemes.

7. Conclusion

Vehicular ad hoc networks (VANETs) are one of the most promising technologies nowadays. For the sake of providing efficient and secure authentication for VANETs, a key-insulated V2I authentication scheme has been constructed in this paper. The core idea of the proposed scheme is dividing private key of the vehicle into two parts which are, respectively, held by a temper-proofing device (TPD) and the vehicle itself, and these two parts of the private key are used to generate a signature. The proposed scheme supports dynamically updating private key in different time periods. For the vehicle, it obtains its updated secret key by the help of TPD before signing. For the RSU, it first checks whether the time stamp is valid before verification, and then it validates the signature from the vehicle. The security analysis manifests that the proposed scheme is secure under the adaptive chosen message attack. The comparison is also conducted among our scheme and other similar schemes. The performance evaluation shows that our bilinear pairing-free scheme harvests a better trade-off between security and efficiency, and it is feasible for VANET environment.

In our proposed scheme, the helper is assumed as a fully-trusted device, and the private key of the vehicle is generated by its helper. However, the helper is actually semitrusted in some situations, which means that the assistant device can generate signature without the user's approval. In this situation, 2-out-of-2 threshold manner should be a considerable method to prevent the misuse of the user's secret key by the helper. The core idea of 2-out-of-2 threshold manner is that the user and the helper device could share the threshold value n using standard threshold techniques, where the user keeps n_1 and the helper keeps n_2 such that $n_1 + n_2 = n$. In addition, because the RSU would receive and verify numerous signatures from the vehicles within its region, this would inevitably cause burden of computational consumption if the RSU proceeds verification on by one. Taken into account the requirements for efficiency and security in the context discussed above, design of an efficient threshold key-insulated authentication scheme is our future work, which aims to achieve feasible secure V2I communication for VANETs.

Acknowledgments

This work was jointly supported by the National Social Science Foundation of China (no. 14CTQ026), the National Natural Science Foundation of China (no. 61702067, no. 61672119, and no. 61472464), the Chongqing Research Program of Application Foundation and Advanced Technology (no. cstc2017jcyjAX0201), the Natural Science Foundation of Shandong Province, China (no. ZR2015FL024), and the Science and Technology Research Project of Chongqing Municipal Education Commission (no. KJ1600445).

References

[1] C. K. Toh, *Ad Hoc Mobile Wireless Networks: Protocols and Systems*, Pearson Education, London, UK, 2001.

[2] S. Biswas, J. Mišić, and V. Mišić, "DDoS attack on WAVE-enabled VANET through synchronization," in *Proceedings of the Global Communications Conference (GLOBECOM)*, pp. 1079–1084, Anaheim, CA, USA, December 2012.

[3] H. Oh, C. Yae, D. Ahn, and H. Cho, "5.8 GHz DSRC packet communication system for ITS services," in *Proceedings of the Vehicular Technology Conference*, pp. 2223–2227, Amsterdam, The Netherlands, September 1999.

[4] Y. Wang, H. Zhong, Y. Xu, and J. Cui, "ECPB: efficient conditional privacy-preserving authentication scheme supporting batch verification for VANETs," *International Journal of Network Security*, vol. 18, no. 2, pp. 374–382, 2016.
[5] Z. Wang, "A privacy-preserving and accountable authentication protocol for IoT end-devices with weaker identity," *Future Generation Computer Systems*, 2017, in press.
[6] J. Shen, S. Chang, J. Shen, Q. Liu, and X. Sun, "A lightweight multi-layer authentication protocol for wireless body area networks," *Future Generation Computer Systems*, vol. 78, no. 3, pp. 956–963, 2016.
[7] C. C. Lee and Y. M. Lai, "Toward a secure batch verification with group testing for VANET," *Wireless Networks*, vol. 19, no. 6, pp. 1441–1449, 2013.
[8] M. Bayat, M. Barmshoory, M. Rahimi, and M. R. Aref, "A secure authentication scheme for VANETs with batch verification," *Wireless Networks*, vol. 21, no. 5, pp. 1733–1743, 2015.
[9] S. Wang and N. Yao, "LIAP: a local identity-based anonymous message authentication protocol in VANETs," *Computer Communications*, vol. 112, pp. 154–164, 2017.
[10] S. Jiang, X. Zhu, and L. Wang, "An efficient anonymous batch authentication scheme based on HMAC for VANETs," *IEEE Transactions on Intelligent Transportation Systems*, vol. 17, no. 8, pp. 2193–2204, 2016.
[11] M. Azees, P. Vijayakumar, and L. J. Deboarh, "EAAP: efficient anonymous authentication with conditional privacy-preserving scheme for vehicular ad hoc networks," *IEEE Transactions on Intelligent Transportation Systems*, vol. 18, no. 9, pp. 2467–2476, 2017.
[12] J. Cui, J. Zhang, H. Zhong, and Y. Xu, "SPACF: a secure privacy-preserving authentication scheme for VANET with cuckoo filter," *IEEE Transactions on Vehicular Technology*, vol. 66, no. 11, pp. 10283–10295, 2017.
[13] Y. Xie, L. B. Wu, J. Shen, and A. Alelaiwi, "EIAS-CP: new efficient identity-based authentication scheme with conditional privacy-preserving for VANETs," *Telecommunication Systems*, vol. 65, no. 2, pp. 229–240, 2017.
[14] N. W. Lo and J. L. Tsai, "An efficient conditional privacy-preserving authentication scheme for vehicular sensor networks without pairings," *IEEE Transactions on Intelligent Transportation Systems*, vol. 17, no. 5, pp. 1319–1328, 2016.
[15] D. He, S. Zeadally, B. Xu, and X. Huang, "An efficient identity-based conditional privacy-preserving authentication scheme for vehicular ad hoc networks," *IEEE Transactions on Information Forensics and Security*, vol. 10, no. 12, pp. 2681–2691, 2015.
[16] Y. Dodis, J. Katz, S. Xu, and M. Yung, "Key-insulated public key cryptosystems," in *Proceedings of the International Conference on the Theory and Applications of Cryptographic Techniques*, pp. 65–82, Amsterdam, The Netherlands, April-May 2002.
[17] Y. Dodis, J. Katz, S. Xu, and M. Yung, "Strong key-insulated signature schemes," in *Proceedings of the International Workshop on Public Key Cryptography*, pp. 130–144, Miami, FL, USA, January 2003.
[18] N. Gonzlez-Deleito, O. Markowitch, and E. Dall'Olio, "A new key-insulated signature scheme," in *Proceedings of the International Conference on Information and Communications Security*, pp. 465–479, Malaga, Spain, October 2004.
[19] Z. Le, Y. Ouyang, J. Ford, and F. Makedon, "A hierarchical key-insulated signature scheme in the CA trust model," in *Proceedings of the International Conference on Information Security*, pp. 280–291, Palo Alto, CA, USA, September 2004.
[20] G. Hanaoka, Y. Hanaoka, and H. Imai, "Parallel key-insulated public key encryption," in *Proceedings of the International Workshop on Public Key Cryptography*, pp. 105–122, New York, NY, USA, April 2006.
[21] D. H. Yum and P. J. Lee, "Efficient key updating signature schemes based on IBS," in *Proceedings of the IMA International Conference on Cryptography and Coding*, pp. 167–182, Cirencester, UK, December 2003.
[22] J. Weng, S. Liu, K. Chen, and X. Li, "Identity-based key-insulated signature with secure key-updates," in *Proceedings of the International Conference on Information Security and Cryptology*, pp. 13–26, Beijing, China, November-December 2006.
[23] Y. Zhou, Z. Cao, and Z. Chai, "Identity based key insulated signature," in *Proceedings of the International Conference on Information Security Practice and Experience*, pp. 226–234, Hangzhou, China, April 2006.
[24] Z. Wan, J. Li, and X. Hong, "Parallel key-insulated signature scheme without random oracles," *Journal of Communications and Networks*, vol. 15, no. 3, pp. 252–257, 2013.
[25] H. Zhao, J. Yu, S. Duan, X. Cheng, and R. Hao, "Key-insulated aggregate signature," *Frontiers of Computer Science*, vol. 8, no. 5, pp. 837–846, 2014.
[26] J. Weng, X. Li, K. Chen, and S. L. Liu, "Identity-based parallel key-insulated signature without random oracles," *Journal of Information Science and Engineering*, vol. 24, no. 4, pp. 1143–1157, 2008.
[27] J. Chen, W. Wang, and K. Yu, "Attribute-based threshold key-insulated signature," *Revista Tecnica de la Facultad de Ingenieria Universidad del Zulia*, vol. 39, no. 5, pp. 378–386, 2016.
[28] C. Zhou, Z. Zhao, W. Zhou, and Y. Mei, "Certificateless key-insulated generalized signcryption scheme without bilinear pairings," *Security and Communication Networks*, vol. 2017, Article ID 8405879, 17 pages, 2017.
[29] H. Hong, Y. Xia, and Z. Sun, "Provably secure key insulated attribute based signature without bilinear pairings for wireless communications," *DEStech Transactions on Computer Science and Engineering*, 2016.
[30] H. Kun, D. Xuefeng, and L. Jing, "ID-based key-insulated encryption with message linkages for peer-to-peer network," *Computer Engineering*, vol. 40, no. 4, pp. 124–129, 2014.
[31] Y. Shi, J. Lin, G. Xiong, X. Wang, and H. Fan, "Key-insulated undetachable digital signature scheme and solution for secure mobile agents in electronic commerce," *Mobile Information Systems*, vol. 2016, Article ID 437507, 18 pages, 2016.
[32] V. Kumar and R. Kumar, "An optimal authentication protocol using certificateless ID-based signature in MANET," in *Proceedings of the International Symposium on Security in Computing and Communication*, pp. 110–121, Kochi, India, August 2015.
[33] Y. Park, C. Sur, C. D. Jung, and K. H. Rhee, "Efficient anonymous authentication protocol using key-insulated signature scheme for secure VANET," in *Proceedings of the International Conference on Mobile Lightweight Wireless Systems*, pp. 35–44, Athens, Greece, May 2009.
[34] P. Manzoni, C. T. Calafate, and J. C. Cano, *Encyclopedia of Information Science and Technology*, Information Resources Management Association, New York, NY, USA, 2nd edition, 2009.
[35] M. Bellare and P. Rogaway, "Optimal asymmetric encryption," in *Proceedings of the Workshop on the Theory and Application of Cryptographic Techniques*, pp. 92–111, Perugia, Italy, May 1994.

[36] D. Pointcheval and J. Stern, "Security proofs for signature schemes," in *Proceedings of the International Conference on the Theory and Applications of Cryptographic Techniques*, pp. 387–398, Saragossa, Spain, May 1996.
[37] E. Okamoto, N. Kanayama, A. Kanaoka, F. Rodrguez-Henrquez, and T. Teruya, "TEPLA (University of Tsukuba Elliptic Curve and Pairing Library) manual," January 2013. http://www.cipher.risk.tsukuba.ac.jp/tepla/install_e.html.
[38] OpenSSL Management Committee, "Cryptography and SSL/TLS toolkit," January 2017, https://www.openssl.org/.

12

Online Signature Verification on MOBISIG Finger-Drawn Signature Corpus

Margit Antal, László Zsolt Szabó, and Tünde Tordai

Faculty of Technical and Human Sciences, Sapientia University, Cluj-Napoca, Romania

Correspondence should be addressed to Margit Antal; manyi@ms.sapientia.ro

Academic Editor: Paolo Bellavista

We present MOBISIG, a pseudosignature dataset containing finger-drawn signatures from 83 users captured with a capacitive touchscreen-based mobile device. The database was captured in three sessions resulting in 45 genuine signatures and 20 skilled forgeries for each user. The database was evaluated by two state-of-the-art methods: a function-based system using local features and a feature-based system using global features. Two types of equal error rate computations are performed: one using a global threshold and the other using user-specific thresholds. The lowest equal error rate was 0.01% against random forgeries and 5.81% against skilled forgeries using user-specific thresholds that were computed a posteriori. However, these equal error rates were significantly raised to 1.68% (random forgeries case) and 14.31% (skilled forgeries case) using global thresholds. The same evaluation protocol was performed on the DooDB publicly available dataset. Besides verification performance evaluations conducted on the two finger-drawn datasets, we evaluated the quality of the samples and the users of the two datasets using basic quality measures. The results show that finger-drawn signatures can be used by biometric systems with reasonable accuracy.

1. Introduction

One of the oldest ways of proving your identity is giving your signature. Many official documents require signatures from agreeing parties. Signature recognition can be divided into off-line (static) and online (dynamic) methods. While off-line systems work with images, only the shape of the signature is available, but online systems use information related to the dynamics of the signature. Due to this additional information, online systems outperform off-line systems [1].

Biometric systems can produce two types of errors: false rejections of genuine signatures (false rejection rate (FRR)) and false acceptance of forged signatures (false acceptance rate (FAR)). The overall system error is usually reported in terms of EER (equal error rate), which is defined as the system error rate when FAR and FRR are equal.

In signature database evaluations, two types of forgeries are considered: skilled and random forgeries. Skilled forgery evaluation is based on using the forgery samples available in the database (forgery samples are provided by forgers who know both the shape and dynamics of the imitated signature). Random forgery (or zero effort) evaluation is based on using random genuine samples from the dataset (corresponding to the case when the forger does not know the signature to be forged, therefore is using his/her own signature). The state of the art in automatic signature verification is presented in a study by Impedovo and Pirlo [1].

Online signature recognition is not a new research area: several online signature corpora have already been collected using digitizer tablets. While PHILIPS [2], SVC2004 [3], and SUSIG [4] databases contain only online signatures, MCYT [5] is a bimodal database containing both fingerprints and online signatures. BIOMET [6] and BioSecurID [7] contain several types of biometric data including online signatures.

Due to the increasing number of touchscreen-based mobile devices and the familiarity of users with using signatures, we consider that signatures are plausible candidates for authentication on mobile devices. A number of researches have already been conducted in this topic, although the signature databases are not publicly available [8–10], except for DooDB database [11]. Compared to the DooDB database, which was collected on a device using a resistive touchscreen, our database was collected on a device with a capacitive touchscreen. Specifically, while DooDB contains only the coordinates of the points of the signature and the corresponding time information, our MOBISIG database contains additional information such as pressure, finger area, and data saved from the accelerometer and gyroscope.

In this paper, we analyze whether signatures can be used for authentication in a mobile device context. Therefore, two state-of-the-art methods, a local- or function-based and a global- or feature-based system, were implemented and evaluated on the MOBISIG database. In order to compare our database to the other publicly available online signature database, we performed the same evaluations for the DooDB database using the same parameters and features. In addition, we present a few basic quality evaluations for both databases.

The main contribution of this paper is the presentation and analysis of MOBISIG signature database containing data from 83 users. The signatures are not the original signatures of the users, but users were assigned a family name and were required to create a signature for that name. Signature collection was performed on a Nexus 9 tablet under supervision; data providers were instructed on how to draw signatures using their finger. The database was collected during three sessions and contains 45 genuine and 20 forged signatures for each user. The database is publicly available at http://www.ms.sapientia.ro/~manyi/mobisig.html.

On the MOBISIG database, the best EER for skilled forgeries was obtained by our function-based DTW system: 5.81% for a posteriori user-specific thresholds and 20.82% for common thresholds. In case we added pressure information to the coordinates and their first- and second-order differences, only the a posteriori user-specific threshold result was improved. When using a 9-inch diameter device for data collection, users tend to put down the device on the table while drawing the signatures. Therefore, we did not use the data obtained from the accelerometer and gyroscope sensors in the computations.

Comparisons with the DooDB database indicate higher signature quality in the case of the MOBISIG database, and correspondingly better performances for the verification methods studied in our paper. Our study is limited by the sample size (83 users) and a slightly unbalanced age distribution (77% of the users aged under 25); in addition, our data providers were not experts in forging signatures.

The rest of the paper is organized as follows. A literature review on signature recognition on mobile devices is presented in Section 2, completed with a review of a few papers on signature quality evaluation. Section 3 presents our MOBISIG dataset, followed by a detailed description of the methods used for signature verification. Experiments and benchmark results are presented in Section 5, whereas Section 6 compares DooDB and MOBISIG datasets along verification system performance and quality measures. Section 7 concludes the paper.

2. Related Work

2.1. Signature Recognition on Mobile Devices. Little research has been carried out in the field of online signature recognition on mobile devices. We have only found six studies [8–13] reporting results obtained on signature databases captured in mobile context. The properties of the databases used in these studies are presented in Table 1.

In most of the studies concerning signature recognition, results are reported using signature databases captured on a pen tablet. However, touchscreens present some drawbacks compared to pen tablets, the most important being the quality of the captured signal. While pen tablets sample the signal uniformly with relatively high frequency, hand-held device sampling is usually event-driven with lower sampling frequency than pen tablets. Moreover, while both touchscreen devices and pen tablets are able to capture trajectory and pressure, the latter can track pen orientation. The only advantage of a touchscreen device is that it allows capturing the signature by fingertip.

One of the objectives of BioSecure Signature Evaluation Campaign (BSEC'2009) was to study the influence of acquisition conditions (digitizing tablet or PDA) on authentication systems' performance [14]. Results are reported using signatures from 382 writers, acquired on a digitizing tablet and on a PDA, respectively. The authors reported a significant quality degradation of signatures acquired in mobile conditions.

The semester thesis of Bissig [12] is the first study reporting results using a signature database captured on a resistive touchscreen with fingertip. Four types of signals were acquired: coordinates $x(t), y(t)$, pressure $p(t)$, and finger area $a(t)$. Both local (function-based: DTW) and global systems (feature-based: one-class SVM and Mahalanobis distance) and the combination of these were evaluated. Unfortunately, neither the number of subjects nor the number of forgeries in the captured database is reported. However, this is the only study reporting the influence of pressure on the performance of a signature verification system.

Houmani et al. [8] report results on a new dataset collected from 64 subjects on a PDA. Unfortunately, neither the number of sessions nor the acquired signals are reported. However, they propose an entropy-based quality metric for selecting reference signatures in the enrollment phase.

Krish et al. [9] collected a new signature database using a Samsung Galaxy Note device from 25 users (20 genuine signatures from a user). Their verification algorithm combines two state-of-the-art algorithms (function-based DTW and feature-based Mahalanobis distance). Due to the missing forgery samples, only the results obtained by random forgery evaluations are reported.

Sae-Bae and Memon [10] collected an online signature dataset from 180 users using HTML5. Users were allowed to enter their signatures using their own iOS devices. The dataset is not publicly available and contains only genuine signatures; therefore, only the random forgery-type evaluation was feasible. They proposed a new histogram-based feature set and reported performance evaluations both on their and MCYT datasets.

The first publicly available database collected on a hand-held device (HTC Touch HD mobile phone) is the DooDB. This database contains data from 100 users, and it also contains doodles besides pseudosignatures. Martinez-Diaz et al. [11] report the database analysis and benchmark results,

Table 1: Mobile signature databases.

Paper	Users	Device	Input Method	#GEN	#FOR	#SESS	Signals	Public
Bissig [12]	NA	HTC Desire 3.7', capacitive	Finger	20	NA	NA	$x(t)$, $y(t)$, $p(t)$, $fa(t)$	No
Houmani et al. [14]	432	PDA HP iPAQ hx2790	Pen	30	20	2	$x(t)$, $y(t)$	Yes
Houmani et al. [8]	64	PDA Qtek 2020 ARM	Pen	30	20	2	$x(t)$, $y(t)$	No
Krish et al. [9]	25	Samsung Galaxy Note	Pen	20	0	2	$x(t)$, $y(t)$, $p(t)$	No
Martinez-Diaz et al. [11]	100	HTC Touch HD mobile, resistive	Finger	30	20	2	$x(t)$, $y(t)$	Yes
Sae-Bae and Memon [10]	180	User-owned iOS devices	Finger	30	0	6	$x(t)$, $y(t)$	No
This paper	83	Nexus 9, capacitive	Finger	45	20	3	$x(t)$, $y(t)$, $p(t)$, $fa(t)$ $vx(t)$, $vy(t)$, $ax(t)$, $ay(t)$, $az(t)$ $gx(t)$, $gy(t)$, $gz(t)$	Yes

#GEN, number of genuine samples; #FOR, number of forgeries; #SESS, number of sessions; NA, unspecified.

using a function-based DTW verification system with several local features. Although the EERs obtained by the random forgery evaluations are low (around 3%), those obtained for skilled forgery evaluations are too high (around 27%). In a later study [13], the skilled forgery result was improved (20.9%) by using the Gaussian mixture method.

2.2. Signature Quality. Quality evaluation of biometric datasets is a difficult problem. A biometric dataset consists of biometric samples from a number of users, usually containing a fixed number of samples from each user collected in a fixed number of sessions. Moreover, signature datasets contain skilled forgery samples for each user.

There are two ways to achieve the quality evaluation of a biometric dataset: (i) evaluating each sample of the dataset and (ii) evaluating each user of the dataset. In each case, we obtain a set of scores, and then an average score can be computed from these scores. Both the samples and the users can be evaluated by using only the genuine signatures or by using both the genuine and forgery signatures. Müller and Henniger [15] proposed two quality metrics for signature dataset evaluation. One of the quality metrics evaluates the samples, while the other one evaluates the users of the dataset. Both metrics use the DTW distance between samples.

Houmani et al. [16] proposed a personal entropy measure for online signatures and showed the existence of a clear relationship between the proposed measure and the verification performance of the user revealed by the signature verification system. This measure allowed them to categorize the users of several signature datasets. In a later study [17], they adapted the measure to the skilled forgery samples of signature datasets. Similar to their previous study, they proved the effectiveness of the quality measure by evaluating several online signature databases by using state-of-the-art signature verification methods.

One of the objectives of the BSEC'2009 competition was the evaluation of online signature algorithms with respect to the quality of the signatures [14]. The personal entropy measure introduced by Houmani et al. was used to group the signatures into different categories. The results of the competition showed that the performance of classifiers varied significantly with respect to the good and bad quality signatures. Houmani and Garcia-Salicetti [18] extended the Biometric Menagerie to online signatures and categorized the users of MCYT database using the Personal Entropy quality measure.

Kahindo et al. [19] proposed a novel signature complexity measure to select reference signatures for online signature verification systems. Guest and Henniger [20] used commercial engines for the assessment of the quality of handwritten signatures. They concluded that predicting the utility of a signature sample using a multifeature vector was possible. More recently, another novel method was proposed for the quality evaluation of off-line signatures [21].

3. The MOBISIG Database

Due to security reasons (people are reluctant to give their own signatures), participants were asked to create a signature for a given family name. Family names were selected from the first 100 most frequent Hungarian family names. Participants were also asked to practice the created signatures by drawing and deleting several attempts. The first five attempts were deleted.

The database contains signatures from 83 subjects: 49 men and 34 women, with the following age distribution: 64 subjects under 25, 12 between 25 and 40, and 7 over 40.

3.1. Data Acquisition Protocol. Data collection was performed using a Nexus 9 tablet. The device has a capacitive touchscreen of $228.2 \times 153.7 \times 8$ mm ($8.98 \times 6.05 \times 0.31$ in.). Signatures were sampled at about 60 Hz (event-driven sampling) when the users pressed the screen. The resolution of the screen is 1536×2048 pixels (approx. 281 ppi pixel density). Each signature was stored as a sequence of discrete values $[x_t, y_t, t, p_t, fa_t, vx_t, vy_t, ax_t, ay_t, az_t, gx_t, gy_t, gz_t]$, where x_t, y_t are the coordinate values, t is the time stamp, p_t, fa_t are the pressure and finger area (these are normalized values [0, 1] and can be obtained through the standard Android API), vx_t, vy_t are the directional velocities, ax_t, ay_t, az_t are the directional acceleration of the device, and gx_t, gy_t, gz_t are the values obtained from the gyroscope

Figure 1: Data collection screen.

sensor. The accelerations and the values obtained from the gyroscope characterize the holding position of the device.

The screen of the device was divided into two sections (Figure 1): the upper section was the replay section, where users were shown the animated signature, and the lower section was designed to draw the signature. The animation functionality was available in both types of signature collections: genuine and forgery. The animation allows participants to recall the shape and the dynamics of genuine and forged signatures. The animation could be replayed any number of times. Before data collection, users were asked to become familiar with the device usage as well as their pseudosignatures. Additionally, signatures were saved after user's acceptance. Any signature could be deleted by the provider if he/she was not satisfied with the result.

The data collection process was divided into three sessions with one week between consecutive sessions. In the first session, each user had to provide 15 genuine pseudosignatures for the assigned name. In the second and third sessions, participants had to provide 15 genuine pseudosignatures and 10 forgeries for two assigned users (two times 5 forgeries). At the end of the data collection process, we had 45 genuine signatures and 20 forgeries for each participant. A few of these signatures are shown in Figure 2.

3.2. Signature Files. Each user has a dedicated folder which contains the 45 genuine signatures of the user and the 20 forgeries made by other users. The naming convention of the files is as follows: SIGN_[T] _USER[SID] _USER[WID] _[NR] , where T is FOR for forgeries and GEN for genuine signatures. SID identifies the user whose signatures are in the folder. WID is the identifier of the user, who gave the signature; SID and WID are equal in the case of genuine signatures, while they are different in the case of forgeries. The NR at the end of the filename is a value from 1 to 45 for genuine signatures and 1 to 20 for forgeries. The first 15 genuine signatures were collected in the first session, the second 15 signatures in the second session, and the third 15 signatures were collected in the third session. The first 10 forgeries were collected in the second session, while the second 10 forgeries in the third session. The naming conventions of the folders is USER[SID]. Each signature is represented as a sequence of points and is stored in a file. Each line of the file represents one point of the signature and consists of the following features: x-coordinate, y-coordinate, time stamp, pressure, finger area, x-velocity, y-velocity, x-acceleration, y-acceleration, z-acceleration, x-gyroscope, y-gyroscope, and z-gyroscope.

4. Methods

In order to assess the authentication performance based on pseudosignatures, both a function-based and a feature-based verification system were implemented. Features used by signature verification systems can be local and global ones. Local features correspond to sample points along the signature's trajectory (e.g.,point-wise pressure). Global features are computed from the signature as a whole (e.g.,duration). Function-based systems use local features and feature-based systems use global features.

4.1. Function-Based Verification. The first system is based on DTW (dynamic time warping), and it compares the captured time sequences $s[1\ldots n]$ and $t[1\ldots m]$ (Algorithm 1). Each signature $s[1\ldots n]$ is represented as a time sequence with $s[j] = (f_1^j, f_2^j, \ldots, f_{N_f}^j)$, $j = 1, \ldots, n$, where N_f is the number of local features. In this work, the following local features were employed: the x, y coordinates, the x^1, y^1 first-order differences, the x^2, y^2 second-order differences, and the p, p^1 pressure and its first-order difference. Before computations, the features of time sequences were standardized $f_i' = (f_i - \mu_{f_i})/\sigma_{f_i}, i = 1, \ldots, N_f$, where μ_{f_i} and σ_{f_i} are the mean and standard deviation for the ith local feature computed over all sampling points of the signature). The Euclidean distance function was used to compute the distance between two elements of the time sequences (distance$(s[i], t[j])$). Finally, the obtained DTW distance was divided by the sum of the time sequence lengths $(n + m)$.

The verification process works as follows: in the enrollment stage, a set of N reference signatures are selected $\{e_1, e_2, \ldots, e_N\}$. In the verification stage, the DTW distances between the test signature and all the reference signatures are computed, and the final score results as the average of these distances. Finally, this distance-based score (Dscore)

$$\text{Dscore} = \frac{1}{N}\sum_{i=1}^{N} d_{\text{DTW}}(\text{test}, e_i). \quad (1)$$

is transformed into a similarity score using (2):

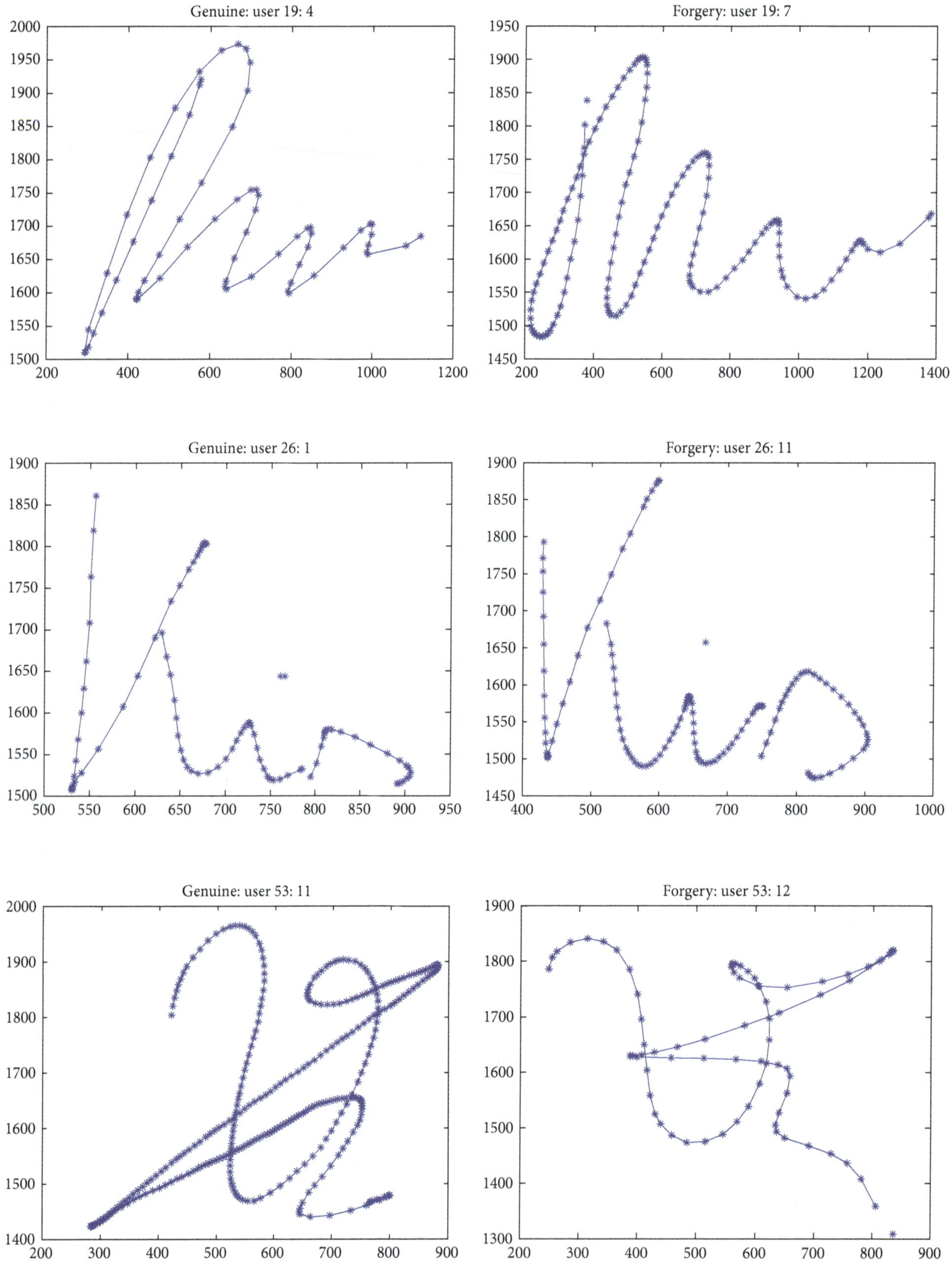

(a)

FIGURE 2: Continued.

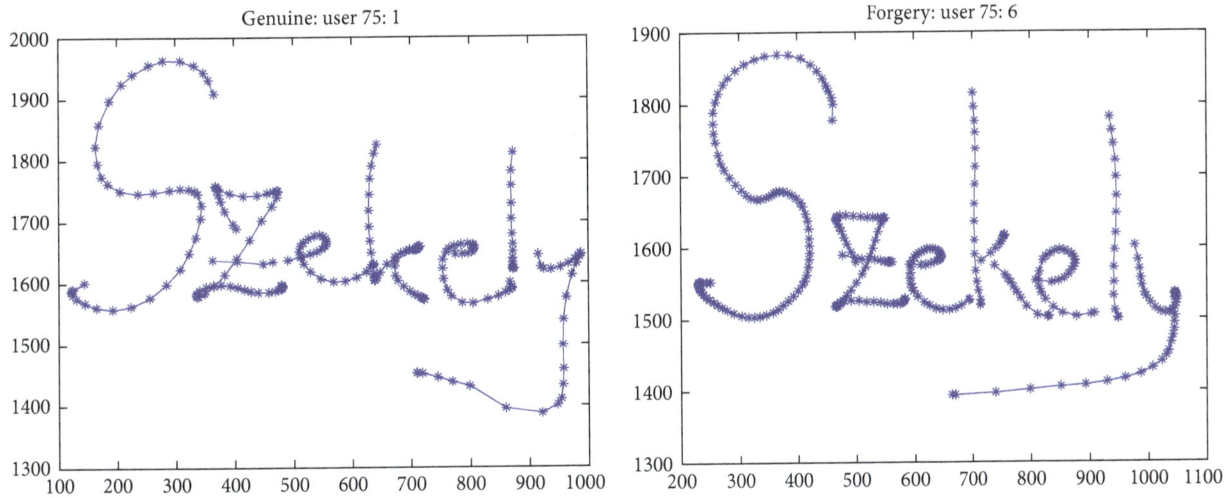

(b)

FIGURE 2: Pseudosignatures from the database. The first column contains genuine signatures and the second column contains forgeries.

$$\text{score} = \frac{1}{(1 + \text{Dscore})}. \tag{2}$$

The architecture of our function-based system is shown in Figure 3.

4.2. Feature-Based Verification. The second type of verification system is a feature-based or a global system, which computes a fixed size feature vector from each signature. Each feature is a global feature related to the signature as a whole. The components of our feature-based system are the following: feature extractor, user template creation, and matcher. No preprocessing was applied. However, before computations, features were scaled (separately for each user). The architecture of our feature-based system is shown in Figure 4.

Our feature extractor computed position- and time-based features [12, 22], such as duration and different types of velocity, as well as different types of sign change, a few sensor-related features (touchscreen pressure and finger area), and 2 types of histogram-based features. In the computation of histogram-based features, we followed the work of Sae-Bae and Memon [10] except that we used fewer features.

Let $X = \{x_1, x_2, \ldots, x_n\}$ and $Y = \{y_1, y_2, \ldots, y_n\}$ be the x, y coordinates of a signature and $P = \{p_1, p_2, \ldots, p_n\}$ the pressure attribute. We compute the first- and second-order differences of these sequences as follows: $X^1 = \{x_i^1 \mid x_i^1 = x_{i+1} - x_i\}$, $Y^1 = \{y_i^1 \mid y_i^1 = y_{i+1} - y_i\}$, and $P^1 = \{p_i^1 \mid p_i^1 = p_i\}$, where $i = 1, 2, \ldots, n-1$, and $X^2 = \{x_i^2 \mid x_i^2 = x_{i+1}^1 - x_i^1\}$, $Y^2 = \{y_i^2 \mid y_i^2 = y_{i+1}^1 - y_i^1\}$, and $P^2 = \{p_i^2 \mid p_i^2 = p_{i+1}^1 - p_i^1\}$, $i = 1, 2, \ldots, n-2$.

Angles $\Theta_i^1 = \tan 2^{-1}(y_i^1, x_i^1), i = \overline{1, n-1}$, were computed in order to derive a uniform-width histogram from this sequence. The tan 2^{-1} trigonometric function is a common variation of the standard arctan function, which produces results in the range $(-\pi, \pi]$. Angles characterize the shape of

```
procedure DTWDist s[1···n], t[1···m]
  DTW[0···n][0···m]
  for i←1,n do
    DTW[i][0]←infinity
  end for
  for i←1,m do
    DTW[0][i]←infinity
  end for
  DTW[0][0]←0
  for i←1,n do
    for i←1,m do
      cost←distance(s[i],t[j])
      DTW[i][j]←cost + min(DTW[i − 1][j]),
      DTW[i][j − 1],DTW([i − 1][j − 1] )
    end for
  end for
  result←DTW[n][m]/(n + m)
end procedure
```

ALGORITHM 1: DTW algorithm.

the signature. This interval of angles was divided into 8 equal bins in order to compute the histogram.

The $r_i^1 = \sqrt{(x_i^1)^2 + (y_i^1)^2}, i = \overline{1, n-1}$, sequence captures the speed distribution, and it is considered very useful in combating skilled forgeries [10]. In the histogram computation, we considered only the values from the interval $[0, \mu + 3\sigma]$, where μ represents the mean and σ the standard deviation of the r sequence obtained from a single signature. This interval was divided into 16 equal bins for the histogram computation. The full list of the features is presented in Table 2.

In feature-based methods, each signature is representedby a D-dimensional feature vector. After selecting the signatures used for template creation, features were scaled. We applied the normalization for each feature ($f_i^{'} =$

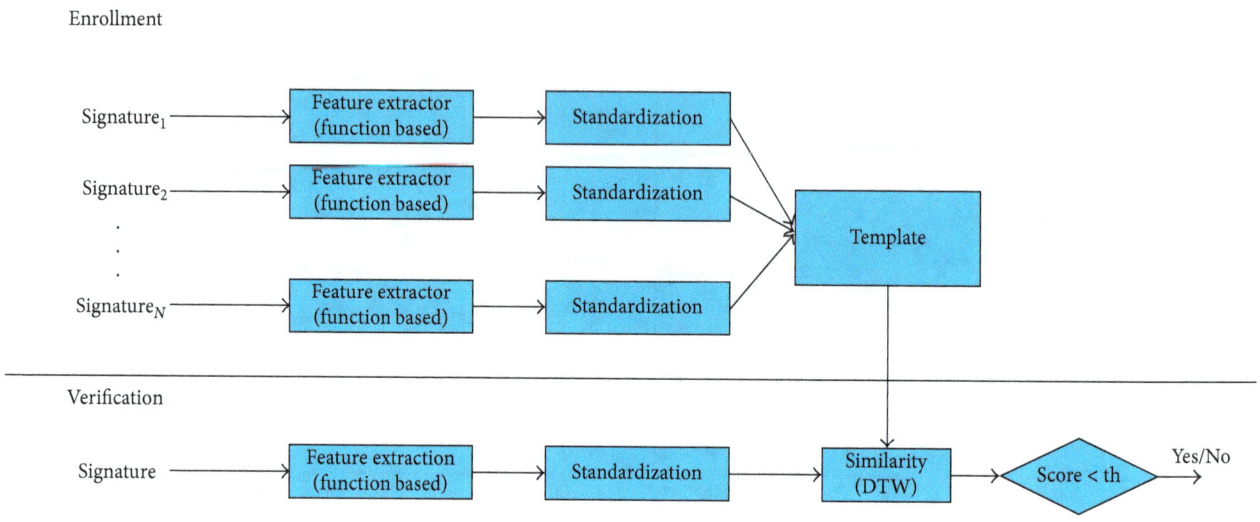

FIGURE 3: System architecture of function-based method.

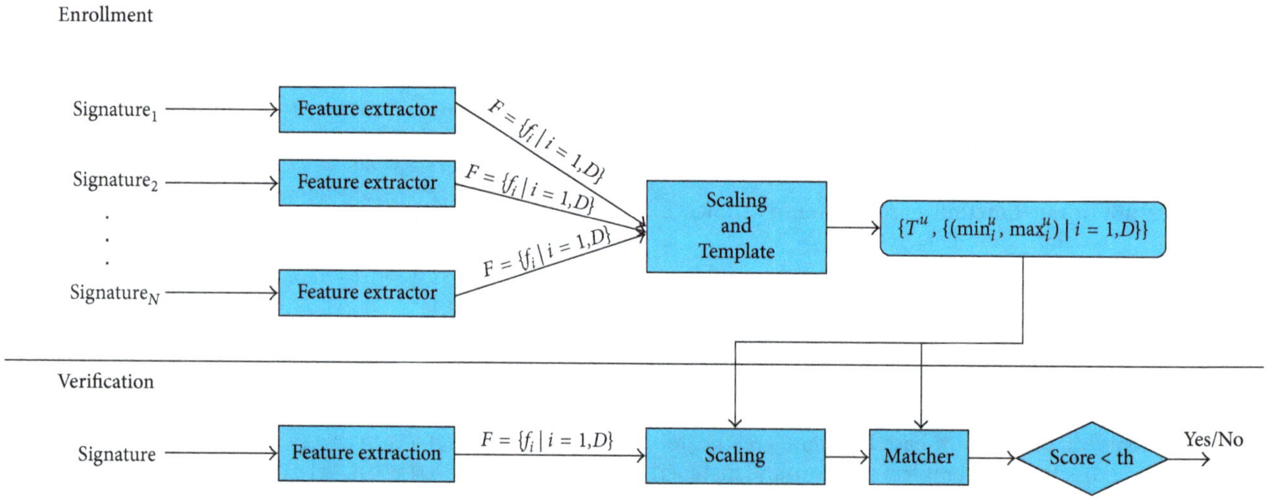

FIGURE 4: System architecture of feature-based method.

$(f_i - \min(f_i))/(\max(f_i) - \min(f_i)),\ i = 1, \ldots, D)$ before each template creation. In order to be able to apply the same scaling to the test signature before matching, the scaling parameters (max and min values for each feature) were stored as part of the user's template. User-based scaling of the data meets real biometric verification systems, because the system does not take into account the data of other users. The performance of the system is increased significantly by normalizing all users' data in a single step. This type of normalization was already evaluated for the DooDB database [23].

In data mining, anomaly detection (also outlier detection) is the identification of observations which do not conform to an expected pattern in a dataset [24] and therefore can be used for impostor detection. Killourhy and Maxion [25] evaluated 14 anomaly detectors for keystroke dynamic biometrics.

Four anomaly detectors were implemented for template creation and matching. Each anomaly detector uses a specific template and score computation (see below). Dissimilarity scores were transformed into similarity scores by using formula (2) as for the DTW algorithm.

In the training phase, user models (templates) are constructed using a fixed number of training samples. Testing or verification refers to an anomaly score computation for the test sample, which in our case is a distance from the user model.

The classic Euclidean anomaly-detection algorithm in the training phase calculates the mean vector of the training samples, while in the test phase it calculates the Euclidean distance between the mean vector and the test sample.

The Manhattan detector is similar to the Euclidean detector except that the Manhattan distance is used in the testing phase. Another variant of the Manhattan detector is the Manhattan scaled detector, described in Araujo et al. [26]. In the training phase, besides the mean vector of the training samples $(m_i, i = \overline{1, D})$, the mean absolute deviation of each feature is also calculated $(a_i, i = \overline{1, D})$. The anomaly

TABLE 2: Global features.

Description	#nf
Total signature duration	1
Number of finger strokes	1
Amount of time the finger touched the display	1
The number of sign changes of $x^1, y^1, x^2, y^2, p^1, p^2$	6
Mean of $x^1, y^1, x^2, y^2, p^1, fa$	6
Standard deviation of x^1, x^2, y^1, y^2	4
Maximum value of p^1	1
Mean velocity	1
Mean acceleration	1
Mean velocity in x direction	1
Mean velocity in y direction	1
Normalized point in time when the maximal value of $x, x^1, x^2, y, y^1, y^2, p^1$ is reached	7
Normalized point in time when the minimal value of x, x^1, x^2, y, y^1, y^2 is reached	6
Starting direction	1
Histogram of Θ sequence	8
Histogram of r sequence	16
Total	**62**

#nf, number of features.

score is calculated using the formula: $\text{score}(x, \text{model}) = \sum_{i=1}^{D} |x_i - m_i|/a_i$, where x is a test sample ($x_i, i = \overline{1, D}$), and the model comprises the mean vector of the features and the mean absolute deviation of the features: $\text{model} = (m_i, a_i), i = \overline{1, D}$.

The k-nearest neighbour (kNN) detector works as follows: in the training phase, the detector saves the training samples (reference-based system). In the test phase, the detector calculates the Euclidean distance between each of the training samples and the test sample. The anomaly score is calculated as the mean of the distances to the k-nearest training samples.

One-class SVM is an outlier detector based on the support vector method. We used the LibSVM implementation [27] provided by e1071 package in R, with the following parameters: type = one-classification (for novelty detection), kernel = "radial," and gamma = 0.04. The SVM parameter nu was set to 0.4, and parameters were not tuned for individual users.

5. Benchmark Results

5.1. Evaluation Protocol. The same evaluation protocol was used for both verification systems, which consists of three measurements. The systems were trained with the first 5, 10, and 15 samples from the first session, respectively. The 15 genuine signatures from the second session were used for positive score computation. All of the 20 available forgeries per user are used for skilled forgery score computations. Random forgery scores were computed for each user by using the first genuine signature from all the other users (Table 3).

TABLE 3: Number of testing samples used in evaluations.

Ev. type	Positive samples	Negative samples
SF	15 (session 2)	20
RF	15 (session 2)	1/each other user

SF, skilled forgeries; RF, random forgeries.

Two types of EERs were computed. The global EER was computed based on a global threshold, which was computed using common genuine and forgery score lists for all users of the database. The second type of EER, the a posteriori user-specific EER, is reported in some recent papers [13, 28]. This type of EER is computed by averaging the user-specific EERs. User-specific EERs are computed independently for each subject of the dataset and therefore are based on user-specific decision thresholds. Martinez-Diaz et al. [13] used the notation of aEER for this type of EER. As we compare our results with their results, it is important to use the same notation. However, it is important to note that this type of EER is based on a posteriori user-specific thresholds [29]. Hence, the corresponding EER (aEER) represents the best global EER that would be obtained if optimal score normalization techniques were known a priori [13].

5.2. Results

5.2.1. Training Set Size. The effect of the number of training samples during enrollment was investigated. Three cases were evaluated for each type of local and global features: using 5, 10, and 15 samples during enrollment.

Table 4 shows the function-based system results for the MOBISIG database. The five types of local feature sets were as follows: (i) xy—the coordinate sequence; (ii) x^1y^1—the first-order differences of the coordinate sequence; (iii) x^2y^2—the second-order differences of the coordinate sequence; (iv)$xyx^1y^1x^2y^2$—coordinate sequence with first- and second-order differences; (v) $xyx^1y^1x^2y^2pp^1$—coordinate sequence with first- and second-order differences and pressure with first-order differences. Both types of evaluations, skilled and random forgeries, are reported.

In the case of skilled forgeries, the best global EER was obtained by using the xy time sequences only (20.82%). However, the best aEER was 5.81% using type (v) local features.

In the case of random forgeries evaluation, the best result was obtained by using the first differences of x and y time sequences (EER: 1.41% and aEER: 0.01%). The very low EERs obtained show that this type of function-based verification system is highly reliable against random forgeries.

As expected, the more samples used in the enrollment phase, the better the verification system performance was. This is true for each type of local feature and for both types of evaluations: skilled and random forgeries. However, the improvements are small, especially between using 10 and 15 samples for enrollment.

The results obtained for the feature-based system are reported in the following. In order to show the contribution of different categories of global features, we formed three

Table 4: Verification performance in terms of EER and averaged individual EER (aEER) for the MOBISIG database using DTW and local features (%).

Features	Skilled forgeries		Random forgeries	
	EER	aEER (std dev.)	EER	aEER (std dev.)
Enrollment: 5 samples				
(i)	25.45	14.72 (14.62)	2.12	0.32 (1.29)
(ii)	31.37	11.42 (14.64)	1.75	0.00 (0.00)
(iii)	42.92	24.72 (21.46)	22.21	3.34 (9.27)
(iv)	32.02	10.28 (14.72)	3.18	0.10 (0.71)
(v)	31.52	8.56 (11.50)	4.51	0.18 (1.00)
Enrollment: 10 samples				
(i)	22.71	12.10 (13.98)	1.91	0.34 (1.31)
(ii)	28.87	8.60 (11.99)	1.41	0.01 (0.07)
(iii)	42.66	20.43 (18.34)	20.61	1.81 (3.52)
(iv)	30.35	6.80 (12.17)	2.55	0.01 (0.09)
(v)	29.97	6.69 (10.05)	4.09	0.16 (0.99)
Enrollment: 15 samples				
(i)	20.82	10.77 (13.29)	1.76	0.27 (1.18)
(ii)	27.15	7.34 (11.53)	1.68	0.01 (0.07)
(iii)	42.13	18.74 (17.49)	20.14	1.70 (3.60)
(iv)	29.79	6.27 (11.54)	2.62	0.01 (0.09)
(v)	29.76	5.81 (9.64)	4.27	0.15 (0.98)

Table 5: Verification performance in terms of EER and averaged individual EER (aEER) for the MOBISIG database using anomaly detectors and 62 features (%).

Detector	Skilled forgeries		Random forgeries	
	EER	aEER (std dev.)	EER	aEER (std dev.)
Enrollment: 5 samples				
Eucl.	22.99	18.95 (15.81)	12.24	7.43 (9.13)
Manh.	19.27	13.57 (12.43)	7.28	3.60 (5.21)
kNN	21.68	17.19 (15.22)	11.85	6.96 (8.46)
SVM	29.81	18.34 (15.66)	25.78	7.61 (9.52)
Enrollment: 10 samples				
Eucl.	20.12	13.78 (12.84)	9.18	5.80 (8.99)
Manh.	16.58	10.82 (11.73)	5.37	2.60 (4.72)
kNN	18.40	13.35 (12.62)	8.63	5.80 (8.90)
SVM	19.93	14.03 (12.98)	10.59	5.80 (8.90)
Enrollment: 15 samples				
Eucl.	17.11	12.14 (12.60)	8.69	5.01 (8.47)
Manh.	14.31	9.35 (11.28)	5.21	2.10 (4.01)
kNN	17.25	12.15 (12.21)	9.14	5.00 (8.87)
SVM	17.35	12.45 (12.84)	9.00	5.00 (8.57)

feature sets from the features shown in Table 2: feat62 : all features; feat56 : feat62\{pressure and finger area features}; and feat32 : feat56\{histogram-based features}.

Table 5 shows the feature-based system results for the MOBISIG database using all features (feat62). As for the function-based system, we report results for using 5, 10, and 15 samples for enrollment. Similar to the function-based system, using more samples for enrollment improves the performance of the feature-based verification system. The performance gaps are higher for cases between 5 and 10 samples than for cases between 10 and 15 samples.

Comparing Table 5 with Table 4, it can be observed that the DTW system achieved far better performances in the case of random forgeries than the anomaly detectors. In the case of skilled forgeries, both the best global EER (14.31%) and the best aEER (9.35%) were obtained by the Manhattan detector. Interestingly, the differences between global EER and aEER are not so large as in the case of DTW system. The ROC curves for the global EERs are shown in Figure 5.

5.2.2. Global Feature Sets. The results obtained for the three global feature sets are shown in Table 6. The best (lowest) error rates were obtained by using the Manhattan detector both for skilled and random forgery evaluations. In the case of this detector, using less features resulted in higher error rates. However, not all detectors were affected negatively by using less features. For example, the SVM detector performance increased by reducing the dimension of the feature set, especially when the pressure-related features were excluded (feat56).

5.2.3. Intersession Variability. We evaluated the intersession variability for the MOBISIG dataset. Two cases were evaluated: (1) using session 1 for training and session 2 for testing and (2) using session 2 for training and session 3 for testing. The results for using 15 samples for enrollment are shown in Table 7. The performance differences between the two cases are between 0.16% and 3.30% and are highly dependent on the used method. For example, verification results for system-wide EER are improved for the DTW (type (ii) features) method in case of session pair 2-3 (compared to session pair 1-2), but this tendency is not similar for aEER value in the case of the Manhattan detector.

According to the results from Table 7, no general tendency of improvement or deterioration of verification results could be stated. The results might suggest similar signature quality in session pairs 1-2 and 2-3 considering verification performance, but further investigations are necessary.

6. DooDB and MOBISIG comparison

In this section, comparative results are presented for our MOBISIG database and the other publicly available DooDB database. The results are presented along (i) their statistical properties, (ii) the verification system performances, (iii) score distributions, and (iv) some signature quality metrics.

6.1. Statistical Properties. Table 8 presents the most important characteristics regarding the data collection process of the two publicly available mobile device context signature databases.

6.2. Verification Performance. In the case of function-based DTW system, only the coordinates and their first- and second-

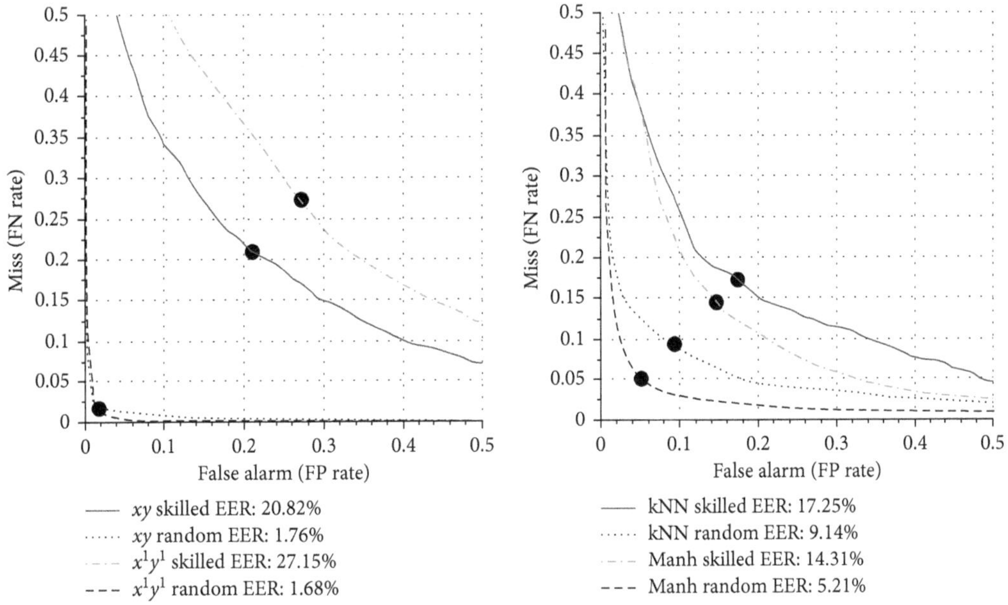

Figure 5: MOBISIG. ROC curves for local and global methods. 15 sample training cases. (a) DTW and (b) anomaly detectors.

Table 6: MOBISIG: EER and aEER (different feature sets).

Detector	Skilled forgeries		Random forgeries	
	EER	aEER (std dev.)	EER	aEER (std dev.)
		feat62		
Eucl.	17.11	12.14 (12.60)	8.69	5.01 (8.47)
Manh.	14.31	9.35 (11.28)	5.21	2.10 (4.01)
kNN	17.25	12.15 (12.21)	9.14	5.00 (8.87)
SVM	17.35	12.45 (12.84)	9.00	5.00 (8.57)
		feat56		
Eucl.	16.00	13.05 (13.55)	7.44	5.17 (8.64)
Manh.	15.45	10.97 (12.75)	5.26	2.32 (4.26)
kNN	16.91	13.48 (13.10)	7.46	5.56 (9.08)
SVM	15.73	13.22 (13.29)	7.39	5.35 (9.07)
		feat32		
Eucl.	17.58	14.87 (14.81)	8.80	5.89 (9.08)
Manh.	17.17	13.32 (13.62)	7.44	3.83 (6.56)
kNN	18.07	14.44 (13.62)	8.83	6.17 (9.09)
SVM	17.76	14.78 (14.86)	8.76	5.83 (9.15)

Note. Training: 15 samples.

Table 7: MOBISIG: EER and aEER (intersession variability).

Method	Skilled forgeries		Random forgeries	
	EER	aEER (std dev.)	EER	aEER (std dev.)
	Training: session 1; testing: session 2			
DTW	27.15	7.34 (11.53)	1.68	0.01 (0.07)
Manh.	14.31	9.35 (11.28)	5.21	2.10 (4.01)
SVM	17.35	12.45 (12.84)	9.00	5.00 (8.57)
	Training: session 2; testing: session 3			
DTW	24.69	9.05 (15.85)	1.78	0.33 (1.98)
Manh.	15.49	10.94 (12.27)	5.05	2.48 (4.33)
SVM	18.24	13.74 (13.66)	8.59	4.84 (6.52)

Note. Training samples: 15. We used type (ii) local features for DTW and 62 features (feat62) for the Manhattan anomaly detector and SVM.

order differences were used (no pressure information is available in the DooDB database). The pressure-based features were omitted for the feature-based system, consequently measurements were performed on feat56 and feat32 feature sets. The DooDB database contains $(0, 0)$ coordinate values in the case of sampling errors. In order to correct this error, coordinate values from the previous sample were assigned to the sample.

The same evaluation protocol was followed as presented in Section 5.1. We present the results obtained by the two verification methods on the DooDB database (Tables 9 and 10), followed by the comparison of the results obtained on the two databases (Table 11).

Comparing the skilled forgery results by the DTW method (Tables 4 and 9), we observe that using more samples for reference signatures decreases both global and user average EERs. Among the xy, x^1y^1, and x^2y^2 features, the first-order differences (x^1y^1) provide the lowest aEER for both databases (7.34% for MOBISIG and 16.67% for DooDB). However, combining all features $(xyx^1y^1x^2y^2)$, the aEERs become slightly better (6.27% for MOBISIG and 15.78% for DooDB). Kholmatov and Yanikoglu [30] also reported that the first-order differences had given the lowest error rates. The worst results were obtained by using thesecond-order differences (x^2y^2) alone, which means that these features do not contain too much user-specific information.

Table 8: DooDB versus MOBISIG data collection.

	DooDB	MOBISIG
Device	HTC Touch DD mobile phone	Nexus 9 tablet
Touchscreen	Resistive	Capacitive
Frequency	100 Hz (periodic sampling)	Approx. 60 Hz (event-based sampling)
Coordinates	x: 0–2000, y: 0–3500	x: 0–1536, y: 0–2048
Raw data	$[x_t, y_t, \Delta_t]$	$[x_t, y_t, t, p_t, fa_t, vx_t, vy_t, ax_t, ay_t, az_t, gx_t, gy_t, gz_t]$
Sessions	2 (two weeks between)	3 (one week between)
Subjects	100 (44 women and 56 men)	83 (34 women and 49 men)
Samples	Session 1: 15 genuine + 10 forgeries Session 2: 15 genuine signatures + 10 forgeries	Session 1: 15 genuine signatures Session 2: 15 genuine signatures + 10 forgeries Session 3: 15 genuine signatures + 10 forgeries

Table 9: Verification performance in terms of EER and averaged individual EER (aEER) for the DooDB database using DTW (%).

Features	Skilled forgeries		Random forgeries	
	EER	aEER (std dev.)	EER	aEER (std dev.)
Enrollment: 5 samples				
(i)	30.57	22.33 (19.60)	4.31	1.71 (8.27)
(ii)	31.93	19.63 (19.36)	4.96	1.37 (3.54)
(iii)	41.73	24.35 (20.19)	20.41	5.86 (9.28)
(iv)	32.15	19.68 (18.82)	4.44	1.14 (3.82)
Enrollment: 10 samples				
(i)	29.48	19.59 (18.73)	3.63	1.51 (7.69)
(ii)	29.69	17.51 (18.35)	4.31	1.14 (2.55)
(iii)	42.25	21.49 (19.32)	19.63	5.05 (7.92)
(iv)	31.09	17.15 (17.74)	3.60	1.06 (3.78)
Enrollment: 15 samples				
(i)	27.41	17.66 (17.57)	2.80	1.40 (7.66)
(ii)	29.27	16.67 (18.32)	3.71	1.12 (2.54)
(iii)	41.51	20.27 (19.02)	18.75	4.74 (7.69)
(iv)	30.12	15.78 (17.54)	2.94	1.15 (4.19)

Table 10: Verification performance in terms of EER and averaged individual EER (aEER) for the DooDB database using anomaly detectors and 56 features (%).

Detector	Skilled forgeries		Random forgeries	
	EER	aEER (std dev.)	EER	aEER (std dev.)
Enrollment: 5 samples				
Eucl.	34.69	30.98 (17.91)	17.98	9.97 (11.95)
Manh.	31.24	25.29 (18.16)	13.38	5.11 (7.29)
kNN	35.33	31.42 (17.76)	18.58	10.29 (12.26)
SVM	35.6	31.73 (17.86)	27.74	10.84 (11.93)
Enrollment: 10 samples				
Eucl.	27.632	24.63 (17.59)	12.20	7.61 (10.81)
Manh.	24.45	20.22 (16.49)	9.80	4.56 (7.08)
kNN	27.44	24.47 (17.74)	11.82	7.55 (10.33)
SVM	27.14	24.66 (17.38)	12.60	7.55 (10.79)
Enrollment: 15 samples				
Eucl.	23.48	20.98 (16.05)	10.32	6.76 (10.28)
Manh.	21.71	18.75 (16.66)	7.82	4.19 (7.24)
kNN	24.25	20.83 (16.19)	10.10	7.07 (10.43)
SVM	23.26	20.85 (16.07)	10.40	6.97 (10.39)

The effect of using a reduced feature set is shown in Table 12. It can be seen that not using the histogram-based features resulted in a very small performance degradation (about 1%).

The ROC curves for some verification method are shown in Figure 6.

6.3. Score Distributions. The score distributions of genuine and impostor samples (skilled forgeries case, 15 training samples) for both verification methods are shown in Figure 7.

6.4. Signature Quality. Alonso-Fernandez et al. [31] reviewed the state of the art in the biometric quality problem. They consider that a biometric sample is of good quality if it is suitable for personal recognition. According to the ISO/IEC 29794-1 standard, biometric quality has three dimensions: fidelity, utility, and character. Fidelity is the degree of similarity between the sample and its source. Utility is related to the impact of the sample on the overall performance of a biometric system, while character indicates the inherent discriminative capability of the source.

Influence of the character of a biometric sample on its utility was investigated by Müller and Henniger [15]. In case of signatures, recognition systems will give different results depending on the reference samples (the sample set used for template creation). Selection of samples used in the template can be controlled by sample quality assessment algorithms. Evaluating the quality of a single sample based only on sample features is difficult, but methods exist to evaluate considering other samples. Some a posteriori methods use only genuine samples, others use skilled forgery samples also. Using a set of genuine samples for quality assessment still restricts selection of reference samples to the genuine user and assures usage of the method in real applications. In consequence, we used two metrics which examine the character of the sample proposed by Müller and Henniger [15].

The first method, the sampleEER, consists of computing the EER for each genuine sample. This is obtained by

Table 11: EER and aEER comparison for DooDB and MOBISIG databases.

Method	Skilled forgeries		Random forgeries	
	EER	aEER	EER	aEER
DooDB				
Manh.	21.71	18.75 (16.66)	7.82	4.19 (7.24)
DTW	29.27	16.67 (18.32)	3.71	1.12 (2.54)
MOBISIG				
Manh.	15.45	10.97 (12.75)	5.26	2.32 (4.26)
DTW	27.15	7.34 (11.53)	1.68	0.01 (0.07)

Note. 15 training samples (session 1); type (ii) local features for DTW and 56 features for the Manhattan anomaly detector (%).

Table 12: Verification performance in terms of EER and averaged individual EER (aEER) for the DooDB database using anomaly detectors (different feature sets).

Detector	Skilled forgeries		Random forgeries	
	EER	aEER (std dev.)	EER	aEER (std dev.)
feat56				
Eucl.	23.48	20.98 (16.05)	10.32	6.76 (10.28)
Manh.	21.71	18.75 (16.66)	7.82	4.19 (7.24)
kNN	24.25	20.83 (16.19)	10.10	7.07 (10.43)
SVM	23.26	20.85 (16.07)	10.40	6.97 (10.39)
feat32				
Eucl.	24.66	22.59 (17.09)	11.20	7.79 (11.00)
Manh.	22.60	19.63 (16.10)	8.64	5.19 (8.20)
kNN	25.43	22.68 (16.81)	11.34	8.21 (10.74)
SVM	24.60	22.43 (17.05)	11.55	7.88 (11.05)

Note. Training: 15 samples.

computing the distances to all the other corresponding genuine and forgery samples. We formed a positive and a negative score lists from the obtained distances (scores) and computed the EER. We computed DTW distances between signatures using type (i) local features.

The second metric, the userAvgDist, was computed as the average of the pairwise distances (DTW distance—type (i) local features) of genuine samples for each user (in the case of having N genuine samples, we computed $N(N-1)/2$ distances). The lower it was, the more similar the samples of the user were, so the more stable the user's signature was. For both databases, we used the first $N = 30$ genuine samples for computations.

After applying the two metrics for each genuine sample (sampleEER) or each user userAvgDist, we obtained two sequences of values. In order to characterize the dataset, both the mean and the standard deviation of each sequence were computed. The obtained results are shown in Table 13.

From the two quality metrics proposed by Müller and Henniger, we favor the userAvgDist because this metric does not rely on forgery samples; therefore, it could be used during data collection. Samples that are not closely similar to those already collected should be discarded.

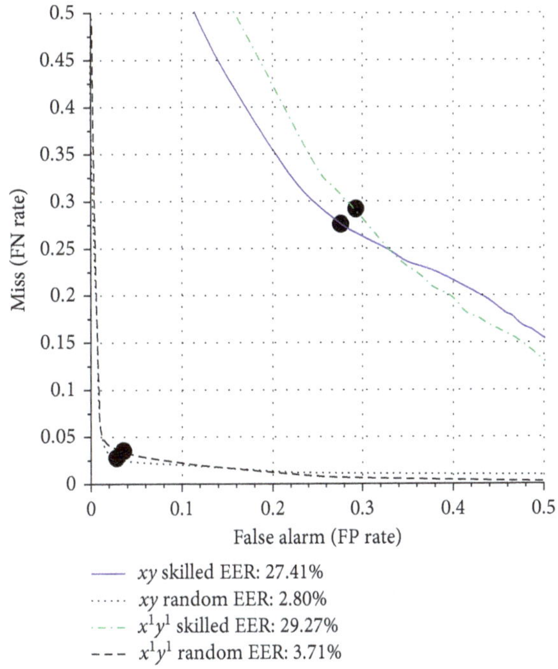

(a)

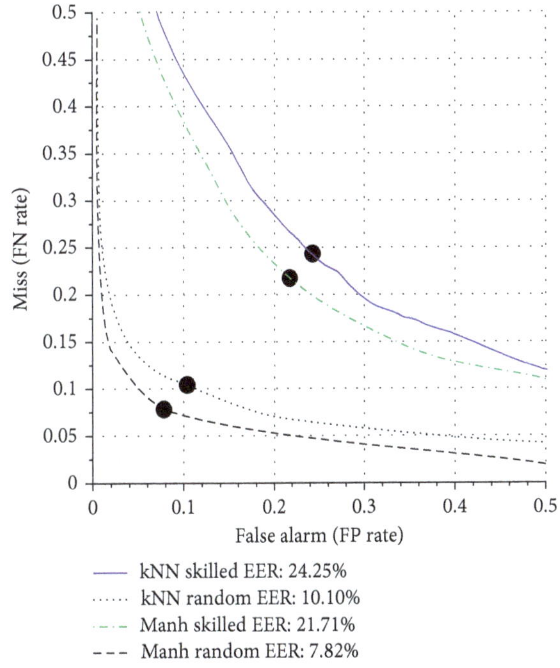

(b)

Figure 6: DooDB. ROC curves for local and global methods. 15 sample training cases. (a) DTW and (b) anomaly detectors.

7. Conclusion

In this paper, the MOBISIG finger-drawn online signature dataset has been presented. The dataset comprises pseudo-signatures from 83 users, both genuine and forgery samples. Benchmark verification experiments have been performed

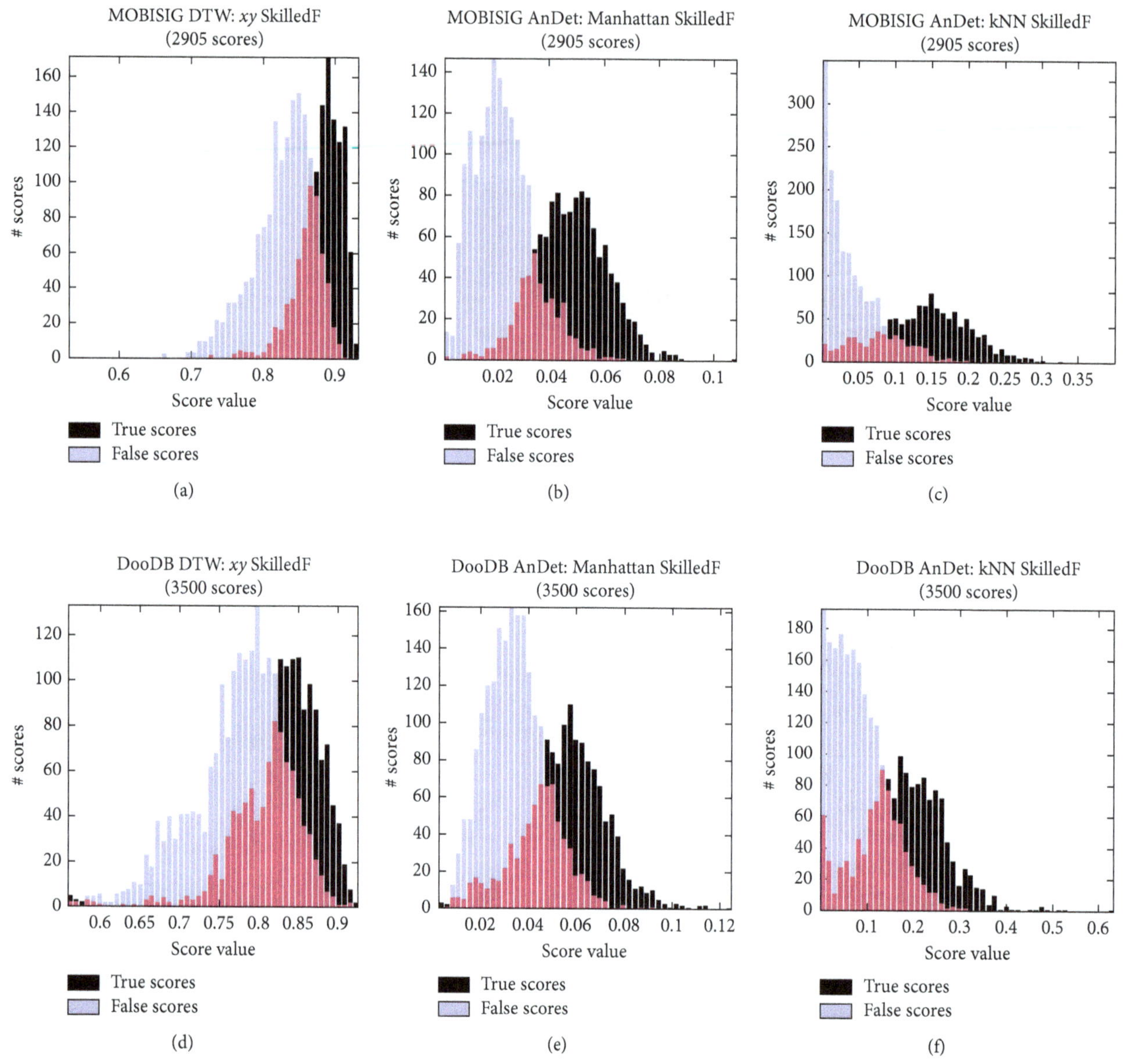

FIGURE 7: Score distributions. Training: 15 samples (session 1). Testing: genuine signatures (15 samples (session 2)) and skilled forgeries (20 samples). (a) MOBISIG-DTW-xy, (b) MOBISIG-Manhattan detector, (c) MOBISIG-kNN detector, (d) DooDB-DTW-xy, (e) DooDB-Manhattan detector, and (f) DooDB-kNN detector.

using both function and feature-based methods. Two types of EERs are reported: one is based on global threshold, while the other one on a posteriori user-specific thresholds. However, the global threshold-based EER results are not outstanding. Nevertheless, further improvement may be obtained by score normalization techniques. Good results were obtained using a posteriori user-specific thresholds (aEER). The lowest aEER for the skilled forgery case was 5.81% and 0.01% for the random forgery case. Both results were obtained by the function-based DTW method. Although feature-based methods offer poor results in the case of global threshold, they are significantly better than using function-based methods (skilled forgery case). The lowest aEERs were 9.35% (skilled forgery) and 2.10% (random forgery), both obtained by the Manhattan outlier detector.

TABLE 13: Signature quality measures for MOBISIG and DooDB databases (%).

Quality method	MOBISIG	DooDB
sampleEER	14.94 (12.58)	16.84 (12.59)
userAvgDist	12.85 (1.57)	19.85 (3.45)

The second objective of this paper was to compare our new dataset with similar publicly available ones. The only publicly available dataset collected on mobile devices and containing finger-drawn signatures was the DooDB. Therefore, we have presented a comparison of our MOBISIG and the DooDB dataset along (i) their statistical properties, (ii) the verification system performances using exactly the same methods and features, and (iii) some signature quality metrics. Signature quality measures indicate slightly better values in the case of the MOBISIG dataset. This difference is also expressed by better results in all evaluations in this study for the MOBISIG dataset compared to the DooDB. These results might be explained by the larger touchscreen for data acquisition in the case of the MOBISIG dataset, furthermore by the average duration of individual signatures in MOBISIG dataset (twice as long as those in the DooDB dataset).

Future work may include the introduction of new signature quality metrics for online signatures, as well as their evaluation on several datasets.

Acknowledgments

The research has been supported by Sapientia Foundation Institute for Scientific Research.

References

[1] D. Impedovo and G. Pirlo, "Automatic signature verification: the state of the art," *IEEE Transactions on Systems, Man, and Cybernetics-Part C: Applications and Reviews*, vol. 38, no. 5, pp. 609–635, 2008.

[2] J. G. A. Dolfing, E. H. L. Aarts, and J. J. G. M. van Oosterhout, "On-line signature verification with hidden Markov models," in *Proceedings of the Fourteenth International Conference on Pattern Recognition*, vol. 2, pp. 1309–1312, Brisbane, Queensland, Australia, August 1998.

[3] D. Y. Yeung, H. Chang, Y. Xiong et al., *SVC2004: First International Signature Verification Competition*, Springer Berlin Heidelberg, Berlin, Germany, 2004.

[4] A. Kholmatov and B. Yanikoglu, "SUSIG: an on-line signature database, associated protocols and benchmark results," *Pattern Analysis and Applications*, vol. 12, no. 3, pp. 227–236, 2009.

[5] J. Ortega-Garcia, J. Fierrez-Aguilar, D. Simon et al., "MCYT baseline corpus: a bimodal biometric database," *IEE Proceedings-Vision, Image and Signal Processing*, vol. 150, no. 6, pp. 395–401, 2003.

[6] S. Garcia-Salicetti, C. Beumier, G. Chollet et al., *BIOMET: A Multimodal Person Authentication Database Including Face, Voice, Fingerprint, Hand and Signature Modalities*, Springer Berlin Heidelberg, Berlin, Germany, 2003.

[7] J. Fierrez, J. Galbally, J. Ortega-Garcia et al., "BiosecurID: a multimodal biometric database," *Pattern Analysis and Applications*, vol. 13, no. 2, pp. 235–246, 2010.

[8] N. Houmani, S. Garcia-Salicetti, B. Dorizzi, and M. El-Yacoubi, *On-Line Signature Verification on a Mobile Platform*, Springer Berlin Heidelberg, Berlin, Germany, 2012.

[9] R. P. Krish, J. Fierrez, J. Galbally, and M. Martinez-Diaz, *Dynamic Signature Verification on Smart Phones*, Springer Berlin Heidelberg, Berlin, Germany, 2013.

[10] N. Sae-Bae and N. Memon, "Online signature verification on mobile devices," *IEEE Transactions on Information Forensics and Security*, vol. 9, no. 6, pp. 933–947, 2014.

[11] M. Martinez-Diaz, J. Fierrez, and J. Galbally, "The DooDB graphical password database: data analysis and benchmark results," *IEEE Access*, vol. 1, pp. 596–605, 2013.

[12] P. Bissig, "Signature verification on finger operated touchscreen devices," Master's thesis, ETH Zürich, Distributed Computer Group, Zürich, Switzerland, 2011.

[13] M. Martinez-Diaz, J. Fierrez, and J. Galbally, "Graphical password-based user authentication with free-form doodles," *IEEE Transactions on Human-Machine Systems*, vol. 46, no. 4, pp. 607–614, 2016.

[14] N. Houmani, A. Mayoue, S. Garcia-Salicett et al., "Biosecure signature evaluation campaign (BSEC'2009): evaluating online signature algorithms depending on the quality of signatures," *Pattern Recognition*, vol. 45, no. 3, pp. 993–1003, 2012.

[15] S. Müller and O. Henniger, *Evaluating the Biometric Sample Quality of Handwritten Signatures*, Springer Berlin Heidelberg, Berlin, Germany, 2007.

[16] N. Houmani, S. Garcia-Salicetti, and B. Dorizzi, "A novel personal entropy measure confronted with online signature verification systems' performance," in *Proceedings of the 2008 2nd IEEE International Conference on Biometrics: Theory, Applications and Systems (BTAS 2008)*, pp. 1–6, Washington, DC, USA, October 2008.

[17] N. Houmani, S. Garcia-Salicetti, and B. Dorizzi, "On measuring forgery quality in online signatures," *Pattern Recognition*, vol. 45, no. 3, pp. 1004–1018, 2012.

[18] N. Houmani and S. Garcia-Salicetti, "On hunting animals of the biometric menagerie for online signature," *PLoS One*, vol. 11, no. 4, article e0151691, pp. 1–26, 2016.

[19] C. Kahindo, S. Garcia-Salicetti, and N. Houmani, "A signature complexity measure to select reference signatures for online signature verification," in *Proceedings of the 2015 International Conference of the Biometrics Special Interest Group (BIOSIG)*, pp. 1–8, Darmstadt, Germany, September 2015.

[20] R. Guest and O. Henniger, "Assessment of the quality of handwritten signatures based on multiple correlations," in *Proceedings of the 2013 International Conference on Biometrics (ICB)*, pp. 1–6, Madrid, Spain, June 2013.

[21] E. N. Zois, L. Alewijnse, and G. Economou, "Offline signature verification and quality characterization using poset-oriented grid features," *Pattern Recognition*, vol. 54, pp. 162–177, 2016.

[22] J. Fierrez-Aguilar, L. Nanni, J. Lopez-Peñalba, J. Ortega-Garcia, and D. Maltoni, "An on-line signature verification system based on fusion of local and global information," in *Audio- and Video-Based Biometric Person Authentication, vol. 3546 of Lecture Notes in Computer Science*, pp. 523–532, Springer Berlin Heidelberg, Berlin, Germany, 2005.

[23] M. Antal and L. Z. Szabó, "On-line verification of finger drawn signatures," in *Proceedings of the 2016 IEEE 11th International Symposium on Applied Computational Intelligence and Informatics (SACI)*, pp. 419–424, Timisoara, Romania, May 2016.

[24] V. Chandola, A. Banerjee, and V. Kumar, "Anomaly detection: a survey," *ACM Computing Surveys*, vol. 41, no. 3, pp. 1–58, 2009.

[25] K. Killourhy and R. Maxion, "Comparing anomaly-detection algorithms for keystroke dynamics," in *Proceedings of the IEEE/IFIP International Conference on Dependable Systems Networks (DSN'09)*, pp. 125–134, Lisbon, Portugal, June–July 2009.

[26] L. C. F. Araujo, L. H. R. Sucupira, M. G. Lizarraga, L. L. Ling, and J. B. T. Yabu-Uti, "User authentication through typing biometrics features," *IEEE Transactions on Signal Processing*, vol. 53, no. 2, pp. 851–855, 2005.

[27] C. C. Chang and C. J. Lin, "Libsvm: a library for support vector machines," *ACM Transactions on Intelligent Systems and Technology (TIST)*, vol. 2, no. 3, pp. 1–27, 2011.

[28] A. Morales, J. Fierrez, R. Tolosana et al., "Keystroke biometrics ongoing competition," *IEEE Access*, vol. 4, pp. 7736–7746, 2016.

[29] J. Fierrez-Aguilar, J. Ortega-Garcia, and J. Gonzalez-Rodriguez, "Target dependent score normalization techniques and their application to signature verification," *IEEE Transactions on Systems, Man, and Cybernetics-Part C: Applications and Reviews*, vol. 35, no. 3, pp. 418–425, 2005.
[30] A. Kholmatov and B. Yanikoglu, "Identity authentication using improved online signature verification method," *Pattern Recognition Letters*, vol. 26, no. 15, pp. 2400–2408, 2005.
[31] F. Alonso-Fernandez, J. Fierrez, and J. Ortega-Garcia, "Quality measures in biometric systems," *IEEE Security and Privacy Magazine*, vol. 10, no. 6, pp. 52–62, 2012.

13

Security and Privacy Issues in Vehicular Named Data Networks

Hakima Khelifi,[1] **Senlin Luo**,[1] **Boubakr Nour**,[2] **and Sayed Chhattan Shah**[3]

[1]*School of Information and Electronics, Beijing Institute of Technology, Beijing, China*
[2]*School of Computer Science, Beijing Institute of Technology, Beijing, China*
[3]*Department of Information Communication Engineering, Hankuk University of Foreign Studies, Seoul, Republic of Korea*

Correspondence should be addressed to Senlin Luo; luosenlin2012@gmail.com

Academic Editor: Mohamed Elhoseny

A tremendous amount of content and information are exchanging in a vehicular environment between vehicles, roadside units, and the Internet. This information aims to improve the driving experience and human safety. Due to the VANET's properties and application characteristics, the security becomes an essential aspect and a more challenging task. On the contrary, named data networking has been proposed as a future Internet architecture that may improve the network performance, enhance content access and dissemination, and decrease the communication delay. NDN uses a clean design based on content names and Interest-Data exchange model. In this paper, we focus on the vehicular named data networking environment, targeting the security attacks and privacy issues. We present a state of the art of existing VANET attacks and how NDN can deal with them. We classified these attacks based on the NDN perspective. Furthermore, we define various challenges and issues faced by NDN-based VANET and highlight future research directions that should be addressed by the research community.

1. Introduction

During the past two decades, research academies and industrials focused their attention on vehicular ad hoc networks (VANETs) [1] in order to provide safety and assistance applications, improve the driving experience, and control the road traffic. Towards this goal, several protocols are proposed such as dedicated short-range communication (DSRC), wireless access in vehicular environment (WAVE), and other protocols that run as an overlay on DSRC/WAVE rather than the IP protocol [2]. In a vehicular environment, a huge amount of data are exchanged under different types of communication such as vehicle-to-vehicle (V2V), vehicle-to-infrastructure (V2I), and with pedestrian, satellite, charging stations, and smart grids (V2U). V2I communication can be used in applications that support Internet access, while V2V is mostly used in real applications that need to send emergency and real-time information about an accident or traffic information [3].

However, exchanging information and sharing data in VANET under the use of IP protocol have been a challenging task [4] due to the nature of this network that frequently and quickly changes under the use of poor-quality wireless link which makes VANET more complex in terms of routing, mobility, and security. Also, due to the fact that vehicles may exchange personal and sensitive information, securing the content and communication and preserving user and data privacy are indispensable [5]. VANET communication must ensure different security requirements including privacy, confidentiality, integrity, and nonrepudiation.

Meanwhile, different solutions have been proposed in the literature as future Internet architectures [6]. Information-centric networking (ICN) [7] has been proposed as one of the promising paradigms to replace the current host-centric network. Hereby, many solutions under the name of ICN have been implemented, such as named data networking (NDN) [8]. NDN replaces the IP address with the name of the content and implements a simple content discovery and data delivery mechanism using an Interest-Data exchange model. NDN aims to improve that data dissemination and facilitates content access. Also, it may enhance mobility support and intermittent connectivity challenge that are difficult to provide through traditional IP networks, by merely re-requesting any unsatisfied request during the mobility and enabling vehicles to retrieve content from the most convenient cache store. Additionally, NDN follows a content-based security concept by protecting the content at the packet level instead of the communication channel level.

Security and privacy are one of the most critical aspects of the whole Internet and not only VANET. Bringing NDN in the vehicular environment has already been introduced in the literature [9]. However, most of the existing works focus on the NDN forwarding plane [10, 11] where the security solutions are rarely elaborated.

1.1. Motivation and Main Contributions. Despite the existing efforts on security, most of them did not consider the nature of VANET communication and attacks in their study [12]; they generally focused on denial of service (DoS) ignoring other attacks. Thus, the main motivation behind this work is to discuss and uncover VANET attacks and security issues from the NDN perspective. Also, as ICN/NDN and VANET merging is still on its first stage and taking a shape, it is quite important to focus on the security and privacy concerns in this phase. It is worth noting that we take VANET as the main network environment, studying its security issues using NDN as a communication model, and not the inverse. To our best knowledge, our work is the first one focusing on VANET security using NDN as a communication plane. Thus, we classify all existing VANET attacks and issues based on the NDN point of view. We overview each one of them and map it to NDN communication. Also, we provide research directions aiming to overcome these attacks and helping the research community to investigate more in this context.

1.2. Organization of the Paper. The rest of the paper is organized as follows: in the following section, we discuss the transaction from current host-centric vehicular networks toward the information-centric paradigm and overview the VANET and NDN architecture, focusing on security issues. Then, in Section 3, we categorize each VANET attack from the NDN perspective and present the existing efforts and summary for each category. Later, we highlight various future research directions in Section 4 and conclude our paper in Section 5.

2. From Host-Centric to Information-Centric Vehicular Networks

ICN is a new communication paradigm that aims to replace the current host-centric model. Shifting from the current IP-based solutions to ICN is not an easy task. Table 1 provides a comparison of host-centric and information-centric paradigms. Regardless of the deployment and transaction methods, mapping ICN communication logic to the existing Internet applications and networks needs more investigation. In this section, we present a quick overview of both VANET and NDN and discuss the mapping between them and the advantages that can be ported by NDN to the vehicular environment. Also, as this work focuses on security, we discuss also some of the security issues related to NDN design.

2.1. Vehicular Ad Hoc Network Overview. Vehicular ad hoc network (VANET) [13] is a part of mobile ad hoc network (MANET) where the node could be a *vehicle* or *roadside unit* (RSU) [14]. Vehicles exchange data with other vehicles (V2V), with RSU (V2I), or with charging stations, personal communication devices, and smart grids (vehicle to uniform) using different VANET applications. Each type of application has its particular features regarding content properties and the way the VANET applications consume it. These characteristics make VANET a challenging environment in terms of security and privacy.

2.1.1. VANET Security and Privacy Problems. As in any communication domain, security requirements in VANET should guarantee authentication, nonrepudiation, integrity, data availability, and confidentiality [15] in order to protect the exchanged messages in the network from modification, deletion, and delay by attackers. The vehicular communication characteristics and application properties have major effects on the security and privacy and make a more challenging environment. The security challenges caused by VANET characteristics are the following:

(1) *Scalability.* VANET is considered as an unbounded network that can be scalable from a small town to a big city until country [16]. It is growing larger and faster with no authority which makes the security enforcement and standardized rules and policies more challenging.
(2) *Mobility.* Due to the high speed of vehicles, the VANET topology can face quick and frequent changes, especially on the highway. Therefore, a very short connection duration and frequently disconnection may occur. Hence, it is very hard to prevent malicious nodes and mitigate attacks in time.
(3) *Time Constraints.* One of the most important of VANET's use cases is safety applications that aim to prevent crashes when an accident happens. These applications require reliable and real-time message delivery. However, due to the possibility of launching denial of service attacks, providing such time constraint services needs more efforts.
(4) *Data Dissemination.* Most of the information and messages in VANET are usually disseminated through vehicles; herein, several applications are vulnerable to attacks to modify, delete, or resend the information at the inappropriate time.
(5) *Privacy.* Preserving data and user privacy is an open issue in the whole Internet including VANET. Vehicles should trust the sender that may have an identity or not, as well as trust the intermediate forwarder vehicles. Thus, trade-off mechanisms are

Table 1: Comparison of host-centric and information-centric models.

Aspect	Host-centric model	Information-centric model
Addressing	(1) Host addresses (2) DNS for host resolution	(1) Content name (2) No DNS required
Routing	(1) Sends packets to the destination address (2) Stateless data plane (3) Point-to-point connectivity (4) Maintains one routing table (5) Routing is based only on the next hop information	(1) Uses Interest packets to fetch the data (2) Stateful data plane (3) Supports multipoint connection (4) Maintains three tables: FIB, PIT, and CS (5) FIB table contains multiple-hop information
Security	(1) Secures the communication channel	(1) Secures the content (content-based security)
Caching	(1) No caching concept	(1) Buffers data packets and reuses them
Mobility	(1) Resends packets to destination addresses	(1) Uses in-network caching and fetches data packets from the most convenient cache point

required between anonymity communication and privacy with the possibility to show real vehicle identity.

2.2. Named Data Networking Overview. Named data networking (NDN) [17] is a promising ICN architecture [18] that follows the content-based paradigm. NDN uses the content name to forward and deliver data between consumers and a producer, with a clean design based on the Interest-Data exchange model. Data packets are sent by a producer/replica node only when receiving an Interest packet triggered by the consumer for the existing content. Both Interest and Data packets contain the same content name.

Every NDN node maintains three data structures: content store (CS), pending interest table (PIT), and forwarding information base (FIB). The CS maintains the locally cached data that can be served for future requests, while the PIT is used to track the received Interest packets, aggregate them, and forward the Data packet downstream. It maintains *content name*, *list of incoming interfaces*, and *nonce* tuple. Similar Interests for the same requests are aggregated within the same PIT entry by pending only the received interface, whereas the FIB table acts as the routing database that contains a list of reachable prefix-names with the outgoing interfaces.

2.2.1. NDN Working Principle. Interest packets are triggered by consumers to discover the content in the network that can be satisfied by either a replica node or the original content producer, where a Data packet is used to deliver the content. Thus, the NDN working principle can be divided into two phases: Interest forwarding and Data forwarding, as illustrated on the right side in Figure 1:

(1) *Interest Forwarding.* When an NDN node receives an Interest packet, it checks its CS whether it already has the requested content. If the content exists on the CS, a Data packet is sent back using the same interface from where the Interest has arrived. Otherwise, PIT exact match is done. When a match is found, it means the same request has already been treated, and the interface name from where the Interest has been received will be appended to the PIT entry; otherwise, a new PIT entry is created for that request by recording the content name and the incoming interface and then performing the FIB longest prefix-name lookup: if a match is found, the Interest will be forwarded to the appropriate interface; otherwise, based on the network policies, the Interest will be either dropped or broadcasted to all interfaces.

(2) *Data Forwarding.* When a Data packet is created by a replica node or the original producer, it carries the same name as in the Interest packet. Hence, the NDN node checks its PIT to verify if the requests have already been treated by him; if a match is found, the Data packet will be sent out to all interfaces listed in the PIT entry (multicast). Otherwise, the packet is considered an unsolicited packet and dropped immediately by the node. During the Data forwarding, the node decides based on its local caching policies, if the content should be cached or not.

It is important to highlight here that all NDN components (CS, PIT, and FIB) are involved in the Interest forwarding phase, while only PIT and CS are used in the Data forwarding.

2.2.2. Security and Privacy in NDN. Preserving data security and preserving user privacy are the most important aspect of any of the today's network architecture and protocol [19]. As the content in NDN is decoupled from its original location, NDN follows a content-based security concept [20] that consists of securing the content among different network elements regardless of the used communication channel:

(1) *Security.* To ensure data authenticity, every Data packet is signed by the original producer using a public key. Thus, any node in the network can verify the data authenticity. Moreover, as Data can be cached in any place in the network, all necessary security-related information is traversed with the Data packets. Furthermore, data confidentiality and access control are supported by content encryption.

(2) *Privacy.* As compared to IP-based networks, any intermediate node can monitor user activities, by knowing who is requesting and what has been requested. However, in NDN, due to the use of

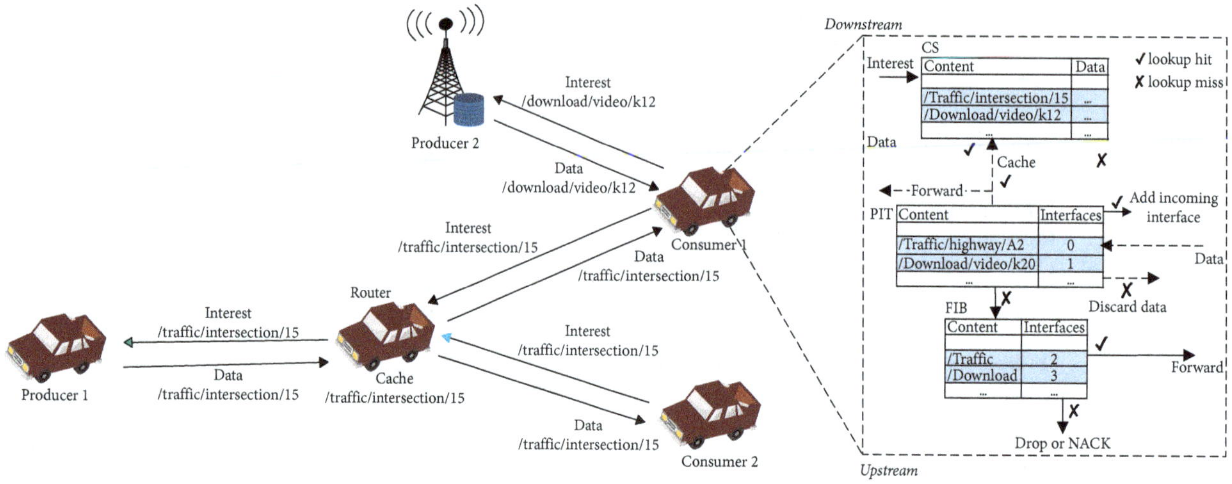

Figure 1: NDN-based VANET.

content names instead of host address, it is not an easy task to know who is requesting, but it is quite easy to know the requested content by monitoring the content names, where various attacks may be launched from different network levels using this vulnerability. Thus, both content and its name should be encrypted to provide high-level privacy.

2.3. NDN Advantages in VANETs. Due to the change of user and application requirements in today's Internet, NDN aims to improve the network scalability and reliability. Towards this, various efforts have been shown tending to bring NDN to vehicular environments [21, 22]. Vehicles in such a merging network can act as data consumers, producers, and intermediates nodes at the same time. Figure 1 shows an example or running NDN architecture on top of the VANET network.

2.3.1. Content Naming. NDN offers a clean, simple, and scale design that can support a large amount of content exchange among vehicles and with infrastructure. Hierarchical NDN names provide a wide addressing range that can be customized by network designers and carry different application semantics [23, 24]. Hence, different application properties can be integrated into the naming scheme to enhance the communication and improve users' needs.

2.3.2. Data Forwarding. The clean content discovery and data delivery mechanisms in NDN make it a suitable solution for VANET [25–27] that can improve the content access among multiple consumers by aggregating the Interests in the first edge node and inherently support multicast communication without the need of extra management protocols. Also, native support of multiple physical interfaces at the same time promotes the necessity of NDN.

2.3.3. In-Network Caching. The time and location decoupling concept in NDN enhances the data availability and improves the overall network performance [28–30]. Data producers are not required to be connected all time to satisfy consumer requests, where the network layer can fulfill different demands especially the hot ones, by retrieving the content from the most closer cache stores.

2.3.4. Data Security. NDN is a session-less architecture, where no session is required to fetch the content from producers or cache store. Also, all security mechanisms are applied to the content itself, by binding the content with its name using the public-private key concept. Hence, securing the communication channel in NDN is not an issue [31].

2.3.5. Mobility Enhancement. By using only the content name to fetch content from the network, NDN aims to enhance node mobility [32]. Mobile nodes are not required to ask for a new address when connecting with the new network. They only need to resend the nonsatisfied Interest by specifying the content name.

3. Security and Privacy Challenges in NDN-Based Vehicular Networks

By moving from a host-centric model to the information-centric concept and using NDN as the primary communication model in VANET, most of the traditional networking aspects will be changed. This also leads to changes in the security and privacy concerns [20]. Also, as ICN/NDN is still taking a shape, more issues will appear when running a large-scale NDN-VANET or fully deployed NDN without overlay protocols. Thus, it is very important to focus on security and privacy in the first design phase. In this section, we first discuss the security and privacy issues in the NDN architecture, and then we classify the existing VANET attacks based on NDN and target only those attacks affecting networking and NDN aspects such as content, routing and forwarding, and caching.

3.1. NDN Security and Privacy Issues. Despite the efforts shown in the NDN project and the research contributions, different security issues still exist in NDN [19]. In the following, we discuss briefly the most critical security issues.

As the network layer takes the responsibilities to satisfy the consumer requests via the ubiquitous in-network caching, different attacks may be launched from different network layers and entities for various goals such as interest flooding attack (due the use of content name), content poisoning attack, and cache poisoning/pollution attack (in-network caching). Hence, the NDN layer should tackle DoS attacks and validate the requested content name.

3.1.1. Content-Name Binding. The content name is the pillar element in ICN/NDN. All network-layer functionalities such as routing, forwarding, mobility, and security are based on the content name. The security of names reflects the network security. In NDN, content and name are bound and validated using the cryptographic function (e.g., public key and signature). Using this secure binding may prove the content ownership. However, in a large-scale network, content may be assigned to different names, which will affect the network scalability especially the routing plane. Thus, a secure control plane to assign names and validate content ownership is required.

3.1.2. Architecture Design. Despite the clean NDN architecture design, some security issues may occur. As NDN communication is based on content name where no host addresses are included, the use of one single interface such as a wireless interface on a vehicle or another device may create a problem of looping in both Interest and Data packets. Even adding a sequence number may solve this problem, however, and according to in-network caching, any node in the network may satisfy the demands. Assuming a malicious node receives all other demands because of the explicit broadcast, it can reply with false content and satisfy these demands, and other nodes in the same rang will receive the Data packets, that by consequence remove the PIT entry where the correct data will be considered unsolicited and dropped by the NDN forwarding plane. In such a scenario, the original requester receives wrong data and will never receive the correct content. Another issue can occur on a node with multiple interfaces: as the Data forwarding plane involves only PIT, a Data packet can be delivered from an interface not indexed in the FIB table but valid on the PIT. Malicious nodes may use this vulnerability to purge all demands on PIT.

3.1.3. Coexistence Issues. NDN can be deployed in three different modes: (a) overlay mode: running NDN on top of the TCP or IP protocol as an overlay layer; (b) coexistence with IP: a node may use the two stacks IP and NDN; and (c) clean-slate mode that consists of running NDN directly. The security of each mode depends on the security of the layer (e.g., IP or TCP). However, to show the real performance of NDN, a clean-slate deployment is required.

In the following sections, we classify VANET attacks into three categories: infrastructure attacks, content protection and access control, and content and user privacy. We overview each attack from both VANET and NDN perspectives and provide a review of existing solutions available.

3.2. Infrastructure Protection. Protecting the infrastructure will by consequence provide high availability and resilience by guaranteeing that only the accurate data are available. Although NDN does not address the hosts directly, securing the infrastructure hosts and endpoints is intuitive, as they are responsible for providing the content. Furthermore, as NDN allows a distributed content caching, this by consequence will increase the content availability and mitigate DoS attacks, which is not always applicable in case of dynamic content that may be generated dynamically only by its original provider.

3.2.1. Denial of Service. Denial of service (DoS) attack [33] is the most famous and dangerous one in the vehicular network, where the attackers send huge requests to the system in order to shut down the network and stop the communication between vehicles and between vehicles and RSUs. The goal of DoS is to stop sending or receiving information to vehicles about the network such as road status.

As NDN deals with content names instead of IP addresses, DoS attacks are based on the use of names [34] and may target consumers, producers, or intermediate nodes. From the NDN point of view, a basic DoS attack can be an *Interest flooding*, where an attacker sends a storm of Interest asking for a different content that may not be available in the cache store.

Figure 2 depicts a simple DoS attack; vehicles, RSUs, and other network infrastructure elements are involved in this scenario. However, due to the *Interest aggregation* feature, intermediate nodes may have more chance to be targeted compared to the content provider, especially when requesting a fake content (i.e., content with a valid name prefix and invalid suffix). The result of that is Interest dropping at the provider level and PIT entries after lifetime expiration at the intermediate nodes level. However, requesting a dynamic content that should be generated by the original content provider upon receiving the Interest (e.g., asking for a fresh patient report) may cause DoS at the provider level due to the fact that dynamic content is not popular and may not be aggregated by intermediate nodes.

A malicious vehicle sends a storm of different Interests asking for different content names, as RSU does not have the content, and it forwards the request and creates a new PIT. Because of the huge number of malicious Interests, the PIT is fitted, where a legitimate vehicle cannot send more requests, and may not benefit of the cache capabilities of the RSU or event forwarding its requests to other nodes. The attack is more severe when it comes to sensitive and urgent communications that may affect people's life.

Various countermeasure solutions have been proposed to overcome and mitigate DoS attacks, by using either rate-limiting mechanisms (e.g., per face or per name-prefix)

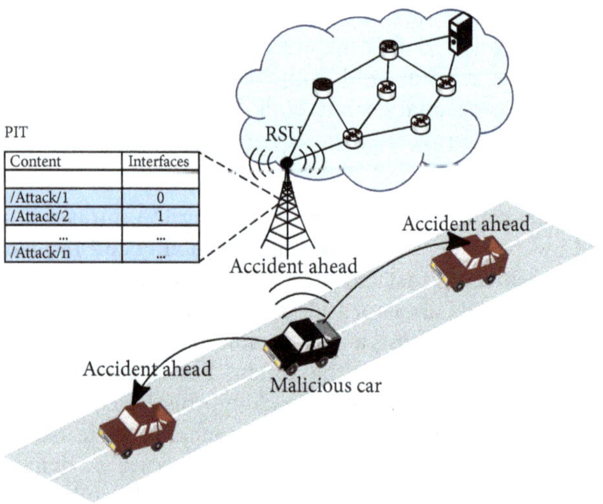

FIGURE 2: Denial of service attack.

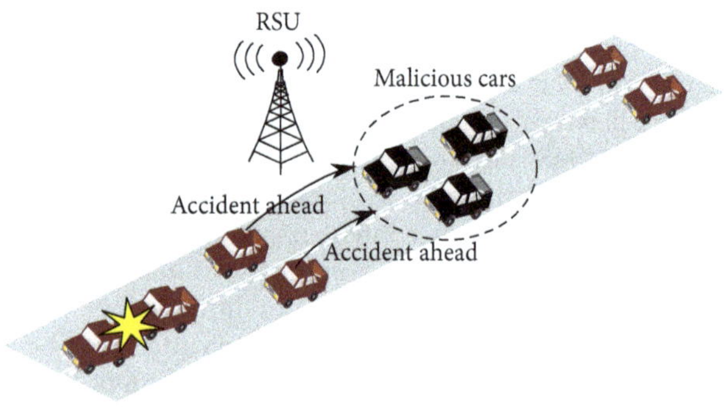

FIGURE 3: Black-hole attack.

[35, 36] or statistical modeling approaches [37, 38]. The former consists of monitoring the face/name-prefix timeout rates and/or the PIT size, when detecting a DoS attack, and the router limits the interest arrival rate (IAR) on the suspicious face, while the latter relies on statistical information about the PIT and interfaces to identify the abnormal traffic pattern. However, these solutions need to make an extensive modification of the regular PIT structure or excessive storage statistics.

3.2.2. Black-Hole and Gray-Hole Attacks. Another dangerous attack, shown in Figure 3, that especially affects safety applications is the black-hole attack [39], where vehicle attackers engage other vehicles by claiming that they have the best route to the destination or have the best position to forward the packet. After the other vehicles send their packet to attackers, the malicious vehicle discards all packets from the network which caused lose of huge packets including critical information and safety messages. Similarly to the black-hole attack, in gray-hole attack, malicious vehicles act as black nodes and misguiding packets, filtering them according to their benefits. A single malicious node or a set of malicious nodes selects some packets to forward and drops others.

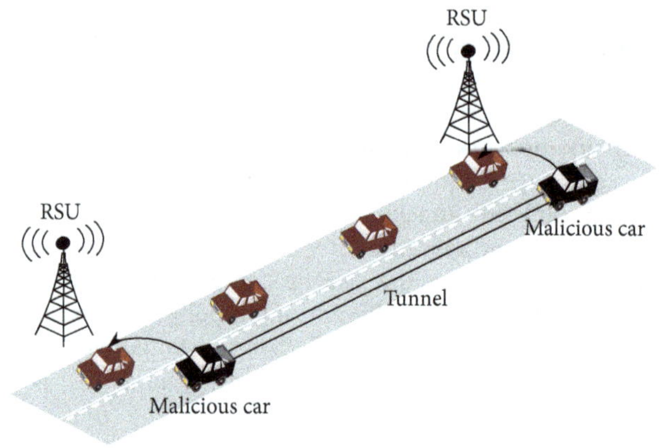

FIGURE 4: Wormhole attack.

In a nutshell, NDN uses a name-based forwarding scheme to forward the requests and deliver data back to consumers. Solving the black-hole and gray-hole attacks can be achieved either by securing the forwarding plane itself or by using secure namespaces to forward name-prefixes that do not exist in the FIB table. Furthermore, as the NDN forwarding plane forwards Interest/Data packets without knowing who is requesting or who will serve, these attacks may not affect VANET-based NDN even by announcing that they have the best route. However, as NDN uses hierarchical names to identify content and services, a malicious node can easily monitor the forwarding system and filter based on content names that allowed and denied packets, which makes these attacks hard to solve in such cases especially when a group of malicious vehicles launches the attack.

3.2.3. Wormhole Attack. The wormhole attack consists of creating a tunnel between two or more collaborative malicious vehicles, aiming to record and transmit data packets between them. Similarly to the black-hole attack, malicious vehicles engage other neighbor vehicles about the link between them as the best path to fetch the data instead of using the original trust path. After malicious vehicles receive packets from victim vehicles (Figure 4), they encapsulate and tunnel to another malicious vehicle, where the latter opens the encapsulated packets and spreads them in the network. The main objective of this attack is to change the network logical topology and make lose the important information that is sent through the tunnel, as well as creating a private network among the malicious vehicles. In an IP-based network, attackers use their IP addresses to create the tunnel. However, due to the use of names instead of addresses and forwarding packets without the need to know who is requesting and to whom should forward, wormhole attack may not be successfully executed in NDN-based networks.

3.2.4. Man-in-the-Middle Attack (MiMA). In man-in-the-middle attack, a malicious vehicle in the communication path keeps listing to all traversed information and injects false information between vehicles. This attack, as shown in Figure 5, has serious effects on the safety applications especially if the injected information is about accidents that may cause life-endangering accidents.

Thanks to the content-based security, all information is signed by the original producer during its creation, and any changes in the data payload during the communication will be exposed to changes in the original signatures [40].

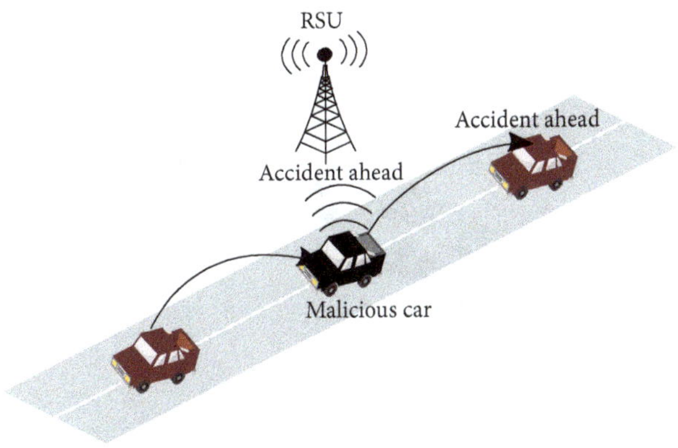

FIGURE 5: Man-in-the-middle attack.

3.2.5. Summary and Insight. In this section, we have reviewed the existing VANET attacks that may affect the network infrastructure including DoS, black-hole, wormhole, gray-hole, and man-in-the-middle attacks. We found that because of using the content name instead of host addresses, many issues can be overcome, especially when binding the content name with the shared information. Also, providing content security at the packet level enhances the communication security.

3.3. Content Protection. As each Data packet is self-signed in NDN, content requesters verify the content signature before consuming the content. Signature verification can also be done by intermediate nodes. However, it will cost more overhead and communication delay. The content signature may ensure *data integrity*, *authentication*, and *correctness*.

3.3.1. Bogus Information. In this type of attack, a malicious vehicle may generate false or wrong information and send it to the network in order to manipulate other vehicles. We find other attacks that can be classified as bogus information attacks, such as the following:

(1) *False Position Information.* Most of the safety applications are based on the particular position, where broadcasting false position information is a hard and critical issue in VANETs. A strong trust and validation model is needed to prevent such false information.

(2) *GPS Spoofing.* A malicious node utilizes the GPS satellite simulator to produce signals which are stronger than the actual satellite signals, tending to deceive vehicles to accept the false position information. This attack is related to physical devices. However, NDN should deal with trust in such data propagation, where collaborative vehicles may detect this information and stop it.

(3) *Illusion Attack.* Attackers disseminate wrong messages to create an illusion to vehicles by exploiting the current road conditions, like a group of cars moves slowly in order to deceive drivers to believe in this wrong information. This attack is hard to detect as the physical vehicle's sensors are used to create and spread the wrong traffic information.

Bogus information attack is usually associated with authentication security conditions, which is an easy task to deal with NDN, as the content is protected and authenticated at the packet level with a secure content-name binding of mechanisms based on hashing techniques and public-private keys.

3.3.2. Replay Attack. In the replay attack, a malicious vehicle saves a copy of the message and resends it later in the network in order to deceive other vehicles, making unnecessary stopping. As NDN is a cache-based network, this attack can be overcome by using the content name and checking the lifetime value in Data packets to know the data freshness, compared with the requested content.

3.3.3. Summary and Insight. Most of the existing NDN attacks related to content in VANETs can be solved by following the content-based security concept. Indeed, securing the content after its creation helps the content security life cycle. Also, when securing the content, access control rules and policies can be used to enforce who can access the content. Moreover, a robust trust model with the validation system is required in NDN to enforce content security and mitigate false content created by malicious nodes.

3.4. Content and User Privacy. Regardless of the content protection level, the user and content privacy still can be compromised in NDN, especially by using plain-text names. Any malicious node may receive the traversing requests and data back. By monitoring the content names, attackers can create a fake content and cache it in any near cache store, that by consequence will be served for future requests.

3.4.1. Sybil and Masquerade Attacks. Sybil attack is considered as one of the most dangerous attacks in VANETs, where the malicious vehicle acts that it is more than a hundred vehicles by creating chaos and a large number of pseudonymous as shown in Figure 6. The goal of this behavior is to deceive other vehicles that there is congestion and force them to change their routes. On the contrary, as the name indicates, in a masquerade attack, a malicious vehicle changes its identity [41] to be another vehicle, trying to produce different messages, alter, and replay with information to deceive other vehicles. For example,

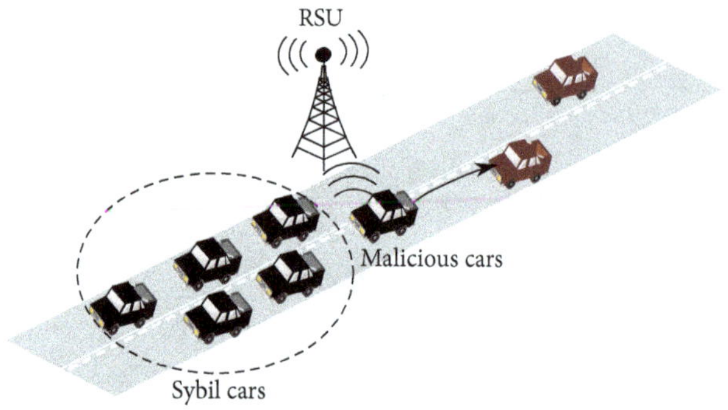

FIGURE 6: Sybil attack.

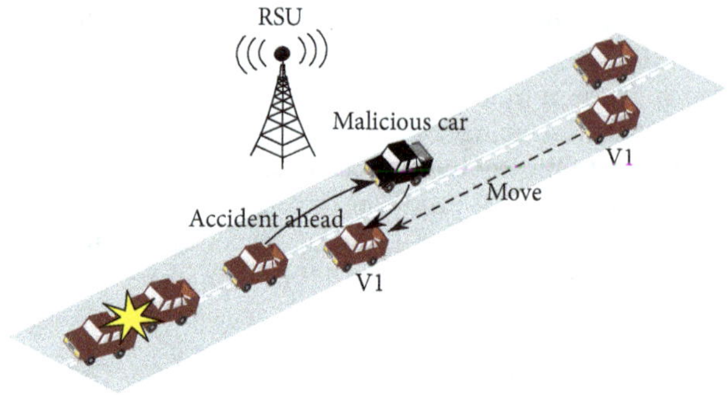

FIGURE 7: Timing attack.

a malicious vehicle can change its identity to be an ambulance and force other vehicles to slow down or change their routes. Works in [42, 43] propose a trust model based on NDN for autonomous vehicular applications in order to prevent bugs information and vehicle tracking. The authors designed a hierarchical naming scheme that composed of four levels: *autonomous vehicle*, *manufacturers*, *vehicles*, and *data*. Furthermore, they used a pseudonym and proxy-based scheme in order to make it difficult for attackers to track vehicles.

NDN binds content names using cryptography algorithms such as public-private keys that may secure the binding and outdo these issues. Furthermore, distributed solutions such as blockchain can be applied to enforce the content-name binding and preserve content and user privacy.

3.4.2. Timing Attack. In timing attacks, the malicious vehicles do not forward the emergency messages and information at the right time (Figure 7) they received it, by creating an explicit communication delay and adding time slots to the received messages. Their neighbor's vehicles receive these messages too late after the time they need it. The timing attack is a critical issue, especially when dealing with time-constraint applications.

3.4.3. Snooping Attack. Snooping is a passive attack, where the malicious vehicle accesses the content and information that traverse it, in order to use it for its benefits without modification. However, as the content is secure and signed, using cryptographic hashing techniques, when it has been created, only legitimate users can access it. Hence, snooping attack may not have an effect on VANET-based NDN.

3.4.4. Summary and Insight. Content and user privacy issues are presented in this part. Content privacy can be preserved using the content-name-binding mechanisms, and more investigation is required in such a context. Also, due to the illumination of host addresses, monitoring attacks can be decreased. The only issue that can occur because of NDN names is the caching-related attacks. Finally, serious solutions and secure forwarding schemes are required to overcome the timing attacks.

4. Future Research Directions

Based on the presented security attacks and issues and their relation with NDN, in the following section, we identify several NDN research directions that may enhance and improve the security of the NDN architecture. Table 2 summarizes this discussion.

4.1. Denial of Service. Solutions based on limiting Interest rate on the malicious interfaces or based on name-prefixes may end by punishing legitimate consumers and authorized requests. Hence, using the software-defined networking (SDN) approach [44] may detect DoS attack in the early stages using the global view of controllers and a per flow rate limiting that may yield highest fairness compared to interface/name-prefix-based solutions. Also, the core routers can collaborate to identify and filter the malicious name-prefixes and malicious traffic pattern using AI-based algorithms.

4.2. Content Poisoning. The purpose of this attacker is to fill the node's cache stores with fake contents. All the existing solutions require performing a data packet signature verification at the intermediate nodes level, that affects the content retrieval delay; comparing content hash with the hash taken from the corresponding Interest may reduce the network scalability, or ranking the content using consumers' feedback, where attackers also have the chance to send malicious feedbacks.

4.3. Naming and Content-Name Binding. Providing a secure naming scheme is still an open research challenge for ICN and NDN. A secure naming scheme should ensure an efficient and scalable binding between the name and the content that may avoid various types of attacks. All the existing binding schemes require signature verification for each and every data packet. This process is costly in terms of resources, as well as affects the data retrieval delay, where an intermediate node cannot perform it at the line speed rate.

4.4. Caching Pollution. The main objective of the caching pollution attack is to diminish the operation of in-network caching and augment content retrieval latency. The existing

TABLE 2: Summary of issues and research directions.

Category	Attacks	Compromised services	Target NDN aspects	Possible directions
Infrastructure protection	DOS	(1) Authentication (2) Availability	(1) Routing and forwarding plane	(1) Interface-based rate limit (2) Name-based rate limit (3) Statistical rate limit
	Black-hole and gray-hole	(1) Availability	(1) Routing and forwarding plane	(1) Securing the forwarding plane (2) Use of secure namespace
	Wormhole	(1) Confidentiality	(1) Data packets	(1) Use of content names instead of device identifications (2) Performing name-based forwarding
	Man-in-the-middle	(1) Authentication (2) Confidentiality (3) Integrity (4) Nonrepudiation	(1) Data packets (2) Forwarding plane	(1) Content-based security mechanism (2) Securing the content during the creation (3) Attaching access control policies with content
Content protection	Bogus information	(1) Authentication (2) Integrity	(1) Data plane	(1) Securing the content using cryptographic hashing techniques and public-private keys
	Replay	(1) Authentication (2) Integrity	(1) Data packet (2) Cache store	(1) Fetching content from the cache store based lifetime (2) Requesting only the fresh content
Content and user privacy	Sybil	(1) Authentication (2) Availability	(1) Routing and forwarding plane	(1) Securing content-name binding (2) Preserving blockchain-based identity
	Masquerade	(1) Authentication (2) Nonrepudiation (3) Integrity	(1) Routing and forwarding plane	(1) Preserving blockchain-based identity
	Timing attack	(1) Availability	(1) Data packets (2) Caching store	(1) Securing the forwarding plane (2) Trust-based forwarding scheme (3) Reputation-based caching and forwarding
	Snooping attack	(1) Authentication	(1) Data packets	(1) Applying content-based security mechanisms (2) Adding access rules within Data packets

solutions have a high computation overhead at intermediate nodes. We consider a collaborative caching scheme a suitable solution to help the core network to mitigate this attack by exchanging feedback between cache stores, keep only the popular content, and reduce the nonpopular ones.

4.5. Secrecy of Correspondence. Preserving the SoC and content copyrights is one of the most critical privacy topics in the whole networking domain and not only ICN [31]. From SoC perspectives, the content owner should state all privacy and content-use policies in the Data packet or in the content itself. We believe that the blockchain-like structure combined with the smart contract can be one of the promising solutions. The content owner specifies different smart contracts depending on the policies such as caching, providing, and consuming the content, that may be executed automatically when the action is triggered, to enforce SoC and content copyright.

4.6. Application Design Patterns. Several application-level security mechanisms have been proposed in ICN such as the following: (i) request filtering that intends to identify and remove the unwanted or forged content from untrusted providers, by using provider's information (e.g., public keys and name-prefix) and consumers' votes for content ranking. (ii) Anomaly detection aims to detect unwanted activities or network misbehavior, using statistical data analyses, fuzzy detection algorithms, and traffic clustering. However, there is no all-in-one scheme that deals with the existing application-level threats or discusses the design patterns for a secure application in ICN, which is a strong future research topic.

5. Conclusion

NDN architecture is a suitable candidate for the future Internet, including vehicular communication. Deploying NDN on top of VANET is still in the early phase. Security and privacy issues have a strong impact on the success of such merging. This article addresses the major networking security and privacy issues in VANET from the perspective of the NDN communication model. The nature of VANET communication and applications changes the way of seeing security; also, adding the NDN model on VANET makes the task more challenging. We categorized VANET security challenges and discussed them from the NDN perspective. Also, we highlighted different NDN research directions and guidelines.

Acknowledgments

This work was supported by the Hankuk University of Foreign Studies Research Fund of 2018 and National Research Foundation of Korea (2017R1C1B5017629).

References

[1] S. Al-Sultan, M. M. Al-Doori, A. H. Al-Bayatti, and H. Zedan, "A comprehensive survey on vehicular ad hoc network," *Journal of Network and Computer Applications*, vol. 37, pp. 380–392, 2014.

[2] F. Cunha, L. Villas, A. Boukerche et al., "Data communication in VANETs: protocols, applications and challenges," *Ad Hoc Networks*, vol. 44, pp. 90–103, 2016.

[3] T. Mekki, I. Jabri, A. Rachedi, and M. ben Jemaa, "Vehicular cloud networks: challenges, architectures, and future directions," *Vehicular Communications*, vol. 9, pp. 268–280, 2017.

[4] T. Mekki, I. Jabri, A. Rachedi, and M. B. Jemaa, "Proactive and hybrid wireless network access strategy for Vehicle Cloud networks: an evolutionary game approach," in *Proceedings of 13th International Wireless Communications and Mobile Computing Conference (IWCMC)*, pp. 1108–1113, IEEE, Valencia, Spain, June 2017.

[5] H. Hasrouny, A. E. Samhat, C. Bassil, and A. Laouiti, "VANet security challenges and solutions: a survey," *Vehicular Communications*, vol. 7, pp. 7–20, 2017.

[6] J. Pan, S. Paul, and R. Jain, "A survey of the research on future internet architectures," *IEEE Communications Magazine*, vol. 49, no. 7, pp. 26–36, 2011.

[7] A. V. Vasilakos, Z. Li, G. Simon, and W. You, "Information centric network: research challenges and opportunities," *Journal of Network and Computer Applications*, vol. 52, pp. 1–10, 2015.

[8] L. Zhang, A. Afanasyev, J. Burke et al., "Named data networking," *ACM SIGCOMM Computer Communication Review*, vol. 44, no. 3, pp. 66–73, 2014.

[9] Z. Su, Y. Hui, and Q. Yang, "The next generation vehicular networks: a content-centric framework," *IEEE Wireless Communications*, vol. 24, no. 1, pp. 60–66, 2017.

[10] M. A. Yaqub, S. H. Ahmed, S. H. Bouk, and D. Kim, "Interest forwarding in vehicular information centric networks: a survey," in *Proceedings of 31st Annual ACM Symposium on Applied Computing-SAC'16*, pp. 724–729, ACM, Pisa, Italy, April 2016.

[11] S. H. Bouk, S. H. Ahmed, D. Kim, K.-J. Park, Y. Eun, and J. Lloret, "LAPEL: hop limit based adaptive PIT entry lifetime for vehicular named data networks," *IEEE Transactions on Vehicular Technology*, vol. 67, no. 7, pp. 5546–5557, 2018.

[12] S. Signorello, M. R. Palattella, and L. A. Grieco, "Security challenges in future NDN-enabled VANETs," in *Proceedings of 2016 IEEE Trustcom/BigDataSE/ISPA*, pp. 1771–1775, IEEE, Tianjin, China, August 2016.

[13] S. Boussoufa-Lahlah, F. Semchedine, and L. Bouallouche-Medjkoune, "Geographic routing protocols for Vehicular Ad hoc NETworks (VANETs): a survey," *Vehicular Communications*, vol. 11, pp. 20–31, 2018.

[14] D. Kim, Y. Velasco, W. Wang, R. Uma, R. Hussain, and S. Lee, "A new comprehensive RSU installation strategy for cost-efficient VANET deployment," *IEEE Transactions on Vehicular Technology*, vol. 66, no. 5, pp. 4200–4211, 2017.

[15] H. Sedjelmaci and S. M. Senouci, "Cyber security methods for aerial vehicle networks: taxonomy, challenges and solution," *Journal of Supercomputing*, pp. 1–17, 2018.

[16] M. Laroui, A. Sellami, B. Nour, H. Moungla, H. Afifi, and S. Boukli-Hacéne, "Driving path stability in VANETs," in *IEEE Global Communications Conference*, Abu Dhabi, UAE, December 2018.

[17] L. Zhang, D. Estrin, J. Burke et al., *Named Data Networking (NDN) Project*, Relatório Técnico NDN-0001, Xerox Palo Alto Research Center-PARC, Palo Alto, CA, USA, 2010.

[18] B. Ahlgren, C. Dannewitz, C. Imbrenda, D. Kutscher, and B. Ohlman, "A survey of information-centric networking," *IEEE Communications Magazine*, vol. 50, no. 7, pp. 26–36, 2012.

[19] T. Chatterjee, S. Ruj, and S. D. Bit, "Security issues in named data networks," *Computer*, vol. 51, no. 1, pp. 66–75, 2018.

[20] R. Tourani, S. Misra, T. Mick, and G. Panwar, "Security, privacy, and access control in information-centric networking: a survey," *IEEE Communications Surveys & Tutorials*, vol. 20, no. 1, pp. 566–600, 2018.

[21] M. Amadeo, C. Campolo, and A. Molinaro, "Content-centric networking: is that a solution for upcoming vehicular networks?," in *Proceedings of Ninth ACM international Workshop on Vehicular inter-networking, systems, and applications*, pp. 99–102, ACM, New York, NY, USA, November 2012.

[22] S. H. Bouk, S. H. Ahmed, D. Kim, and H. Song, "Named-data-networking-based ITS for smart cities," *IEEE Communications Magazine*, vol. 55, no. 1, pp. 105–111, 2017.

[23] S. H. Bouk, S. H. Ahmed, and D. Kim, "Hierarchical and hash based naming with compact trie name management scheme for vehicular content centric networks," *Computer Communications*, vol. 71, pp. 73–83, 2015.

[24] B. Nour, K. Sharif, F. Li, H. Moungla, and Y. Liu, "M2HAV: a standardized ICN naming scheme for wireless devices in internet of things," in *Proceedings of International Conference on Wireless Algorithms, Systems, and Applications*, pp. 289–301, Springer, Guilin, China, June 2017.

[25] X. Liu, Z. Li, P. Yang, and Y. Dong, "Information-centric mobile ad hoc networks and content routing: a survey," *Ad Hoc Networks*, vol. 58, pp. 255–268, 2017.

[26] M. F. Majeed, S. H. Ahmed, and M. N. Dailey, "Enabling push-based critical data forwarding in vehicular named data networks," *IEEE Communications Letters*, vol. 21, no. 4, pp. 873–876, 2017.

[27] S. H. Ahmed, S. H. Bouk, M. A. Yaqub, D. Kim, and H. Song, "DIFS: distributed interest forwarder selection in vehicular named data networks," *IEEE Transactions on Intelligent Transportation Systems*, 2017.

[28] M. A. Yaqub, S. H. Ahmed, and D. Kim, "Asking neighbors a favor: cooperative video retrieval using cellular networks in VANETs," *Vehicular Communications*, vol. 12, pp. 39–49, 2018.

[29] H. Khelifi, S. Luo, B. Nour, A. Sellami, H. Moungla, and F. Naït-Abdesselam, "An optimized proactive caching scheme based on mobility prediction for vehicular networks," in *Proceedings of IEEE Global Communications Conference (GLOBECOM)*, pp. 1–6, Abu Dhabi, UAE, December 2018.

[30] B. Nour, K. Sharif, F. Li, H. Moungla, A. E. Kamal, and H. Afifi, "NCP: a near ICN cache placement scheme for IoT-based traffic class," in *Proceedings of IEEE Global Communications Conference (GLOBECOM)*, pp. 1–6, Abu Dhabi, UAE, December 2018.

[31] C. Ghali, G. Tsudik, and C. A. Wood, "When encryption is not enough: privacy attacks in content-centric networking,"

in *Proceedings of 4th ACM Conference on Information-Centric Networking*, pp. 1–10, ACM, Berlin, Germany, January 2017.

[32] J. M. Duarte, T. Braun, and L. A. Villas, "Source Mobility in Vehicular Named-Data Networking: An Overview," in *Proceedings of 9th EAI International Conference on Ad Hoc Networks*, pp. 83–93, Springer, Niagara Falls, ON, USA, March 2018.

[33] S. M. Specht and R. B. Lee, "Distributed denial of service: taxonomies of attacks, tools, and countermeasures," in *Proceedings of 17th International Conference on Parallel and Distributed Computing Systems*, pp. 543–550, San Francisco, CA, USA, September2004.

[34] M. Aamir and S. M. A. Zaidi, "Denial-of-service in content centric (named data) networking: a tutorial and state-of-the-art survey," *Security and Communication Networks*, vol. 8, no. 11, pp. 2037–2059, 2015.

[35] A. Afanasyev, P. Mahadevan, I. Moiseenko, E. Uzun, and L. Zhang, "Interest flooding attack and countermeasures in named data networking," in *Proceedings of 2013 IFIP Networking Conference*, pp. 1–9, IEEE, Brooklyn, NY, USA, May 2013.

[36] H. Dai, Y. Wang, J. Fan, and B. Liu, "Mitigate ddos attacks in NDN by interest traceback," in *Proceedings of 2013 IEEE Conference on Computer Communications Workshops (INFOCOM WKSHPS)*, pp. 381–386, April 2013.

[37] K. Wang, J. Chen, H. Zhou, and Y. Qin, "Content-centric networking: effect of content caching on mitigating dos attack," *International Journal of Computer Science Issues*, vol. 9, no. 6, pp. 43–52, 2012.

[38] T. Nguyen, R. Cogranne, and G. Doyen, "An optimal statistical test for robust detection against interest flooding attacks in ccn," in *Proceedings of 2015 IFIP/IEEE International Symposium on Integrated Network Management (IM)*, pp. 252–260, IEEE, Ottawa, Canada, May 2015.

[39] F.-H. Tseng, H.-P. Chiang, and H.-C. Chao, "Black hole along with other attacks in MANETs: a survey," *Journal of Information Processing Systems*, vol. 14, no. 1, 2018.

[40] C. Ghali, A. Narayanan, D. Oran, G. Tsudik, and C. A. Wood, "Secure fragmentation for content-centric networks," in *Proceedings of IEEE 14th International Symposium on Network Computing and Applications (NCA)*, pp. 47–56, IEEE, Cambridge, MA, USA, 2015.

[41] A. Boualouache, S.-M. Senouci, and S. Moussaoui, "A survey on pseudonym changing strategies for Vehicular Ad-Hoc Networks," *IEEE Communications Surveys & Tutorials*, vol. 20, no. 1, pp. 770–790, 2017.

[42] M. Chowdhury, A. Gawande, and L. Wang, "Anonymous authentication and pseudonym-renewal for VANET in NDN," in *Proceedings of 4th ACM Conference on Information-Centric Networking*, pp. 222-223, ACM, Berlin, Germany, January 2017.

[43] M. Chowdhury, A. Gawande, and L. Wang, "Secure information sharing among autonomous vehicles in NDN," in *Proceedings of Second International Conference on Internet-of-Things Design and Implementation*, pp. 15–25, ACM, Pittsburgh, PA, USA, April 2017.

[44] A. Alioua, S.-M. Senouci, S. Moussaoui, H. Sedjelmaci, and A. Boualouache, "Software-defined heterogeneous vehicular networks: taxonomy and architecture," in *Proceedings of Global Information Infrastructure and Networking Symposium (GIIS)*, pp. 50–55, IEEE, Saint Pierre, Reunion Island, France, October 2017.

14

Interaction Modes for Emergency Mobile Apps

Claudia Nass, Jessica Jung, Eduard C. Groen, Karina Villela, and Konstantin Holl

Fraunhofer Institute for Experimental Software Engineering (IESE), Fraunhofer-Platz 1, 67663 Kaiserslautern, Germany

Correspondence should be addressed to Konstantin Holl; Konstantin.Holl@iese.fraunhofer.de

Academic Editor: Sergio Mascetti

While seemingly irrational behaviors such as panicking or displaying antisocial behavior are the responses to emergency situations the media and movies lead us to believe, several studies show that people rather react based on decision-making such as acting altruistically and protectively. However, what can we really expect from people in a crowd in terms of participation in an emergency response system? In this paper, we present a mobile application called the RESCUER App, which allows civilians to participate in the emergency response process by providing information about the emergency to a command center and to receive instructions from this command center. We developed a human reaction model for emergencies to better understand the human–computer interaction capabilities of people in an emergency situation. Based on this model, we defined three different interaction modes: one-click interaction, guided interaction, and chat interaction. These interaction modes were implemented in an interactive prototype and evaluated in an experiment in which high cognitive load was induced to simulate a stress situation, similar to the stress experienced in an emergency. The experiment results showed that the three predefined interaction modes enabled people to interact with the RESCUER App even though they were in a stress situation.

1. Introduction

Every year, the German state of Rhineland-Palatinate organizes a large festival in the region, with more than 200,000 visitors over the course of the three-day event. People can enjoy local food, buy craftwork, or attend concerts by national artists. Now imagine that a market booth in the middle of the festival grounds catches fire. Some festivalgoers close to the booth will call the emergency number as they cannot quickly find members of the event's safety personnel. They will try to describe the place of the incident, but they may not know exactly where they are as the event location is not familiar to them. They will also try to explain what they see around them, which is not easy for them. On the other side of the line, the command center will try to guide the callers through some questions to obtain relevant information from them, while also alerting the security staff and firefighters at the festival venue. The exact site of the incident will still be unknown at this time, and the situation description will be very imprecise. After a few minutes, the number of calls to the emergency number will increase rapidly. In the meantime, the firefighters present at the venue will try to reach the exact location of the incident and start fighting the fire, but when they reach the fire, it may already have spread rapidly so they will need to call for additional firefighters via the command center. Communication via radio is very short and not persistent, so the command center will still lack vital information about the emergency.

This adverse scenario is exemplary for various problems that may occur in an emergency in a given location with high density of people, that is, that involves a crowd [1]. Typical problems for this scenario include the overloaded communication lines of the command center, which prohibits people from calling for help or providing information about injured people at the site of the incident, and the delay in identifying the exact position of the incident, which may lead to an expansion of the fire and the situation to run out of control. Additionally, the lack of information about injured people, the exact number of people affected (i.e., the people close to the site of the incident who are likely to be affected by the adverse consequences of the fire), or even the physical situation around the fire itself prevent the seamless execution of the emergency response plan and put in danger the success of the operation.

In our work, we try to solve these problems through the use of a mobile application called the RESCUER App, which supports the notification and characterization of an emergency situation by involving the crowd at the place of an incident. Our goal is to stimulate crowd participation just after the incident has occurred when they are still close to the site of the incident, so they can inform the command center about the incident and describe its main characteristics or provide essential details.

The scenario presented above is a typical emergency involving a crowd. An emergency is an event or situation that threatens to inflict serious damage on human welfare (e. g., loss of human life, illness, injury, homelessness, damage to property, or disruption of supplies) or the environment (e. g., contamination of land, water, or air with biological, chemical, or radioactive matter) [2]. Past views assumed that people act out of a reflex in emergencies, thereby showing behaviors that may seem irrational to other persons, such as antisocial behavior or panicking. News broadcasts and the media in general have programmed our minds with this picture [3]. However, recent research suggested that this kind of reaction is a myth without empirical basis. According to Quarantelli [4] and Drury et al. [5], studies and analyses of past emergency situations show that people make decisions during an emergency, for example, the decision of behaving altruistically. They behave calmly and know how to protect themselves and others in their social sphere. However, part of the decision-making process is influenced by the presence of others; in a crowd, people defer responsibility more easily and assume others will act, leading to phenomena like the bystander effect or pluralistic ignorance [6].

The collaborative behavior of people in a crowd involved in an emergency, which was described by Drury et al. [5], is essential for the emergency response process. The use of technologies such as smartphones and other mobile devices increases the resilience of people when responding to an emergency [7]. The point, however, is to know what we can really expect from people in terms of how they interact with a mobile application just after an incident, while they are still in the immediate vicinity of the site of the incident. In particular, we focused our study on the interaction with a mobile application for smartphones, as smartphones are much widely used than other mobile devices such as smart watches. In addition, smartphones allow a richer interaction, as people can take photos and make videos.

Currently, several mobile crowdsourcing-based platforms are available that support people involved in disasters as well as the workforces that respond to such emergencies. The Ushahidi crisis map platform [8] and iLigtas [9] are examples of free general public license software for information collection and visualization, and interactive mapping. Both platforms provide a mobile application that allows sharing incidents via SMS, geotagging, and social media feeding. The platforms create infographics visualized on maps for resource distribution, including medical care, shelter, and food. Both platforms support emergency task coordination through social networking and virtual teaming, geospatial visualization for analysis, and decision-making support.

The platforms InSTEDD [10, 11] and Sahana [12] are open source software platforms that support collaboration and improve the information flow to better deliver critical services to vulnerable populations. Both platforms allow a quick adaptation of the emergency response process and tools to the current emergency scenario, and they additionally offer tools to assist groups in analyzing and visualizing multiple streams of information (SMS, geotagging, and social media) and postdisaster tools for tracking projects and resources.

All these solutions are designed to support long-term interactions especially in the context of natural disasters, where the response process might span days or weeks, and the occurrence of immediate stressors might have already decreased. In contrast, RESCUER (an acronym for "Reliable and Smart Crowdsourcing Solution for Emergency and Crisis Management") focuses on the immediate process running concurrently with the identification of an incident. Specifically, the RESCUER project aims at developing a complete platform to support command centers in quickly handling emergencies and managing crises based on the reliable and intelligent analysis of crowdsourcing information [13]. The application scenarios it focuses on are incidents during large-scale events and in industrial areas. The RESCUER App is one the main components of the RESCUER platform, which also includes Data Analysis Solutions, the Emergency Response Toolkit, and the Ad-hoc Communication Solution.

For the design of the RESCUER App, we modeled human reactions in emergency situations and derived three interaction modes that support user interaction with a mobile device in such circumstances: one-click interaction, guided interaction, and chat interaction. These interaction modes provided concrete requirements for the user interface that were implemented in the RESCUER App accordingly. During an earlier stage of the project, we performed an experiment with a prototype of the RESCUER App in which stress was induced in participants to simulate that they found themselves in an emergency situation. The results indicate that the three predefined interaction modes enable people to interact with the RESCUER App even when they are in a stressful situation.

In Section 2, we present the related work that supported the definition of the RESCUER Human Reaction Model. The model itself is presented in Section 3. Based on that model, we derived the requirements for the interaction design of the RESCUER App, which led to the three concrete interaction modes described in Section 4. Section 5 presents the resulting RESCUER App. In Section 6, we describe the experiment that we conducted, and in Section 7, we discuss how it confirmed that the three predefined interaction modes support the interaction of the crowd with the mobile application even in a stressful situation. Finally, we discuss the main outcomes in Section 8. Conclusions and future work are addressed in Section 9.

2. Related Work

In this section, we present relevant work and definitions that formed the basis for the specification of the human reaction model used for the development of the RESCUER App. Pan [14] presents a human and social behavioral model for typical emergency situations in large-scale event scenarios that involve overcrowding and crushing, which among others occur at sports stadiums, schools, or social gathering places such as nightclubs. His goal was to build a multiagent simulation system for egress analysis. He identified five typical evacuation behaviors, which were modeled in a computational framework: queuing, competitive, leader-following, altruistic, and herding behavior. These behaviors are related to the output of the decision-making process that is responsible for identifying and properly responding to an emergency. Factors influencing this process and thereby influencing human behavior include

(i) human physical characteristics (i.e., body size, mobility, age, and gender),
(ii) environmental characteristics (i.e., geometric constraints, emergency type, and egress systems),
(iii) psychological and sociological characteristics (i.e., individual, interaction among individuals, and group).

Gannt and Gannt [15] propose that in order to make emergency plans and processes in industrial area scenarios more efficient, they should also be based on typical human cognitive capabilities and behaviors. They argue that most emergency plans and systems ignore such human capabilities and, consequently, expose the employees and the environment to higher risks. They paid special attention to the cognitive decision-making process that people execute during emergencies, the influence of the changes in the environment, and especially the social behavior that emerges during an emergency. The authors argue that social bonds between people involved in an emergency are solidified and even newly created.

Similarly, Perry and Greene [16] describe a decision-making process that is executed by individuals who identify and properly respond to an emergency. They characterize human behavior in a disaster as "non-traditional behaviour in response to a changing or changed environment" [16]. This means that individuals who face unexpected environmental changes reexamine their own behavior in relation to the altered conditions and adapt their behavior in order to protect themselves and minimize harmful consequences. This process of behavior adaptation consists of three phases: risk identification, risk assessment, and risk reduction. In the case of an emergency, the changes in one's immediate environment are recognized as a threat to the individual, and the individual takes protective actions accordingly.

According to Leach [17], the decision-making process is managed in the human working memory as a neurocognitive function. The processing of such information involves several steps in order to turn the perception of a threat into an appropriate action. Under optimal conditions, this takes a minimum of eight to ten steps, but this number can increase due to higher task complexity and physiological factors that influence the individual. The processing of an emergency often results in "freeze" behavior experienced by the individual just after he or she has perceived the threat. Leach [17] classified three typical behaviors in response to threats:

(i) The first group (10–15% of the people involved in an emergency) will remain calm, think quickly, and retain situational awareness, hardly affecting their judgment and reasoning abilities.
(ii) The second group (about 75% of the people involved) will be stunned and show impaired reasoning and sluggish thinking.
(iii) The third group (10–15% of the people involved) will show confusion, paralyzing anxiety, and exhibit counterproductive behavior that adds to their risk.

The term "stress" is frequently used in the context of emergencies such as disasters. However, stress is an ambiguous term, as it can have different meanings. Broadly speaking, stress is the biophysiological reaction to a stressor, which in the case of an emergency is the threat. Stressors (i.e., that what causes stress) are "circumstances that threaten a major goal, including the maintenance of one's physical integrity or one's psychological well-being" [18]. The psychological response to a stressor is called distress, which can manifest itself through affective states, such as anxiety, frustration, euphoria, or sadness.

Kemeny [18] proposes an integrated specificity model of stress. In this model, the exposure to a specific stressor evokes an integrated psychophysiological response (which includes emotion, motivation, and physiological responses) that intends to generate a protective action in an individual in reaction to the threat. In this model, the resources available for reacting to a threat moderate the relationship between a stressor and its psychophysiological response. This means that when faced with a threat, the level of stress is regulated by one's own motivation, the physiological responses to the threat, and the individual's resources for reacting. Resources, in this case, may include time, tools, abilities, and space.

Finally, Staal [19] explores the interaction between stress, cognition, and human performance. He found that various stressors can affect cognition in which case these stressors contribute to a reduction in the working memory's performance, especially in combination with a low self-efficacy to averting a threat. Under stress, the resources available for the working memory are reduced, in part due to the physiological changes related to the fight-or-flight response. Considering that when faced with a threat in an emergency, the decision-making process takes place in working memory, stress can consequently affect the decision-making process and its execution. Thus, the level of stress experienced by a person in the face of a threat can diminish his or her cognitive capabilities and performance.

3. RESCUER Human Reaction Model

The RESCUER Human Reaction Model is based on the models presented above and uses the ABC Model of Attitudes [20] as a basis for distinguishing between various

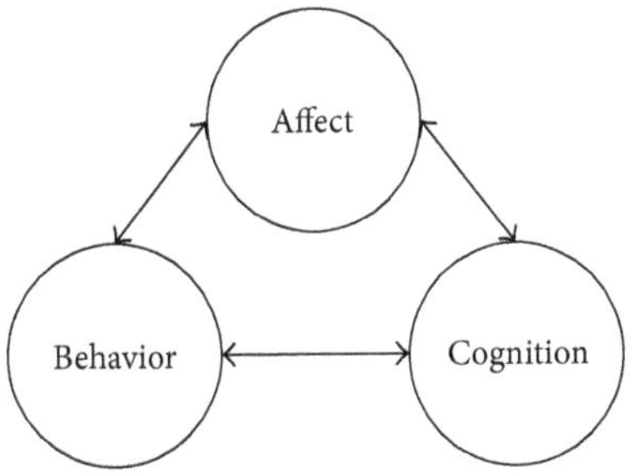

Figure 1: Interplay between affect, behavior, and cognition.

human manifestations. The model is a simplified representation of what happens within the human organism, separated into three components: affect, behavior, and cognition. The affect component describes the emotions that individuals experience during an emergency, that is, what people feel. The behavior component presents different possible behaviors during an emergency, that is, what people do. The cognitive component presents the decision-making process that people execute during an emergency, that is, what people think. As Figure 1 illustrates, these three components bilaterally influence each other.

The result of this interplay characterizes the human reactions in the face of an emergency on a high level of abstraction. In practice, the three components are not easily distinguishable because humans are unified organisms with complex internal interrelationships. However, separating what is essentially a united system into three integrated components helps us approach human reactions in emergencies from multiple points of view and beyond the "black-box" approach in which behavior is the only measurable and perceivable output.

Figure 2 shows the RESCUER Human Reaction Model. Each of the components will be described in the following sections. We will start with the cognitive component, as this component is responsible for the basal thinking process [20]. For a better understanding of Figure 2, we also recommend reading the model starting from the cognitive component at the bottom of the picture.

3.1. Cognitive Component. The cognitive component represents the decision-making process of people in an emergency situation [16]. This process takes place in the human working memory [17] and can be subdivided into three phases: (1) risk identification, (2) risk assessment, and (3) risk reduction.

During risk identification, people assess whether there is a real threat. Factors that influence this decision are the credibility of the authority that is issuing a warning, environmental cues, and the reaction of the social group. In large-scale event scenarios, information coming from authorities (e.g., police or security personnel) is interpreted as being more trustworthy than information from civilians. Environmental cues are a stimulus for gathering information to select the appropriate response. For example, when hearing a fire alarm, people tend to look for a sign of fire, the smell of smoke, or other perceivable indicators of a risk. The third factor is the social group. People look for confirmation that there is a threat by observing the behavior of other people nearby who would theoretically also be affected by the threat. If other people have identified a threat and react accordingly, this will increase people's personal confidence in the severity of the threat. For example, in the case of a fire alarm at a company, employees will observe their colleagues' response. Do they leave the workplace in the direction of a meeting point, or is it a false alarm? If there are inconsistencies between the three factors, people become uncertain and may remain in a closed loop of risk identification until the inconsistencies are resolved. In this state of uncertainty, people tend to follow others until they are able to determine if there is a risk for themselves.

If people recognize that the threat is real, they initiate a risk assessment to determine their individual risk [16]. This involves analyzing the content of the emergency warning, past experiences, and their own knowledge about the threat and its expected consequences. The issued emergency messages should provide information about the probability of impact, the proximity of such an impact, and the severity of the impact. The more detailed the emergency warning, the higher its credibility, whereas the more inconsistent or unclear the message, the harder it is for people to assess the real risk [16]. Past experiences with emergency situations may influence the overall risk assessment positively or negatively. For example, people who have frequently experienced false alarms tend to take alarms less seriously. The third factor influencing the risk assessment of an emergency situation is previous knowledge about a specific threat. For example, if people are well prepared for abandoning the building in the case of a fire alarm and know what the available resources are for reacting to a fire, they will assess the situation to be less severe compared to an emergency they have no knowledge about or for which they have no resources.

Based on the severity level of the emergency, people determine the effectiveness and their chances of responding to a threat in a particular way. If they think that they stand a good chance, they will contemplate the possible action scenarios to reduce their personal risk and/or minimize the impact of the threat and choose one plan. This could, for example, be an official plan, that is, a plan recommended by authorities in charge of security and safety. Once a plan is selected, the feasibility of its execution is considered against factors such as the available time and other resources. People then display the behavior that corresponds to the plan they have in mind. People who decide that a reaction is impossible or insufficient do not take any protective action against the threat at a cognitive level.

3.2. Behavior Component. This component describes how people act during an emergency. Environmental characteristics influence the action plan chosen by the people involved in an emergency. Table 1 presents typical behaviors in emergencies [14].

Figure 2 presents typical relations between the decision-making process in the cognitive component and the resulting behavior during an emergency with the evacuation of people in large-scale event scenarios. Particularly when

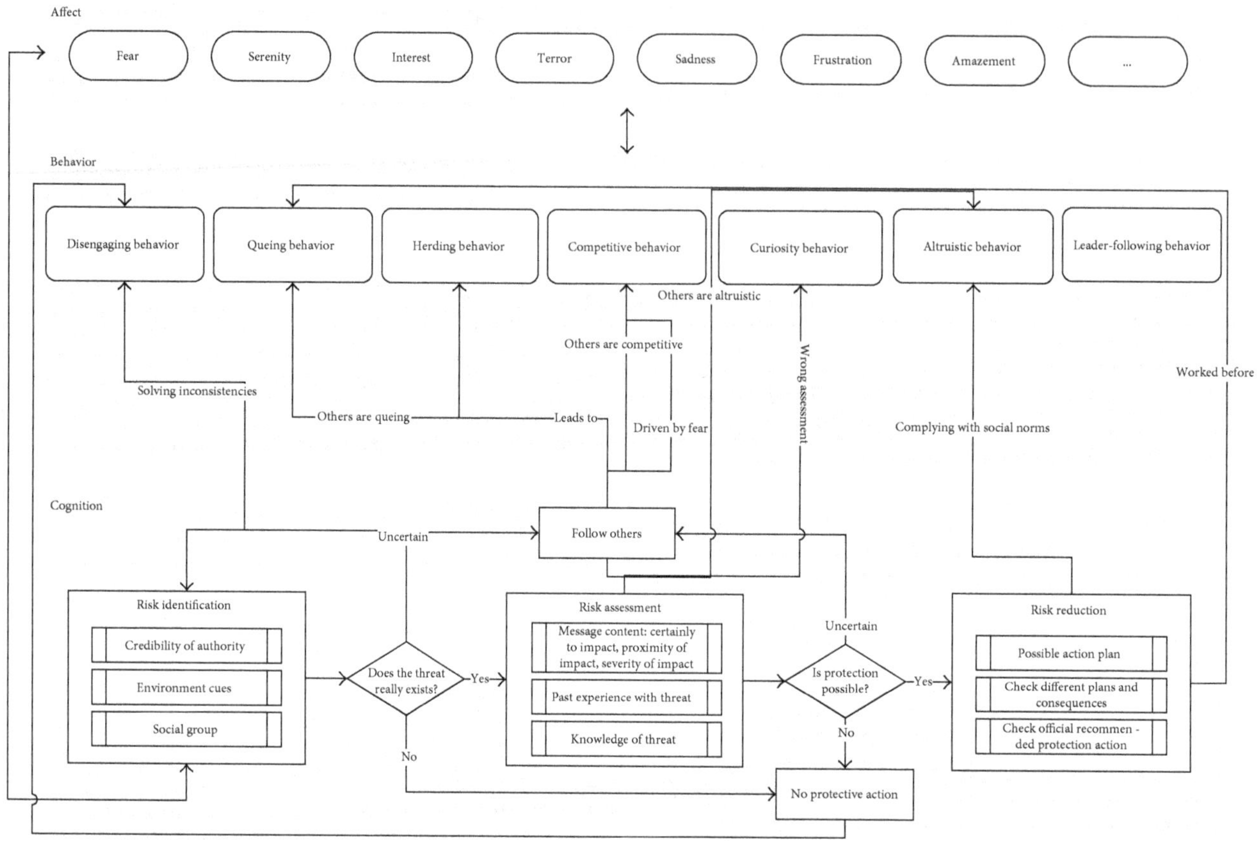

Figure 2: RESCUER Human Reaction Model.

Table 1: Typical behaviors in emergencies.

Behavior name	Behavior description
Disengaging behavior	Individuals are withdrawn and inactive. They do not look for an exit, adding to their both own risk and that of others, as they block exits and escape ways and interrupt the evacuation flow.
Queuing behavior	People self-organize queues to get to the exits; the crowd streams out in an orderly way. Queuing behavior does not lead to blockages, but to a more effective evacuation.
Herding behavior	People evacuate a room through one obstructed exit, while other exits are not fully utilized. This is typical behavior of individuals who are uncertain about what to do and decide to follow others.
Competitive behavior	Individuals compete to exit, violate the personal space of others, and do not respect social structures and rules.
Curiosity behavior	People move closer to the threat, do not take any protective action, take photos and discuss the situation with other curious individuals. This phenomenon is known as “disaster tourism.”
Altruistic behavior	People help other individuals during the evacuation, especially when the other person is impaired. People observing the willingness of others to help also tend more towards helping. People believe they are following social norms or are being ethical by helping others in need.
Leader-following behavior	People tend to follow others of a higher hierarchical rank during an evacuation. Sometimes a leader emerges when the situation is uncertain to many people.

people should leave the site of an incident but encounter inconsistencies while identifying the risk, they may tend to disengage and fail to react to the threat until the inconsistencies are resolved. In the case of high uncertainty, people tend to follow others, exhibiting queuing, herding, or competitive behavior. If people decide to take protective actions, they will most likely follow an assumed leader, for example, a family member who is deemed to be more rational or an officer of an operational force.

3.3. Affect Component. The affect component refers to a person’s feelings and the intensity of those feelings, that is, the emotions triggered during an emergency in response to physiological and cognitive factors [21]. Plutchik [21] describes eight basic human emotions, consisting of four opposite pairs at different degrees of intensity, which are presented in Figure 3.

Negative emotional states are strongly associated with emergencies involving stress. Feelings such as fear, anxiety,

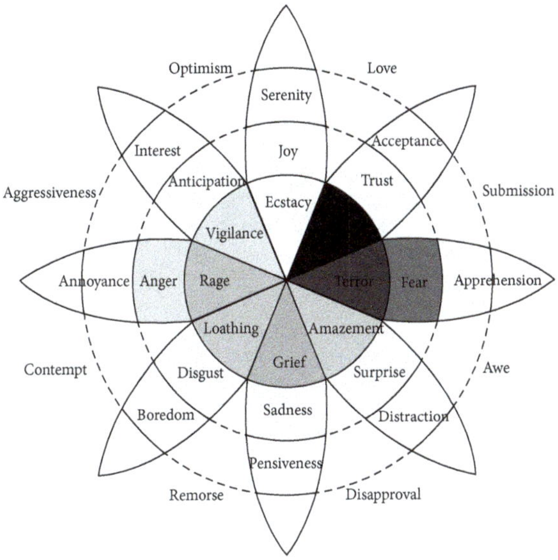

FIGURE 3: Plutchik's wheel of emotional states [21].

irritability, embarrassment, depression, helplessness, euphoria, frustration, and hostility are the most common feelings associated with emergencies [20]. Some of these typical feelings are associated with certain behaviors, but the direct correlation between specific feelings, behavior, and cognition is very complex and difficult to characterize in detail.

Emotions are among the most obvious (and best visible) aspects of human reactions. Humans are capable of perceiving the physiological response of others, such as variations in heart rate, respiration, and facial expressions, and thereby interpret their emotional state. This, in turn, may influence their own emotions. This is why the human reaction in emergencies is often described in terms of affect, and why the emotions experienced and observed in a crowd form an important stimulus for individuals in determining how to relate to the situation [22]. For example, if people in the crowd feel anxiety, they show it through their expressions and behavior, which may influence others who perceive it.

3.4. Reaction Changes, Stress Factors, and Interaction Capabilities. Even when an individual has developed a certain reaction in response to an emergency, this reaction can still be subject to change. The cognitive decision-making process is continuous and capable of adapting quickly. New or changed environmental stimuli received are used to reassess the severity of the situation and adapt the action plan accordingly.

Stress is one of the key factors affecting human reactions in an emergency situation. According to Staal [19], the definition of stress as the interaction between the following three elements includes the contemporary assumptions about the concept of stress:

(i) Perceived demand is the requirement for executing the plan elaborated by the individual to protect him/herself and to minimize the threat.

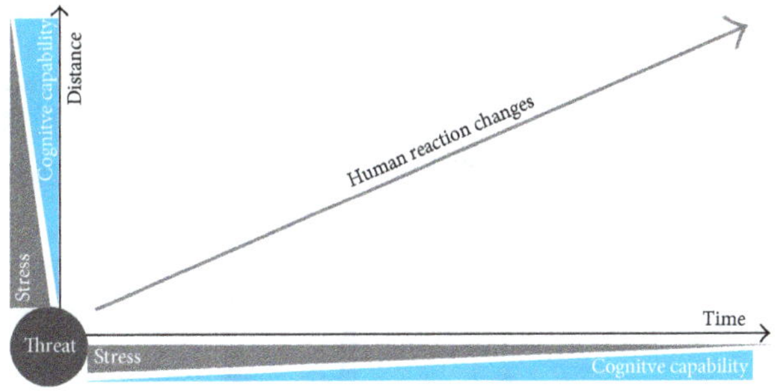

FIGURE 4: Changes in stress, human reaction, and cognitive capabilities.

(ii) Perceived ability to cope with the demand is the subjective assessment of the resources the individual has available to execute the plan.

(iii) Perception of the importance of being able to cope with the demand is the affect-based motivation of the individual to execute the plan and achieve goals.

This definition implies that as long as an individual asserts that he or she can respond to the stressor using the currently available resources and has the internal motivation to do so, the stress level will be low. In contrast, if the resources or the motivation cannot fulfill the demand, the perceived self-efficacy will be low, causing the stress level to be high. In this context, the distance and time from the threat affect the three elements that describe stress. The closer an individual is to the threat, the higher his or her level of stress will be. Similarly, right after the perception of a threat, the level of stress will be heightened by the individual's state of uncertainty and anxiety. Figure 4 illustrates this influence of distance and time on the level of stress.

(i) *The time span* between the occurrence of the threat that triggered the emergency and the first moment the person is confronted with the situation is a factor that influences the stress level. The more recent the threat occurred, the less information about the emergency is available and the fewer actions have been taken to control the situation. This causes stress, because people at the site of the incident causing a threat have not had enough time and information to assess the threat and they are primarily concerned with protecting themselves and possibly others.

(ii) *The distance* between the exact place of the incident causing a threat and the person has a direct impact on the stress level. The closer the victim's position to the epicenter or danger zone of the threat, the higher the physical threat (e.g., heat, noise, and water).

Because stress impairs decision-making [19], fewer stimuli are attended to. Particularly affected are the encoding, rehearsal, and retrieval of information. This means that individuals are less capable of understanding and recognizing information when under stress, especially the closer they are to a threat in terms of time and distance. When the stressor is of an emotional nature (e.g., related to a state of anxiety), peripheral events are remembered even

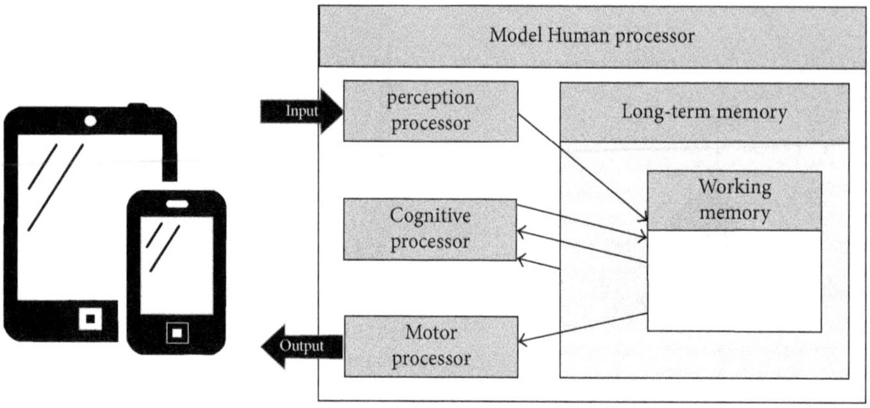

FIGURE 5: Model human processor adapted from Card et al. [23].

less often [19] because individuals tend to use as many mentally available resources as possible on solving the primary task, which is the decision-making process. The reduced attention for peripheral events affects the perception and processing of the perceived stimuli, causing changes in the environment that are not directly related to the primary task to be overlooked or to be registered without processing.

The effects of stress on the cognitive capabilities of a person have major implications on the interaction concept for software systems designed to interact with persons in an emergency situation. Interacting with a software system during an emergency situation requires focusing one's attention and dedicating resources from the working memory to the interaction, especially if the system asks a person to report on the perceived environmental cues and characteristics. Figure 5 presents a model human processor, which reflects how the processing of information perceived during the interaction with a software system occurs in working memory [23]. As part of the interaction with the software system, not just the perception plays a role but also the way of providing input, for example, by operating a touch screen. This is why the preparation of the motor processor to respond to the human–computer interaction stimuli is also required [24] and included in the model human processor accordingly. Breiner [25] describes this interaction process as follows:

> *All information that is perceived by a human (stimulus, e.g., visual, auditory, tactile) acts as input for the perception processor. After the execution phase of this processor, the information is available in the working memory. Based on the long-term memory, the information of the working memory is processed by the cognitive processor. The result of this process is again stored in the working memory. The outcome of the human computation in the working memory then triggers the motor processor to execute the desired action (e.g., press a button, make a statement, etc.). Elements of the working memory are linked to the corresponding elements in the long-term memory. Based on previous experience, data and actions that will be executed depending on the content of the working memory are stored in the long-term memory.*

The stress level may impair people in their interactions with software systems—in our case, a mobile application—because the available resources are being used for the decision-making process. Interaction with mobile devices therefore requires the stress level to be sufficiently reduced to have enough information processing resources available for the interaction.

Additionally, several authors [19] have demonstrated that well-learned tasks tend to be more resistant to the effects of stress. The practice and frequent activation of an operation tend to commit this operation to the long-term memory, which is less susceptible to being influenced by stress. The activation of these memories would allow the user to perform actions with a kind of automaticity, for which fewer mental resources are required during planning and execution [19]. Hence, if the required interaction in an emergency situation is practiced when no emergency is taking place, this may make the interaction easier, or even make it possible for an interaction to take place that would not have been possible without practice.

4. Interaction Modes and Requirements

Considering the RESCUER Human Reaction Model presented in Section 3, concrete requirements for the interaction design of the RESCUER App were defined in order to support the participation of the crowd in the response process of an emergency. As the level of stress in an emergency has an impact on people's ability to interact with mobile devices, our solution strategy provides an interaction spectrum for information gathering with three interaction modes: one-click interaction, guided interaction, and chat interaction. According to our theory, the closer the distance to the exact place of the incident and the more recent the incident, the more information has to be gathered automatically, because the cognitive capabilities of the people present will still be limited so that they cannot provide all the information a command center requires for the emergency response process. Figure 6 depicts this relationship between the available cognitive capabilities and the level of stress in an emergency, indicating when each of the aforementioned interaction modes is most appropriate.

The current characterization of the three modes is as follows:

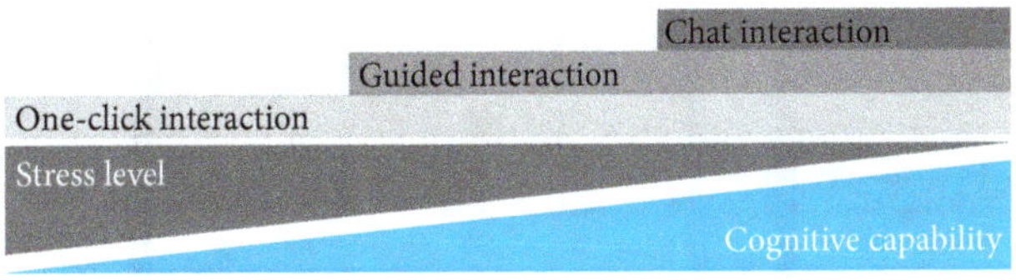

FIGURE 6: Stress, cognitive capabilities, and interaction modes.

(i) *One-click interaction*: Since people's cognitive capability is mainly consumed by their decision-making processes in response to the emergency situation itself, they are unable to respond to any request or complete any task through the user interface. Therefore, any mobile application to be used in an emergency situation has to be able to automatically gather all the information about the situation that is relevant for the workforces.

(ii) *Guided interaction*: When people are still not able to respond to complicated interaction demands but can at least provide a little information about the emergency, a guided approach ensures that the interaction is kept as simple as possible. These people are likely still near the site of the incident and are too distracted to execute a large set of actions with their mobile devices. Therefore, it is advisable to pose precise questions (e.g., by using multiple-choice selections) and ask them to execute easy and fast actions (such as sending them to specific areas of the venue).

(iii) *Chat interaction*: When eyewitnesses of an emergency are safe, they may become observers and can contribute information about the situation more intensively and actively, and at a greater level of detail. Their cognitive load is no longer fully consumed by the situation, and their ability to interact with their mobile devices the way they usually do has increased. Due to the physical distance and the elapsed time, the information that can be gathered from the physical sensors of the mobile device might not be as important as in the other modes.

As people using an interactive system in an emergency situation cannot be pressed into a concrete interaction mode as it is difficult to know their specific level of stress, the proposed interaction spectrum spans an escalation order of the three interaction modes. Interaction through in-air gestures and device shaking are not part of any proposed interaction mode. First, they would require users of the mobile application to previously learn the expected movements. Second, the user must be protected from entering wrong input or input that may lead to dramatic consequences. When the level of stress is high and the cognitive capabilities are consequently low, the use of these interaction means is too error-prone. On the other side, when the level of stress reduces and thus the cognitive capabilities increases, users are expected to provide more information than is possible through in-air gestures and device shaking.

TABLE 2: Requirements of the interaction modes.

Interaction mode	UI requirements
One-click interaction	As users are highly impacted by stress or cognitive overload, a single click is the most complex interaction that can be expected. This can be taken as a confirmation of intent, resulting in fully automated information gathering. This means that all mechanisms responsible for gathering data can be activated.
Guided interaction	As users are not directly impacted by the incident anymore and cognitive workload is available again, more focused interaction can be expected (but not too complex). This can be implemented through simple multiple-choice selection, where the user is guided to a possible answer by being confronted with various answers. By recognizing the correct (or most suitable) answer, the user is able to provide more precise information about the incident.
Chat interaction	Unstructured/open information. Having almost their complete cognitive power available, users can execute more complicated tasks. This means reacting to individually defined tasks or providing a description in plain text. Here, basically all interaction possibilities can be integrated into the app.

Table 2 presents concrete user interface requirements that were elaborated on the basis of the different interaction modes introduced above.

5. The RESCUER App

Figure 7 illustrates a prototypical implementation of the three interaction modes in the RESCUER App. We assert that, as long as people have enough cognitive capabilities, they will interact with the application and move from one interaction mode to the next. If the working memory can no longer support the interaction with the RESCUER App, people will naturally stop using it and focus on the main decision-making task, for example, fleeing from the site of the incident.

Figure 8 presents the navigation in the case of a fire emergency. When touching the fire icon (orange), the user is taken to a confirmation screen and subsequently to a screen where he or she can describe the characteristics of the fire. It is also possible to take pictures and submit the report. Next, the chat screen appears, where the user can provide additional information as well as send more photos and videos to the command center. This means that the interaction modes appear in the same order as the model describes: the easiest screen first, that is, one-click interaction, followed by the more complex screens, namely guided interaction and chat interaction, which is the last screen.

In the one-click interaction mode, people have to select which type of emergency they want to report. They are presented with a set of predefined options of typical incident types (fire, explosion, gas leakage, and environmental emergency) in the application context of the RESCUER App. In addition to the emergency type, the RESCUER App

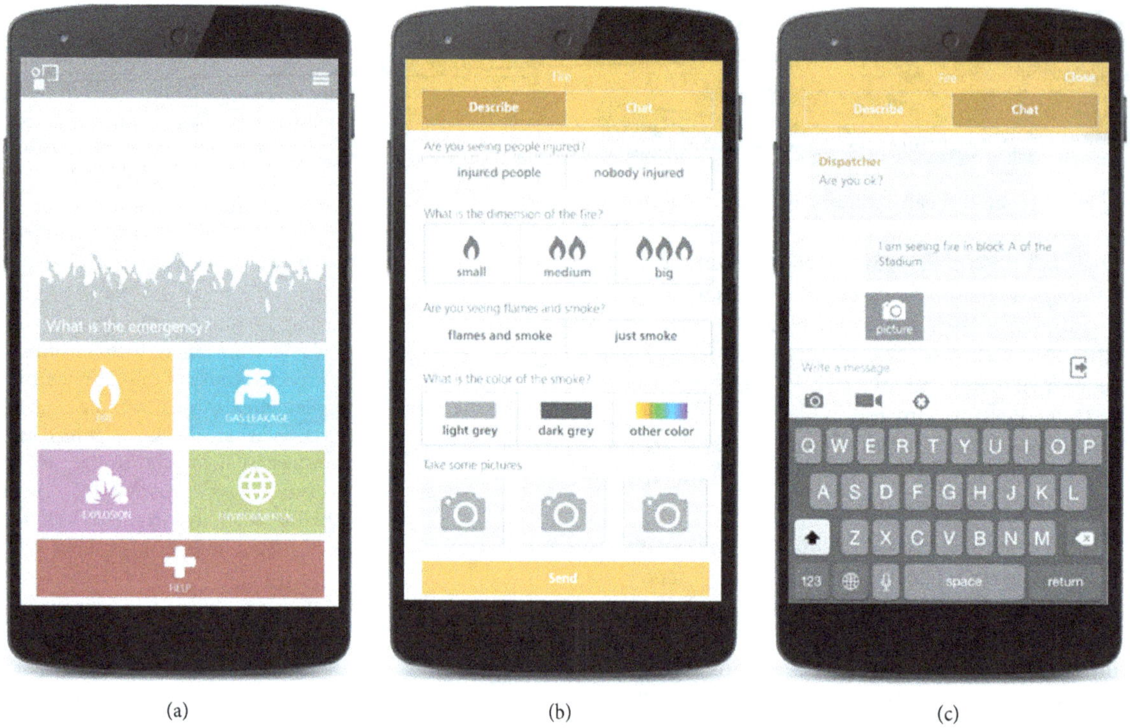

FIGURE 7: The RESCUER App and interaction modes. (a) One-click interaction. (b) Guided interaction. (c) Chat interaction.

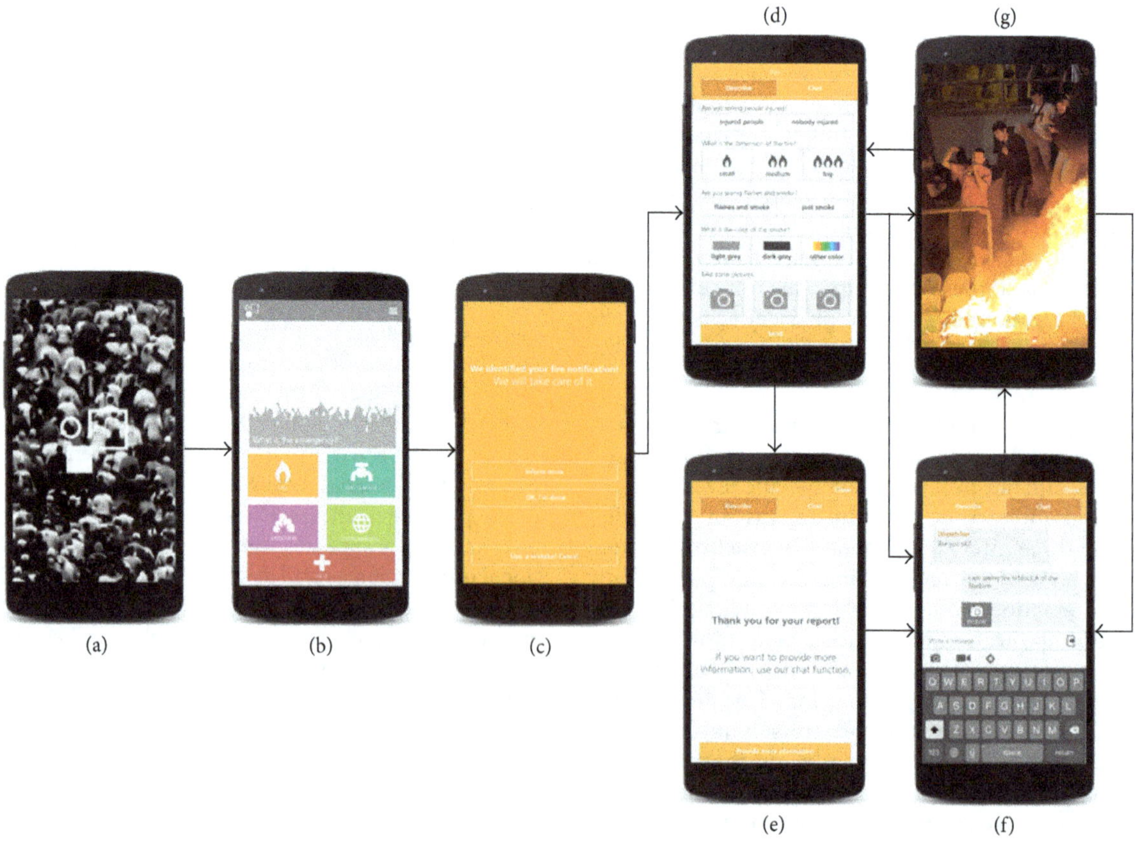

FIGURE 8: Navigation of the RESCUER App through the interaction modes.

automatically provides information about the position of the person sending the report and the time at which the report was sent. Following this one click, the RESCUER App can automatically collect relevant data from users located close to the site of the incident with the help of the physical sensors of the users' mobile devices.

In the guided interaction mode, people answer typical questions that characterize the emergency situation by reporting what they see and send pictures. During the development of the RESCUER App, workforce representatives have helped to select the most relevant questions to be answered by either the general crowd or the workforces at the site of the incident (see screen (b) in Figure 7).

The chat interaction mode screen is a typical chat functionality as encountered in typical mobile communication tools. It allows people to write free text, dictate text, and send pictures and videos to provide information on their own initiative, or in response to specific questions asked by the command center.

6. Experiment

To validate the defined interaction modes and to identify improvement potential for the RESCUER App, we performed an experiment in a laboratory environment with 26 participants. The aim was to explore different aspects of the theory that supports the definition of the interaction modes as well as to observe how people interact with the design conceived for the RESCUER App when exposed to different degrees of stress. Four main research questions guided the experiment with the RESCUER App.

6.1. Research Questions

> RQ1. Mental effort increases from one mode to the next in the following order: one-click interaction, guided interaction, and chat interaction. This increase becomes more intense under high cognitive load conditions.

We assumed that users would perceive low mental effort when interacting with the screen representing the one-click interaction, that the perceived mental effort would increase to a medium level when interacting with the screen representing the guided interaction mode, and finally, that people would perceive high mental effort when interacting with the screen representing the chat interaction mode. This increasing perception of mental effort was expected to be further intensified during the execution of an additional task they must perform in parallel, that is, when participants experience a higher cognitive load. In this research question, we have two independent variables, namely, interaction mode and cognitive load. The dependent variable, which we measured with the help of the Subjective Mental Effort Questionnaire (SMEQ) [26], was the perceived mental effort. The analysis method we used was a repeated measures ANOVA with two within-subject factors.

> RQ2. Previous involvement with an emergency situation does not affect the perceived mental effort when using the RESCUER App under different levels of cognitive load.

The aim of the RESCUER App's design was to develop a mobile application that is easy to use in cognitively demanding situations regardless of a person's previous experience with emergencies. We assumed that having previous emergency experience results in no differences in the SMEQ ratings for the different interaction modes and the different levels of cognitive load, which means that it should be possible to use the RESCUER App equally comfortably, regardless of previous experience with emergencies. To analyze this research question, we used the repeated measures ANOVA model from RQ1 and included a third independent variable relating to previous emergency experience with two conditions (no previous experience versus previous experience) as a between-subject factor. The dependent variable is the perceived mental effort (SMEQ) as in RQ1.

> RQ3. The duration of the interaction with the RESCUER App increases from one mode to the next in the following order: one-click interaction, guided interaction, and chat interaction. This increase in duration is larger in high cognitive load conditions.

With each new screen in the RESCUER App, additional interaction possibilities arise. Therefore, we assumed that the time spent on each screen would increase with the added functionalities. We expected the shortest time span for screen 1 (one-click interaction), a medium time span for screen 2 (guided interaction), and the longest time span for screen 3 (chat interaction). Furthermore, if asked to execute an additional parallel task (an n-back task as presented in Section 6.2) while using the mobile application, participants would need even more time to interact with the screens in the three interaction modes. Again, we used interaction mode and cognitive load as independent variables. As the dependent variable, we measured the time in seconds spent on each screen. This results in a repeated measures ANOVA with two within-subject factors.

> RQ4. Even under high cognitive load conditions, participants are able to use the mobile application and provide correct answers for a parallel task, the n-back task.

The design of the RESCUER App is optimized for usage in cognitively demanding and stressful situations. Consequently, it should be no problem to use the app and still react to external stimuli. As external stimuli, we used the n-back task, specifically as one-back task (see Section 6.2 for a description). We assumed that when using the RESCUER App, it is still possible to provide mainly correct answers in the one-back task. In this research question, we investigated the participants' answers in the one-back task, that is, correct answers, incorrect answers, and omitted answers in response to the stimuli (used as dependent variables). We assume that incorrect answers and omitted answers do not differ significantly from zero. Therefore, we use the one-sample t-test statistic.

In addition to the above research questions, we wanted to identify existing usability problems and opportunities for improving the usability and user experience (UX) of the

Table 3: Experiment design with interaction mode (factor 1) and cognitive load (factor 2).

		Factor 1: interaction mode		
		(A) One-click interaction	(B) Guided interaction	(C) Chat interaction
Factor 2: cognitive load	No/low cognitive load (task 1)	X_{11}	X_{12}	X_{13}
	High cognitive load (task 2)	X_{21}	X_{22}	X_{23}

RESCUER App. Especially in the context of an emergency situation, it is very important that the RESCUER App is without usability problems so that a fast and precise interaction of the crowd with the command center is guaranteed. Therefore, we performed a descriptive analysis of items from a usability questionnaire and implemented one-sample t-tests to check for significance.

6.2. Study Design and Independent Variables. The study design comprised a 2×3 within-subject design, that is, each participant used the RESCUER App twice to report an emergency under different levels of stress caused by cognitive load: the first time with no additional parallel task, resulting in a no/low cognitive load condition (task 1), and the second time with high cognitive load imposed by a parallel task (task 2). Table 3 shows the experimental combination of the independent variables *interaction mode* (factor 1) presented by the RESCUER App in three modes (one-click interaction versus guided interaction versus chat interaction) and *cognitive load* (factor 2) with two levels (no/low cognitive load versus high cognitive load). All participants did the experiment under all experimental conditions.

To induce a higher cognitive load in task 2, we used an audio-based n-back task. An n-back task is a memory task that requires the participant to remember whether the presented stimulus is the same as a stimulus presented n iterations previously [27]. The n-back task traditionally contains spoken word stimuli, but such stimuli could interfere with the language processing required for operating the mobile application [28]. In our study, the auditory stimuli were tones presented for 1000 ms, followed by a 2000 ms interval before the next tone. N-back approaches usually assume 3000 ms trials, consisting of a stimulus and an interstimulus interval. These trials may vary in setup; some use a 1500 ms stimulus with a 1500 ms interstimulus interval (e.g., Schuhfried [29]), while others use a 500 ms stimulus with a 2500 ms interstimulus interval (e.g., Beavon [27]). The stimuli in our study were three sonorous signals in the chromatic scale, with sine wave tones for "low" (261.63 Hz; middle C), "middle" (440 Hz; concert A), and "high" (698.46 Hz, F), generated and sequenced in Audacity (a free audio editor and recorder) [30]. The tones were placed in a randomized order according to a numeric sequence (1 for "low," 2 for "middle," and 3 for "high") generated at http://www.random.org, which was modified to repeat the same tone three times at most. During the execution of the experiment, the tones were played as 320 kbps MP3 files from the moderator's computer. In our experiment, the participants had to perform a one-back task, meaning they had to remember whether the last tone played

Figure 9: Stimulus 1 used for the description of the incident (photo taken from Turner [31]).

Figure 10: Stimulus 2 used for the description of the incident (photo taken from Wikimedia [32]).

was the same as the tone played right before it. If the tones were identical, the participants were to say "yes," and if they differed, they were to say "no."

6.3. Materials. To support the variation of factor 2, two pictures were prepared to support the user in describing a fire situation in a soccer stadium (Figures 9 and 10). The order of the pictures was balanced across the participants over task 1 (no/low cognitive load) and task 2 (high cognitive load). This means that half of the participants were presented with the picture in Figure 9 when performing task 1 and the picture in Figure 10 when performing task 2, and vice versa for the other half of the participants. By balancing the fire pictures, we wanted to reduce the effects that the specific style and event of each picture might cause in the description of the incident when using the RESCUER App.

Because the experiment was conducted in an early phase of the project's development process, we implemented an interactive prototype [33] of the RESCUER App using the prototyping tool Axure Pro 7. The participants were able to navigate between the screens, answer the questions of the guided interaction mode, and type information in the chat interaction mode. The camera function was simulated. The participants had to simulate taking a picture during the study, that is, by pointing the camera at the picture of the event and clicking on the trigger on the screen, following which a sample picture was loaded into the RESCUER App.

6.4. Measurement of the Dependent Variables. Several resources were prepared for the measurement of the dependent variables. In this section, we describe the objective and subjective methods used to investigate our research questions.

For the characterization of the participants, a demographic questionnaire was prepared with questions related to their age, gender, highest level of education, musical skills, and previous experience with emergencies (in terms of whether they have previously been involved in an emergency and how often they have called the emergency number). The participants were also asked to describe positive and negative aspects of their previous experience with making an emergency call.

For measuring the mental effort, we used the Subjective Mental Effort Questionnaire (SMEQ) [26]. The SMEQ is a rating scale ranging from 0 (*not at all hard to do*) to 220 (*tremendously hard to do*). Nine textual items are distributed on a vertical scale at irregular distances. Schuette [34] describes the SMEQ scale as having a good implementing cost for the measurement of subjective mental effort as an indicator for cognitive load. Immediately after using the RESCUER App (in task 1 or task 2), the participants were asked to mark on the vertical line the perceived difficulty of the overall task and of the interaction modes (*one-click interaction*, *guided interaction*, and *chat interaction*). Scoring the paper version of the SMEQ during analysis requires measuring the distance in millimeters from the bottom of the vertical line to the marking made by the participants. The millimeter measurements are converted back to the rating scale of the SMEQ, that is, from 0 to 220.

The eye-tracking system ETG 18-1316-361 of SensoMotoric Instruments (SMI) was used for recording objective data. Eye-tracking systems are used to analyze the eye movements of individuals when they are interacting with a system [35]. Eye tracking allows measuring the degree of interest in certain areas of a user interface, the sequence of fixations on different areas of the user interface, as well as revisits or "stickiness" in certain areas [35]. Pupillary dilations and contractions can also be recorded using SMI's eye-tracking system. In our study, the eye-tracking system helped to precisely measure the time spent in each of the interaction modes in both tasks, namely no/low cognitive load (task 1) and high cognitive load (task 2).

In order to ensure that the participants were executing the *n*-back task correctly, an observer logged the answers provided by the participants when performing the *n*-back task. In order to do so, the observer checked the answers given by the participant against a template with the correct answers and marked the correct and wrong answers as well as omitted answers on the template.

Finally, to measure the perceived usability, a selection of items of the ISONORM 9241/10 questionnaire were used. The ISONORM 9241/10 questionnaire deals with general ergonomic principles that apply to the design of dialogs between humans and information systems [36]. It considers seven aspects of software ergonomics: *suitability for the task*, *suitability for learning*, *suitability for individualization*, *conformity with user expectations*, *self-descriptiveness*, *controllability*, and *error tolerance*. In the ISONORM 9241/10 questionnaire, each aspect of software ergonomics is explored by several questions. In our study, we selected the most relevant aspects for the RESCUER App and also selected the most relevant questions of each selected aspect in order to have a reasonable balance between the time required for answering the questionnaire and the time spent on using the mobile application itself. The participants assessed nine items from the following selected aspects: *suitability for the task*, *suitability for learning*, *conformity with user expectations*, and *self-descriptiveness*. One statement of the item pair was presented to the participants, who could assess this statement by using a 7-point scale from 1 (*does not apply at all*) to 7 (*fully applies*).

6.5. Experimental Setup. The study was performed over the course of three days in a laboratory-like environment. Each experimental run lasted between 30 minutes and one hour. No run exceeded one hour. The participants were invited individually into a meeting room with a table holding the experiment material, a large display with the pictures of the incident, and chairs. The participants performed the study in a sitting position. Figure 11 shows a schema of the experimental setup and pictures of the room where the experiment was performed.

Table 4 presents the step-by-step process of the experiment, which introduces the instruments used for inducing the independent variables and measuring the dependent variables. After the participants arrived in the experiment room, they received a short explanation about the experiment. They signed the informed consent form, and the eye-tracking system was calibrated. In order to get used to wearing the eye-tracking glasses, the participants completed the prequestionnaires while already wearing the eye-tracking glasses. After completing the prequestionnaires, the participants were given the instructions to the task and shown the picture of the fire. The instructions were as follows: "Please imagine being in the soccer stadium of Kaiserslautern. You see a fire on the opposite side of the stadium. You want to report this to the fire department using the mobile application. To give the fire department a better idea of the emergency, you decide to also send a picture of the fire. In the mobile application, you are going to see different screens. Please go through the application and try to use all features: forms, camera, chat function, location, etc." After giving the instruction, the moderator handed a smartphone with the RESCUER App installed on it to the participant, who could then immediately start to report the fire with the RESCUER App.

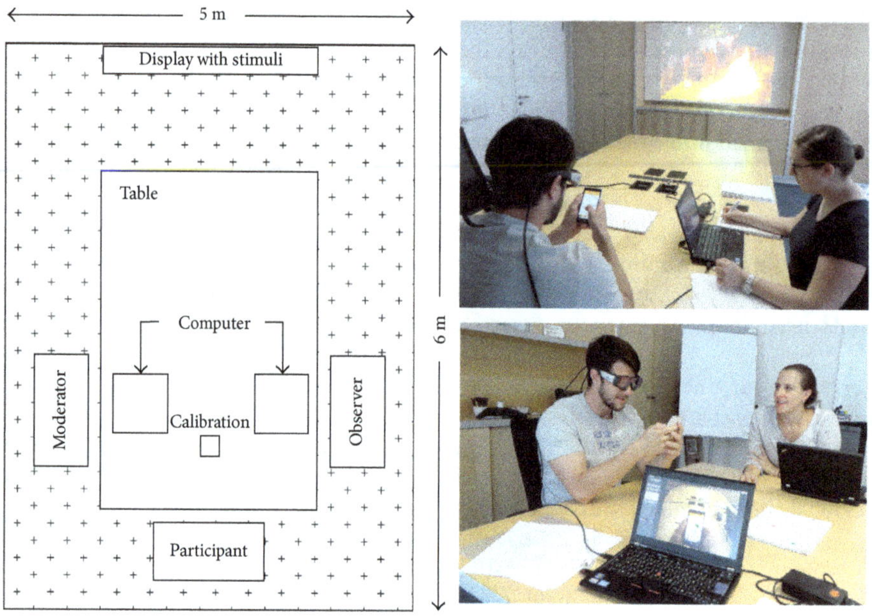

Figure 11: Experiment room and experimental setup.

Table 4: Experimental procedure.

Step	Stage	Activity	
Step 1	Welcome	(i) Introduction to the experiment (ii) Participation consent (iii) Eye-tracker calibration	—
Step 2	Prequestionnaire	(i) Demographic information (ii) Experiences with previous emergencies	
Step 3	Task 1	(i) Use of the RESCUER App (ii) Use of the eye tracker	
Step 4	Post-task 1 questionnaire	(i) SMEQ for each screen (ii) SMEQ overall	Eye tracker recalibration (if necessary)
Step 5	Task 2	(i) Use of the RESCUER App (ii) Use of the eye tracker (iii) Performance of the n-back task	
Step 6	Post-task 2 questionnaire	(i) SMEQ for each screen (ii) SMEQ overall	
Step 7	Postquestionnaire	(i) ISONORM 9241/10	—
Step 8	End	(i) Debriefing (ii) Explanation of study purpose	

After performing task 1, the participants completed the SMEQ for each interaction mode and also provided an overall assessment of task 1 using the SMEQ. As soon as they had completed the questionnaires, they received follow-up instructions for task 2 and the audio-based one-back task was demonstrated. The instructions for task 2 were the same as for task 1, only with the addition of "Do not forget to perform the one-back task while reporting the emergency shown in the picture with the help of the RESCUER App." After receiving the instructions, the participants performed task 2 describing a different picture of an emergency than the one they had described in task 1 (Figures 9 and 10). After completing this task using the RESCUER App, SMEQ questionnaires for each screen and an SMEQ questionnaire for the overall task were once again completed by the participants. If the stress level was too high for the participants, they were allowed to interrupt the task. All participants were able to perform both tasks completely and without interruption.

Finally, the participants completed the ISONORM 9241/10 questionnaire and answered open-ended questions to provide feedback about their experience with RESCUER App. A detailed explanation about the study was offered to participants who wanted to know more about it.

6.6. Subjects. The participants of this study consisted of 19 male and 7 female employees from Fraunhofer IESE ($N = 26$). The age of the participants ranged from 20 to 53 years

Table 5: Mean estimates of the SMEQ for all six experimental conditions ($n = 26$).

Interaction mode (factor 1)	Level of cognitive load (factor 2)	Mean (M)	Standard error (SE)	95% confidence interval	
				Lower bound	Upper bound
One-click interaction	No/low load	19.50	3.18	12.96	26.04
	High load	44.23	6.46	30.92	57.54
Guided interaction	No/low load	69.15	8.81	51.00	87.30
	High load	122.19	9.73	102.15	142.23
Chat interaction	No/low load	105.58	10.94	83.04	128.11
	High load	166.27	7.51	150.80	181.74

($M = 32.77$, SD = 7.63). According to the Shapiro–Wilk test, age was distributed normally: $W(25) = 0.93$, $p = 0.09$. However, the distribution is skewed to the right, with skewness of 0.85 (SE = 0.46) and kurtosis of 1.27 (SE = 0.89). All the participants volunteered for the study and received no compensation.

The participants had a high level of education: 77% of the participants were graduates, 8% had a Ph.D., 11% had a degree in applied sciences, and 4% a secondary-level certificate. All the participants in the experiment possessed a smartphone: 53.84% used an iPhone, 34.62% were Android users, and a minority of 11.54% used Windows Phone.

With respect to prior experience with emergency situations, 42.31% of the participants had never been involved in an emergency situation, while 57.69% of the participants had already experienced an emergency situation (e.g., as an eye witness). The most typical emergencies were traffic accidents (40%), followed by fire (24%), medical emergencies (16%), and gas leakage (8%). Other kinds of incidents totaled 12% of the incidents. Of the 26 participants, 16 had never made an emergency call, six participants had made an emergency call twice, and three participants once. One of the participants was a member of a volunteer workforce and had performed an emergency call four times on duty.

7. Results

In this section, we report on the results separately for each of the research questions described in Section 6.1. First, we will analyze how the perceived mental effort when using the RESCUER App changes, taking into consideration the different interaction models and the levels of externally induced cognitive load (Section 7.1). Secondly, we will then check whether previous experience with emergency situations had an influence on the participants' experience of mental effort (Section 7.2). Thirdly, we will examine the duration of using the RESCUER App under different levels of cognitive load (Section 7.3). Fourthly, we will investigate the results of the one-back task, and how this objective measure relates to the subjective measure of perceived mental effort (Section 7.4). Lastly, we will examine the participants' overall usability evaluation of the RESCUER App based on items from the ISONORM 9241/10 questionnaire (Section 7.5). All statistical analyses were performed using SPSS 19. Unless otherwise stated, all statistical tests assume a significance level of $\alpha = 0.05$.

7.1. RQ1: Subjective Mental Effort of the RESCUER App under Different Levels of Cognitive Load. To analyze RQ1, we performed a 2×3 repeated measurement ANOVA with two within-subject factors (interaction mode and cognitive load) for the dependent variable SMEQ. Table 5 shows the mean estimates for each combination of the levels of the two factors (interaction mode $\times$ cognitive load).

Mauchly's test indicated that the assumptions of sphericity had been violated for the interaction effect of interaction mode (factor 1) and cognitive load (factor 2), $\chi^2(2) = 10.31$, $p = 0.006$, while for the main effects the assumption for sphericity was met (interaction mode: $\chi^2(2) = 0.65$, $p = 0.72$). Therefore, degrees of freedom were corrected for the interaction effect interaction mode $\times$ cognitive load using Greenhouse–Geisser estimates for sphericity ($\varepsilon = 0.74$).

There was a significant main effect of the interaction mode on the rating of the SMEQ, $F(2, 50) = 71.14$, $p < 0.001$, $\eta_p^2 = 0.83$. Contrasts revealed that the ratings for screen 2 (guided interaction) were significantly higher than for screen 1 (one-click interaction), $F(1, 25) = 52.48$, $p < 0.001$, $r = 0.82$, and ratings for screen 3 (chat interaction) were significantly higher compared to screen 2, $F(1, 25) = 24.40$, $p < 0.001$, $r = 0.70$. There was also a significant main effect of the administered level of cognitive load on ratings of SMEQ, $F(1, 25) = 94.08$, $p < 0.001$, $\eta_p^2 = 0.79$. The contrast shows a significantly higher rating of SMEQ in the high cognitive load task compared to the low/no-load task, $F(1, 25) = 94.08$, $p < 0.001$, $r = 0.89$.

There was a significant interaction effect between the interaction mode and the level of cognitive load applied, $F(1.48, 37.06) = 7.11$, $p = 0.005$, $\eta_p^2 = 0.22$. This indicates that the interaction mode used by participants has different effects on participants' rating of SMEQ depending on the cognitive load a person experiences. Figure 12 shows an ordinal interaction between interaction mode (a) and level of cognitive load (b) for ratings of SMEQ. To break down this interaction, contrasts were performed that compared each interaction mode against its preceding mode with fewer interaction possibilities (one-click interaction versus guided interaction, and guided interaction versus chat interaction) for each level of cognitive load. The first contrast revealed a significant difference when comparing no/low cognitive load to high cognitive load for one-click interaction compared to guided interaction, $F(1, 25) = 18.76$, $p < 0.001$, $r = 0.65$. The remaining contrast shows no significant difference when comparing no/low cognitive load to high cognitive load for guided interaction versus chat interaction, $F(1, 25) = 0.42$, $p = 0.53$, $r = 0.13$. However, this contrast yields a small effect size.

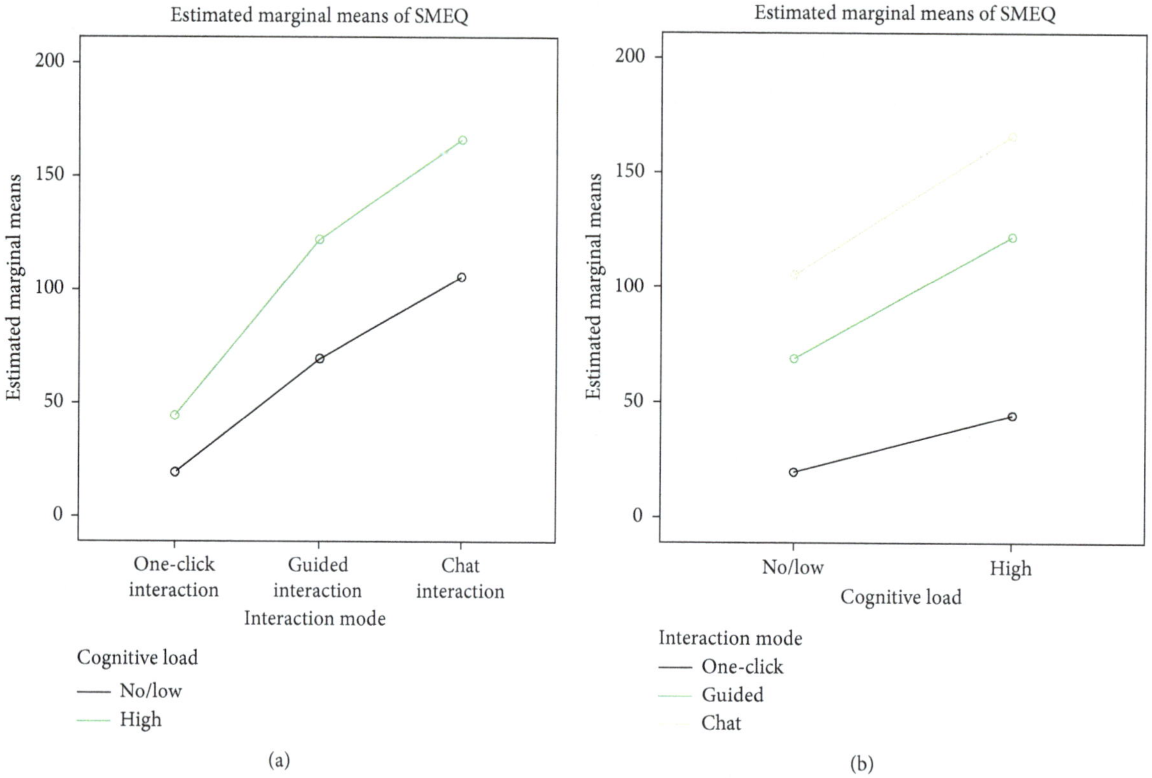

FIGURE 12: Interaction graphs for SMEQ. (a) Interaction mode. (b) Cognitive load.

TABLE 6: Mean estimates of SMEQ for all six experimental conditions separated according to previous experience with emergency situations.

Previous experience with an emergency (factor 3)	Interaction mode (factor 1)	Level of cognitive load (factor 2)	Mean (*M*)	Standard error (SE)	95% confidence interval	
					Lower bound	Upper bound
No ($n = 11$)	One-click interaction	No/low load	18.36	4.51	8.32	28.41
		High load	32.64	4.86	21.80	43.47
	Guided interaction	No/low load	52.64	9.82	30.75	74.52
		High load	103.27	13.89	72.33	134.21
	Chat interaction	No/low load	85.91	14.21	54.26	117.56
		High load	170.55	8.51	72.33	134.21
Yes ($n = 15$)	One-click interaction	No/low load	20.33	4.53	10.63	30.04
		High load	52.73	10.23	30.80	74.67
	Guided interaction	No/low load	81.27	12.85	53.71	108.82
		High load	136.07	12.63	108.97	163.16
	Chat interaction	No/low load	120.00	15.16	87.49	152.51
		High load	163.13	11.62	138.22	188.05

7.2. RQ2: Effect of Previous Emergency Experience on SMEQ while Using the RESCUER App under Different Levels of Cognitive Load. In the analysis of this research question, we were interested in learning how previous experience with emergency situations might alter the effects of the interaction mode in SMEQ and experimentally induced cognitive load. Therefore, we included a third factor in the model, in which previous experience with emergency situations served as an additional independent variable (as a between-subject factor) in the ANOVA.

Of the 26 participants, 11 stated that they had no prior experience with an emergency situation, while 15 participants had already been involved in an emergency situation. Table 6 shows the mean estimates for SMEQ for the six experimental conditions, separately for participants with previous emergency experience (lower part of the table) and without previous emergency experience (upper part of the table). Interestingly, participants with previous emergency experience rated SMEQ in all experimental conditions higher than participants without previous emergency experience, with one exception: for the chat interaction under high cognitive load, participants without emergency experience rated SMEQ higher.

Mauchly's test indicated that the assumptions of sphericity had been violated for the interaction effect of interaction mode (factor 1) and cognitive load (factor 2),

TABLE 7: Test results of the within-subject effects.

Source	F	df	df_{error}	p	η_p^2
Interaction mode	68.19	2	48	<0.001	0.74
Interaction mode × emergency experience	0.72	2	48	0.49	0.03
Cognitive load	91.69	1	24	<0.001	0.79
Cognitive load × emergency experience	0.43	1	24	0.52	0.02
Interaction mode × cognitive load[1]	10.03	1.52	36.41	0.001	0.30
Interaction mode × cognitive load × emergency experience[1]	5.56	1.52	36.41	0.01	0.19

[1]Corrected degrees of freedom using Greenhouse–Geisser estimates for sphericity.

$\chi^2(2) = 8.82$, $p = 0.012$. Thus, degrees of freedom were corrected for the interaction effect interaction mode × cognitive load and interaction mode × cognitive load × emergency experience using Greenhouse–Geisser estimates for sphericity ($\varepsilon = 0.76$).

The between-subject effect for emergency experience showed no significant result, $F(1, 24) = 3.36$, $p = 0.08$, $\eta_p^2 = 0.12$, indicating that participants with previous emergency experience and participants without previous emergency experience gave generally the same SMEQ ratings. The results for the within-subject and interaction effects are listed in Table 7.

The results show a significant main effect of interaction mode. Similarly to the previous analysis, contrasts show a significantly higher rating of the SMEQ for guided interaction compared to one-click interaction, $F(1, 23) = 49.28$, $p < 0.001$, $r = 0.83$, and a higher rating for chat interaction compared to guided interaction, $F(1, 23) = 25.55$, $p < 0.001$, $r = 0.73$. In addition, there is a significant main effect of cognitive load. Again the contrast shows that high cognitive load conditions yield a higher rating of SMEQ compared to no/low cognitive load conditions, $F(1, 24) = 91.69$, $p < 0.001$, $r = 0.89$.

There was no significant interaction effect for both interaction mode × emergency experience and cognitive load × emergency experience. However, the interaction mode × cognitive load interaction effect was significant, $F(1.52, 36.41) = 10.03$, $p = 0.001$, $\eta_p^2 = 0.30$. Correspondingly, this result indicates that the participants' SMEQ ratings differed for the three interaction modes depending on a person's cognitive load. To break down this interaction, contrasts were performed that compared each interaction mode against its preceding mode with fewer interaction possibilities (one-click interaction versus guided interaction and guided interaction versus chat interaction) for each level of cognitive load. The first contrast revealed a significant difference when comparing no/low cognitive load to high cognitive load for one-click interaction compared to guided interaction, $F(1, 24) = 19.83$, $p < 0.001$, $r = 0.67$. The second contrast shows no significant difference when comparing no/low cognitive load to high cognitive load for chat interaction when comparing it to guided interaction, $F(1, 24) < 1$, $p = 0.34$, $r = 0.19$. However, this contrast yields a small effect size.

Finally, the interaction mode × cognitive load × emergency experience interaction was significant, $F(1.52, 36.41) = 5.56$, $p = 0.01$, $\eta_p^2 = 0.19$. This indicates that the interaction mode × cognitive load interaction described previously was different for participants without previous emergency experience and for participants with previous emergency experience. As can be seen in Figure 13, participants with previous emergency experience yielded a higher SMEQ rating in the chat interaction mode under high cognitive load, while for all other combinations of interaction mode and cognitive load, the SMEQ rating is lower compared to participants without previous experience. However, analysis of the contrasts shows that this effect might be overestimated because no contrast shows a significant result. The first contrast shows no significant difference between the one-click interaction mode and the guided interaction mode between the different levels of cognitive load for the two experience groups, $F(1, 24) = 1.12$, $p = 0.30$. The second contrast analyzes the difference between the guided interaction mode and the chat interaction mode between the different levels of cognitive load for participants with previous emergency experience versus participants without previous emergency experience, $F(1, 24) = 4.04$, $p = 0.06$. This contrasts shows no significant result.

7.3. RQ3: Effect of Interaction Mode and Cognitive Load on Time Spent on Each Screen of the RESCUER App. In this research question, we were interested in investigating the participants' usage duration of each screen in seconds in order to describe the emergencies depicted in the stimulus pictures (Figures 9 and 10). To examine RQ3, we performed a repeated measures ANOVA with two within-subject factors (interaction mode and cognitive load). Due to technical problems with the eye tracker, three participants were eliminated from this analysis, resulting in $n = 23$.

Mauchly's test showed that the assumptions of sphericity had been violated only for the interaction effect of interaction mode (factor 1) and cognitive load (factor 2), $\chi^2(2) = 28.78$, $p < 0.001$. Consequently, degrees of freedom were corrected for the interaction effect interaction mode × cognitive load using Greenhouse–Geisser estimates for sphericity ($\varepsilon = 0.57$).

There was a significant main effect of interaction mode on usage duration, $F(2, 44) = 216.73$, $p < 0.001$, $\eta_p^2 = 0.91$. Contrasts revealed that usage duration for screen 2 (guided interaction) was significantly longer than for screen 1 (one-click interaction), $F(1, 22) = 71.83$, $p < 0.001$, $r = 0.87$, and usage duration for screen 3 (chat interaction) was significantly

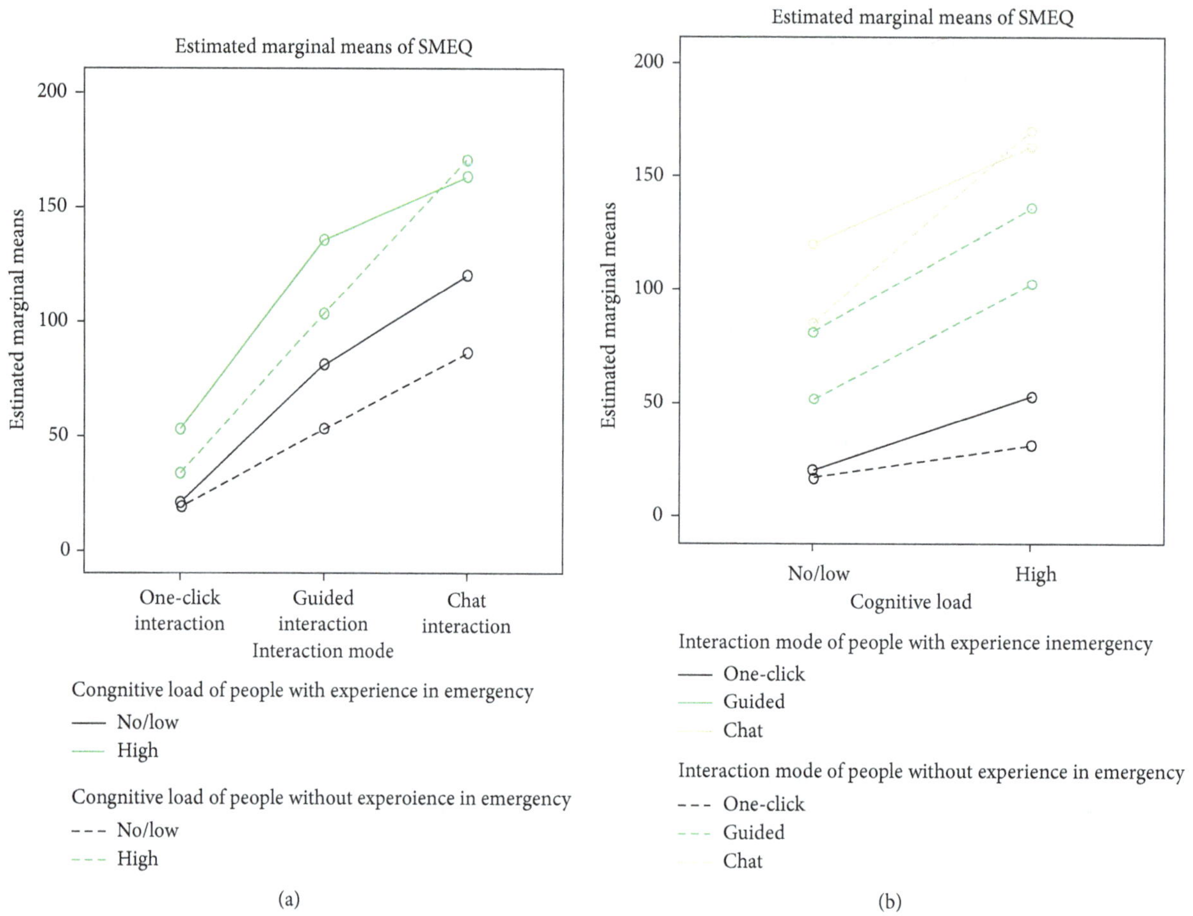

FIGURE 13: Interaction graphs for SMEQ for participants without previous emergency experience (dashed lines) and with previous emergency experience (continuous lines).

longer compared to screen 2, $F(1, 22) = 170.28$, $p < 0.001$, $r = 0.94$. There was also a significant main effect of the level of cognitive load applied on usage duration, $F(1, 22) = 54.38$, $p < 0.001$, $\eta_p^2 = 0.71$. Surprisingly, the contrast shows a significantly lower usage duration in the high cognitive load task compared to the low/no cognitive load task, $F(1, 22) = 54.38$, $p < 0.001$, $r = 0.84$.

Furthermore, there was a significant interaction effect between the interaction mode and the level of cognitive load applied, $F(1.15, 25.20) = 37.38$, $p < 0.001$, $\eta_p^2 = 0.63$. This indicates that the interaction mode used has different effects on the usage duration depending on a person's cognitive load. Figure 14 shows the interaction graphs for usage duration. To break down this interaction, contrasts were performed comparing each interaction mode against its preceding mode with fewer interaction possibilities (one-click interaction versus guided interaction and guided interaction versus chat interaction) separately for both levels of cognitive load. Only the contrast guided interaction versus chat interaction yielded a significant result, $F(1, 22) = 35.11$, $p < 0.001$, $r = 0.78$.

The results reflect that the interaction duration increases along the interaction modes, that is, screens with a larger number of functionalities logically have a longer usage duration. However, in contrast to our assumption, we also see that usage duration significantly drops from task 1 (low/no cognitive load) to task 2 (high cognitive load) with the highest change in the chat interaction mode, possibly due to practice effects with the RESCUER App in the preceding task.

7.4. RQ4: Results of the n*-Back Task.* In the task 2 of the experiment, the participants had to perform an auditory n-back task conceived as a one-back task in parallel to using the mobile application. From the video recordings, we were able to extract the following variables to characterize the participants' performance with the parallel task:

(i) The number of one-back stimuli a participant received in task 2, which depended on the interaction duration

(ii) The number of correct responses to the one-back task (answering with "yes" when a stimulus was similar to the previous one or answering with "no" if two consecutive stimuli differed)

(iii) The number of incorrect responses

(iv) The number of omissions (when a participant skipped an answer).

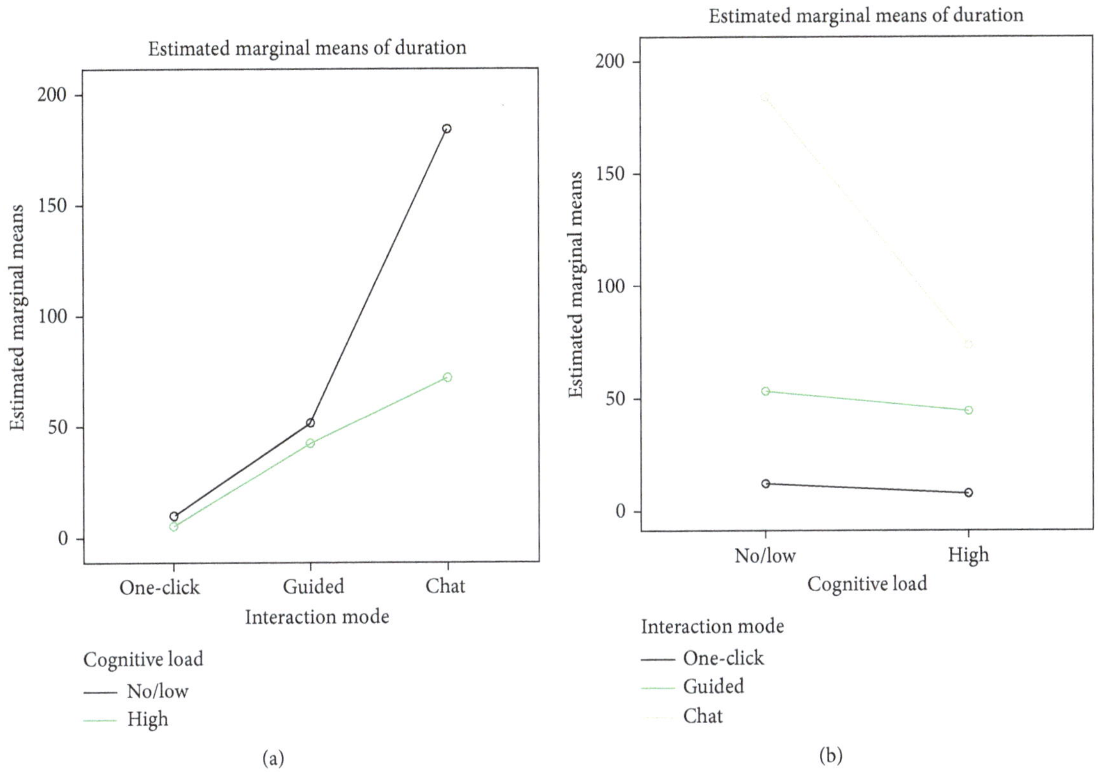

FIGURE 14: Interaction graphs for duration (in seconds).

Because the overall usage duration of the RESCUER App differed between participants, we calculated the ratio of the number of responses (correct, wrong, and omitted) and the number of the presented n-back stimuli for each participant. With this relative score, it is possible to perform a comparison across the study participants.

Overall, the participants provided an average of 86.61% correct responses (SD = 13.67%) with a minimum score of 50% and a maximum score of 100% correct responses to the n-back task. This number significantly differs from the assumed 100% correct responses, $t(25) = -4.99$, $p < 0.001$. The remaining 13.39% are divided into incorrect responses (on average, 3.50% of all stimulus responses were incorrect, SD = 4.47%) and omitted responses (on average, 9.89% of all stimulus responses were omitted responses, SD = 13.37%). Therefore, a larger proportion of the noncorrect responses can be attributed to omitted responses, whereas incorrect responses have the smallest proportion. Still, both categories of noncorrect stimulus responses differ significantly from zero (incorrect responses: $t(25) = 3.99$, $p < 0.001$ and omitted responses: $t(25) = 3.77$, $p < 0.001$).

These results show that a large proportion of the given stimulus responses were correct answers, but a recognizable part of the stimulus responses were not correct, either because a wrong response was given or because no response was given at all because a participant missed a stimulus, possibly due to cognitive overload.

7.5. Usability of the RESCUER App. To analyze the participants' view on the usability of the mobile application, we used nine items from the ISONORM 9241/10 questionnaire (Section 6.4). Table 8 lists the descriptive results, the one-sample t-test results (test value = 4; i.e., middle of response scale), and the frequency distribution for the answer categories for each item.

The participants rated items regarding learnability of the mobile application favorably. Ratings for item 7 ("The app requires just a little time to learn how to use it"), $t(25) = 15.01$, $p < 0.001$, item 8 ("The app is easy to learn without further assistance or a user manual"), $t(25) = 14.88$, $p < 0.001$, and the reversed item 9 ("The app requires that you have to remember many details"), $t(25) = -6.06$, $p < 0.001$, significantly differ from the middle of the response scale (the "neutral middle").

The participants strongly disagreed with the statement that the app "is complicated to operate," $t(25) = -4.57$, $p < 0.001$. Regarding the self-descriptiveness of the RESCUER App, on the one hand, the participants significantly consented that the app uses "understandable terms, abbreviations, or symbols in the screens and menus" (item 2), $t(25) = 7.99$, $p < 0.001$, but on the other hand, they disagreed in their opinion as to whether the app "provides information in adequate extent which entries are permitted or required" (item 3), $t(25) = 1.56$, $p = 0.50$. This apparent contradiction is in line with items regarding conformity with user expectations of the app, which include the ratings of items 4, 5, and 6. All three items measure the user's orientation in the app and feedback of the app to the users' input and revealed no significant positive or negative ratings. We assert that the lack of confirmation screens due to technical problems may have contributed to this dispersed rating.

Overall, the subjective ratings of items from the ISONORM questionnaire show a positive to neutral opinion on

Table 8: Results of the item selection from the ISONORM 9241/10 questionnaire.

Items	Descriptives			One-sample t-test (two-tailed; test value = 4)						Distribution
	N	M	SD	T	df	p	Mean difference	95% confidence interval of the difference		
								Lower	Upper	(--- / -- / - / -/+ / + / ++ / +++)
(1) The app is complicated to operate.	26	2.85	1.29	−4.57	25	0.000	−1.154	−1.67	−0.63	3 / 9 / 7 / 4 / 2 / 1 / 0
(2) The app uses easily understandable terms, names, abbreviations, or symbols in the screens and menus.	26	5.65	1.06	7.99	25	0.000	1.654	1.23	2.08	0 / 0 / 2 / 1 / 5 / 14 / 4
(3) The app provides information in adequate extent which entries are permitted or required.	26	4.50	1.63	1.56	25	0.131	0.500	−0.16	1.16	0 / 5 / 2 / 5 / 5 / 7 / 2
(4) The app facilitates orientation through a consistent look.	26	4.73	1.51	2.47	25	0.21	0.731	0.12	1.34	1 / 1 / 4 / 3 / 8 / 7 / 2
(5) The app lets one in the dark about whether an entry was successful or not.	25	4.04	1.88	0.11	24	0.916	0.040	−0.74	0.82	2 / 5 / 3 / 4 / 5 / 3 / 3
(6) The app sufficiently informs about what it is doing.	26	3.58	1.47	−1.46	25	0.156	−0.423	−1.02	0.17	2 / 4 / 6 / 9 / 2 / 2 / 1
(7) The app requires just a little time to learn how to use it.	26	6.15	0.73	15.01	25	0.000	2.154	1.86	2.45	0 / 0 / 0 / 0 / 5 / 12 / 9
(8) The app requires that you have to remember many details.	26	2.38	1.36	−6.06	25	0.000	−1.615	−2.16	−1.07	8 / 8 / 5 / 3 / 1 / 1 / 0
(9) The app is easy to learn without further assistance or a user manual.	26	6.23	0.77	14.88	25	0.000	2.231	1.92	2.54	0 / 0 / 0 / 1 / 2 / 13 / 10

M, mean; SD, standard deviation; ---, Does not apply at all; --, does not apply; −, rather does not apply; −/+, undecided; +, rather applies; ++, applies; +++, applies fully.

the usability of the RESCUER App, and no explicit strongly negative usability assessment was observed.

8. Discussion

After having performed the user experiment and the statistical analyses, we obtained several insights regarding the interaction modes model as well as their underlying theoretical background and findings on how to improve the usability of the RESCUER App in later iterations. In this chapter, we will discuss the results presented above and the threats to validity that may have an influence on the main outcomes.

8.1. RQ1: Subjective Mental Effort of the RESCUER App under Different Levels of Cognitive Load. The analysis of the data for the first research question revealed that the participants reported higher mental effort (i.e., more difficulty) in operating the app under cognitive load than when they were not under cognitive load. In both conditions, the guided interaction (screen 2) was perceived as being much more difficult than the one-click interaction (screen 1) with the perceived difficulty from the guided interaction to the chat interaction (screen 3). This is surprising, as both screens 1 and 2 ask for simple input by clicking on buttons with predefined answers, whereas screen 3 requires formulating an answer to a question in natural language and could be hypothesized to be much more difficult. On the other hand, it is understandable that answering multiple questions in one interface is more difficult than answering one open question, but answering multiple questions by clicking is still perceived as being easier than answering open questions through a chat interface, as it requires even more effort by the cognitive and motor processors [24].

This is also supported by two additional findings. First, the differences between the reported difficulty of the screens were smaller for the high cognitive load conditions, revealing that the mental effort required to perform simple tasks is impaired more than that for more difficult tasks. Second, the interaction between cognitive load and the difference between screens 2 and 3 did not reach significance, suggesting that performance is impaired equally for the more complex tasks, regardless of their form (e.g., answering multiple questions by clicking versus answering an open question through chatting). This is likely because more complex tasks already require more attention even under less cognitive load, so that other factors demanding a person's attention interfere but do not add to the complexity of the task itself. This suggests that people are generally capable of performing more complex tasks under stress without being impaired too much in executing these tasks. In other words, it is possible to use the RESCUER App in an emergency situation.

8.2. RQ2: Effect of Previous Emergency Experience on SMEQ while Using the RESCUER App under Different Levels of Cognitive Load. Comparing the SMEQ results of participants with or without prior experience in emergency situations revealed that participants with such an experience showed a similar reporting pattern as participants without such an experience, but usually rated the screens more difficult in both types of mental load conditions. Although this difference did not reach significance, it did reveal a trend towards this difference. In addition, a significant interaction mode × cognitive load × emergency experience interaction showed that there is a difference in behavior depending on experience with emergencies, but an absence of other significant interactions with emergency experience reveals that this is based on a more complex interplay of factors. The notion that this might be caused by different expectancy of persons with experience regarding how the process is shaped can be refuted by the finding that the differences held true even in the second condition. Instead, the participants may be primed by their previous experience, and perhaps hindered in their performance by remembering a previous emergency situation they were in. According to the affect infusion model by Bower and Forgas [37], affective states influence cognitive processes such as judgments. Therefore, it is possible that the question about previous emergency experience in the prequestionnaire primed an affective reaction, which in turn activated associations, evaluations, and judgments related to past emergencies that influenced the SMEQ rating in the experiment. However, these effects are small and did not dramatically reduce a participant's performance. This shows two things: people without prior experience with an emergency situation are no less able to successfully report this incident than people who have such an experience, and these people may in fact benefit from not being hindered by such an experience, even though this hindrance is fairly small. Furthermore, it shows that the RESCUER App can be used by anyone, regardless of prior experience.

8.3. RQ3: Effect of Interaction Mode and Cognitive Load on Time Spent on Each Screen of the RESCUER App. Similar to the results on reported effort (RQ1), the time spent per screen increases from screen 1 (one-click interaction) to screen 2 (guided interaction), and from screen 2 to screen 3 (chat interaction). On the SMEQ scale, the effort was perceived as higher under high cognitive load. The performance results in terms of task duration show a different pattern. The participants were quicker on each screen, especially on the third screen, for which they required a lot more time in the first condition. This similarity also becomes obvious from the difference between reported effort and screen duration. On the SMEQ scale, screen 2 was perceived as requiring much more effort than screen 1, more than the difference between screen 2 and screen 3. The results of the actual duration show that under no/low cognitive load conditions, the time spent on screen 3 was considerably higher than that of screen 2, but under high cognitive load conditions, the pattern of screen duration resembled that of reported effort, while the time spent on screens 1 and 2 slightly decreased and showed no significant interaction in terms of spending less time on either screen 1 or screen 2. These results can be explained by the strong effect of practice in task 1 on the performance in task 2, from which it became clear that despite the impairment of additional effort due to a parallel task to increase cognitive load, the performance by participants was still better than in the first condition, where the participants were also still exploring and learning how to work with the app. This suggests that it would be advisable to encourage users to familiarize themselves with the app outside of an emergency context in order to enable them to perform well even when they are under stress in an actual emergency situation.

The results also show that the participants were surprisingly capable of typing under high cognitive load in the chat interaction (screen 3), experiencing an apparently low interference of the n-back task, where the participants had to indicate whether two subsequent tones were different or similar by verbal reporting (saying "yes" or "no," resp.). This means that under stress, multimodal processing is still possible and provides further evidence for the ability to operate the RESCUER App under high cognitive load.

8.4. RQ4: Results of the n*-Back Task.* Under high cognitive load conditions, the participants were instructed to focus on operating the RESCUER App while making as few errors in the parallel one-back task as possible. The results showed that the participants performed well. They navigated swiftly through the app, while still correctly responding to on average 86.61% of the stimuli. The participants who were able to respond correctly to 100% of the stimuli were musically trained and were able to both operate the app and perform the one-back task without much effort. The results of the participants' performance on the one-back task show that people are still able to process multiple sensory inputs (e.g., photo, app interface, and auditory stimuli) and outputs (e.g., tapping and typing) without too much impairment. Some information will be processed slower and some small lapses may occur, but the participants showed a remarkable ability to recuperate and resume both operating the app and attending to the stimuli.

8.5. Interaction Modes' Model and Theoretical Background. The results show that the effort experienced under high cognitive load can be significantly reduced if the number of questions is reduced and users can select preformatted answers. Persons under lower cognitive loads may be able to answer more questions, open questions, and more complex questions, but those directly involved in an emergency should be provided with simpler and preformatted questions. This could be achieved by distributing the questions that responders would like to ask over different persons within the crowd, for example, based on their distance from the site of the emergency or even on sensors in the smartphone. For example, movement sensors could help to identify whether a person is currently fleeing or shaking due to stress, in which case it would be better to provide clear instructions to the person on what to do rather than asking the person to answer a few questions. Furthermore, environmental factors and already collected answers could be used to limit the questions to those that still require answers, thus further reducing the effort required from persons near an emergency site. People who have prior experience with emergencies do not have an advantage in using the RESCUER App. According to Breiner [25], a difference in use could be expected only if people are highly trained in using the RESCUER App so that they have the interaction flow stored in their long-term memory.

8.6. Usability Issues and Improvements. The outcomes of the ISONORM 9241/10 questionnaire are corroborated by the participants' verbal reports. Most of the participants indicated that they were missing feedback from the prototype of the RESCUER App during their interaction. The interactive prototype did not show a confirmation screen to the participants after each screen, such as those presented in Figure 8, screens (c) and (e). This means that the participants did not know if their report was sent. The final implementation of the RESCUER App includes such feedback to inform the users that their messages have been sent.

Another problem identified through the usability measurements was that sometimes the participants wanted to enter some information that was not covered by the multiple-choice answers. They were not sure if answering the question was mandatory or optional, and in that case, they mostly provided their best possible answer. In this respect, the RESCUER App could be improved by clarifying that there is no obligation to answer all questions, for example, by including evasive answer possibilities such as "I do not know," "I cannot say," "I am not sure," or "Pass," so that the user is freed from answering all questions presented to them.

The metaphor used for describing the magnitude of the fire (small, medium, and big) was often criticized by the participants because it left too much room for interpretation. Our recommended way of confirming whether the metaphor corresponds to reality is to perform a user study that compares a set of metaphors with the standard classification and checks whether users perceive them in the same way.

Finally, some users were bothered by some interaction concepts and the reaction of the RESCUER App's prototype if they did not conform to mobile platform interaction standards. The RESCUER App was designed to be a multiplatform mobile app, and the prototype was implemented as a web app. However, users of operating systems such as iOS or Android should be able to expect the RESCUER App to always behave in a similar way as other apps on their smartphone.

8.7. Threats to Validity. Although the study was prepared carefully, not all variables could be controlled for. Some aspects that might influence the degree to which the results and conclusions of this study can be confirmed with certainty and extrapolated to settings outside of this experiment were only recognized afterwards. In this section, we present the main threats to validity of our results.

8.7.1. Statistical Conclusion Validity. Before we conducted the study, we performed a power analysis with G*Power [38] to calculate the optimal sample size so that effects could be detected. For each statistical analysis, the associated assumptions were tested (e.g., sphericity in repeated measures ANOVA) and, where necessary, the appropriate correction (e.g., Greenhouse–Geisser estimates) was used. However, the sample in our study was rather heterogeneous (e.g., wide age range) and therefore this might have had a negative influence on the detection of statistical effects and the interpretation of the results.

8.7.2. Internal Validity. The primary threat to the study's internal validity is certainly the effect of testing between the conditions no/low cognitive load and high cognitive load. Working with the app under the first condition meant that the participants had already been trained in the use of the app when they used it under the second condition. In this experiment, these conditions were not balanced across the participants, as the participants would likely have been hindered in their ability to learn how to use the app when under high cognitive load, and following this up with the low/no cognitive load condition would not have made any sense. It was of more interest to test whether the impairment under high cognitive load causes a person to work slower with the app than when first working with the app.

The use of the eye-tracking system, although non-obtrusive, somewhat narrows the visual field, which may have affected the participants' judgment in a negative sense [39]. Participants with glasses required greater adaptation to the eye-tracking system than users without glasses, which may have caused them to see the task as more difficult because they additionally had to strain their eyes. Although the eye-tracking system occasionally required recalibration, the analysis of the data included controlling for deviations in following the eye movements, meaning that the effect of instrumentation on validity was low.

The prototype used in the experiment was not fully functional and sometimes did not behave like a fully implemented mobile app. For example, clicking on text fields in order to input text required much more precision, and the map image on which participants could indicate their current position did not always appear. This behavior might have influenced the users' judgment regarding the usability of the RESCUER App, and the unexpected app behavior may have increased the participants' stress level during the use of the application.

8.7.3. Construct Validity. Stress was induced and cognitive capability reduced through the n-back task, causing multiple sensorimotor channels to require resources from working memory for the execution of the primary task in combination with the n-back task. Although this was an intended effect, it is an abstraction from real-world settings. When people are confronted by an actual threat, the load on their working memory may be different from that of the n-back task, as the stimuli will not always be auditory and hence may lead to different mental ability.

The presence of two experimenters may have caused an experimenter effect, meaning the participants may have behaved differently than if they had used the app by themselves. In addition, as both the experimenters and the participants are employees of the same institute, they often knew each other well. Although the participants were motivated to help improve the app and the institute has an open and honest culture in which criticism may be expressed, the familiarity may have given rise to participants filling out the questionnaire in a more positive or socially desirable way. We believe this effect to be stronger than the possible negative mood induced by the use of an eye-tracker system, as the participants could nevertheless consciously choose to answer questions in a certain way, regardless of their actual mood.

Although we expected the requirements specified for the interaction modes to be appropriate for an emergency in which the difficulty of an interface increases gradually, adapting to the current state of the individual, it was uncertain whether the sequence from guided interaction to chat interaction was appropriate, especially considering the widespread use of mobile apps for chatting. As a consequence, the participants were highly trained in this kind of interaction so that the chat interaction could perhaps be perceived as being easier than the guided interaction. However, based on the results, we can assume that the content provided by the users during the chat interaction also influenced their cognitive capability. Nevertheless, putting the screens in the same order could have influenced the judgment of the participants regarding the difficulty of each screen and consequently may have confounded the results of the user experiment.

8.7.4. External Validity. As all participants were employees of the institute where the system was conceived, and as most had a background in computer science, there was a selection bias. Most participants are used to working with prototypes and thinking critically about interaction design and do not shun technology. This could have interacted with the experimental variable of the time required to understand the application and solve (intended and unintended) problems that occurred. None of the participants were directly involved in the development, but a primary identification with the institute can also influence how people assess the RESCUER App. The number of men participating was much higher than the number of women, which could have biased the processing of the interface in favor of the way men process information. A more balanced sample with respect to educational level might, among other things, has caused the effects of cognitive load during the guided interaction to be more severe, and the interpretation of metaphors and the description of the fire might have been different.

Moreover, this study did not evaluate the RESCUER App in a real emergency situation, but rather in a laboratory setting. Simulating a stress situation by increasing the cognitive load of the users is far different from the stress caused by a real emergency. A lot of factors related to an emergency situation were not replicated in this study, such as the crowd of people, the fact that people are moving while interacting with the app, and the threat itself. As a result, the stress that was induced may be limited to only working under high cognitive load (or occasionally cognitive overload) and might not have led to a fight-or-flight reaction, as could be expected to occur during an actual emergency situation. To increase the external validity of our user experiment, a more realistic emergency setting would be required, although the development of an experiment in which life and property could be endangered in order to prove our hypotheses would be subject to many ethical concerns.

9. Conclusion and Future Work

This paper has described a human reaction model for the event of an emergency. Three interaction modes were derived from this model: one-click interaction, guided interaction, and chat interaction. The requirements defined for these interaction modes were used to develop a prototype of a mobile application, called the RESCUER App. The RESCUER App supports a command center in gathering information about an emergency with the help of the crowd that is involved in the emergency, as well as in giving instructions to the crowd during the emergency. The RESCUER App is a crowdsourcing-based emergency mobile application that is especially designed for emergency situations in which a large number of people are involved, such as at a festival or in a large industrial park. In particular, the RESCUER App was designed to allow reporting of incidents at the incident site when people are still under stress.

An experiment was performed to verify whether the interaction modes and the concrete mobile app design support the interaction of people with the RESCUER App when they are placed under stress. In the experimental design, the participants first used an interactive prototype of the RESCUER App with no/low cognitive load and a second time under high cognitive load. This high cognitive load was induced by having the participants perform an n-back task in addition to using the app.

The results show that people are remarkably capable of operating the app under high cognitive load, and that this ability does not depend on prior experience with emergency situations. For the three modes of interaction, the results showed that the one-click interaction mode demands very low cognitive load from people, the guided interaction mode generates moderate cognitive load by people, and the chat interaction causes a high mental effort for people. This indicates that the one-click interaction mode and an improved version of the guided interaction mode may be good ways to gather information from the crowd in situations where they experience higher levels of stress. Whenever users have a lower level of stress and require fewer resources for dealing with the emergency, the chat interaction mode can also be used and would allow sharing more detailed information such as the potential cause of the incident and personal impressions. Performance in all of these modes, even under stress, is better when people have familiarized themselves with the reporting process in advance, so it would be helpful to provide a simulation mode in which a user can test what it is like to report an emergency using the RESCUER App.

Improvement of the guided interaction could be achieved by reducing the number of questions presented to the user and spreading subsets of questions over individuals in the crowd, either randomly or according to some metrics. In addition, the use of validated metaphors for the multiple-choice questions (e.g., unambiguous icons) could facilitate easy and quick use of the app and lead to a reduction of cognitive load in the guided interaction. The design of these multiple-choice questions should be validated through one or more studies to compose a question catalog for use by operational forces and presentation to the crowd during an emergency. Other ways to improve the interaction are to make it more self-evident that answering questions is not mandatory and to improve the app's overall usability.

This study has the potential for improving crowd safety in emergencies to the extent that people in the crowd using a mobile application as the one proposed in this paper can notify a command center of an emergency at the first opportunity (as soon as they are able to press a button) without exposing them to further risks (as they can go away from the emergency site and use the application according to their cognitive capabilities), and without any unnecessary delays (without the need for looking for security staff at the venue or for waiting their phone call to be answered). The command center in turn can react faster and better to the situation due the richer set of information available at an earlier point in time. This includes the possibility of promptly sending messages to the crowd to guide them in how to behave or where to go, which again has a positive impact on the crowd's safety in emergencies.

Acknowledgments

This work is supported by the RESCUER project, funded by the European Commission (Grant no. 614154) and by the Brazilian National Council for Scientific and Technological Development (CNPq/MCTI; Grant no. 490084/2013-3). The authors thank all participants in the experiment for their contributions in this project. Furthermore, the authors would like to thank Sonnhild Namingha for proofreading an earlier version of this paper.

References

[1] W. Challenger, W. C. Clegg, and A. M. Robinson, *Understanding Crowd Behaviours: Guidance and Lessons Identified*, UK Cabinet Office, London, UK, 2009.

[2] Association of Chief Police Officers (ACPO), "Guidance on emergency procedures," 2009.

[3] D. A. McEntire, *Wiley Pathways Disaster Response and Recovery*, Wiley, Hoboken, NJ, USA, 2006.

[4] E. L. Quarantelli, "Sociology of panic," in *International Encyclopedia of the Social and Behavioural Sciences*, N. J. Smelser and P. B. Baltes, Eds., pp. 11020–11023, Elsevier: Q5 Pergamon Press, New York, NY, USA, 2001.

[5] J. Drury, C. Cocking, and S. Reicher, "Everyone for themselves? A comparative study of crowd solidarity among emergency survivors," *British Journal of Social Psychology*, vol. 48, no. 3, pp. 487–506, 2009.

[6] S. M. Garcia, K. Weaver, G. B. Moskowitz, and J. M. Darley, "Crowded minds: the implicit bystander effect," *Journal of Personality and Social Psychology*, vol. 83, no. 4, pp. 843–853, 2002.

[7] E. Riddell, A. Rock, P. Clothier, and J. London, *Strengthening Youth Resilience to Natural Disaster with Smartphone Technology: An Exploration of the Communication Needs, Priorities and Preferences of Queensland Youth for Improved Resilience to Natural Disasters and Recommendations for an Appropriate App*, p. 79, University of Queensland, School of Journalism and Communication and the Centre for Educational Innovation and Technology, Queensland, Australia, 2011.

[8] N. Morrow, N. Mock, A. Papendieck, and N. Kocmich, "Independent evaluation of the Ushahidi Haiti project," *Development Information Systems International*, vol. 8, p. 2011, 2011.

[9] Gordon College CCS, "Mitigating disaster risks using a crowd sourcing cross platform mobile apps," 2014, https://www.climatecolab.org/contests/2014/crowdsourcing-for-disaster-risk-management/c/proposal/1306406.

[10] T. A. Kass-Hout and N. di Tada, "International system for total early disease detection (InSTEDD) platform," *Advances in Disease Surveillance*, vol. 5, no. 2, p. 108, 2008.

[11] InSTEDD, "InSTEDD innovative support to emergencies," 2014, http://instedd.org.

[12] K. N. Duc, T. T. Vu, and Y. Ban, "Ushahidi and Sahana Eden open-source platforms to assist disaster relief: geospatial components and capabilities," in *Geoinformation for Informed Decisions*, pp. 163–174, Springer International Publishing, Berlin, Germany, 2014.

[13] K. Villela, K. Breiner, C. Nass, M. Mendonca, and V. Vieira, "A smart and reliable crowdsourcing solution for emergency and crisis management," in *Proceedings of 22nd Interdisciplinary Information Management Talks (IDIMT 2014)*, Networking Societies–Cooperation and Conflict, Podebrady, Czech Republic, September 2014.

[14] X. Pan, "Computational modeling of human and social behaviors for emergency egress analysis," Ph.D. thesis, Stanford University, Stanford, CA, USA, 2006.

[15] P. Gantt and R. Gantt, "Disaster psychology dispelling the myths of panic," *Professional Safety*, vol. 57, no. 8, pp. 42–49, 2012.

[16] R. Perry and M. Greene, "The role of ethnicity in the emergency decision-making process," *Sociological Inquiry*, vol. 52, no. 4, pp. 306–334, 1982.

[17] J. Leach, "Why people 'freeze' in an emergency: temporal and cognitive constraints on survival responses," *Aviation, Space and Environmental Medicine*, vol. 75, no. 6, pp. 539–542, 2004.

[18] M. E. Kemeny, "The psychobiology of stress," *Current Directions in Psychological Science*, vol. 12, no. 4, pp. 124–129, 2003.

[19] M. A. Staal, *Stress, Cognition, and Human Performance: A Literature Review and Conceptual Framework*, NASA Technical Memorandum 2004-212824, Washington, DC, USA, 2004.

[20] M. M. Parker, *Understanding Psychology*, BVT Publishing, Redding, California, USA, 2010.

[21] R. Plutchik, "Emotion: theory, research, and experience," in *Theories of Emotion 1*, Vol. 1, Academic Press, New York, NY, USA, 1980.

[22] L. Minh, C. Adam, R. Canal, B. Gaudou, H. Tuong Vinh, and P. Taillandier, "Simulation of the emotion dynamics in a group of agents in an evacuation situation," in *Principles and Practice of Multi-Agent Systems*, N. Desai, A. Liu, and M. Winikoff, Eds., pp. 604–619, Springer, Berlin, Germany, 2012.
[23] S. K. Card, T. P. Moran, and A. Newell, "The model human processor: an engineering model of human performance," in *Handbook of Perception and Human Performance*, K. R. Boff, L. Kaufmann, and J. P. Thomas, Eds., p. 45, Wiley and Sons, New York, NY, USA, 1986.
[24] W. B. Verwey, "Processing modes and parallel processors in producing familiar keying sequences," *Psychological Research*, vol. 67, no. 2, pp. 106–122, 2003.
[25] K. Breiner, "AssistU–a framework for user interaction forensics," Ph.D. thesis, University of Kaiserslautern, Kaiserslautern, Germany, 2013.
[26] F. Zijlstra, "Efficiency in work behavior. A design approach for modern tools," Ph.D. thesis, Delft University of Technology, Delft, Netherlands, 1993.
[27] P. Beavon, "Improving memory through N-back training," Edith Cowan University, School of Psychology and Social Science Presentations, Queensland, Australia, 2012.
[28] A. Baddeley, "Working memory and language: an overview," *Journal of Communication Disorders*, vol. 36, no. 3, pp. 189–208, 2003.
[29] G. Schuhfried, *Wiener Testsystem*, Vienna Reaction Unit, Basic Program. Development and Production of Scientific Equipment, Molding, Austria, 1992.
[30] D. Mazzoni, M. Brubeck, and J. Haberman, "Audacity: free audio editor and recorder," 2005, http://audacity.sourceforge.net.
[31] L. Turner, "A flare for soccer: open flames in soccer stadiums worldwide," 2009, http://www.boston.com/sports/blogs/bigshots/2009/05/a_flare_for_soccer_open_flames.html.
[32] Wikimedia, "Wikimedia Commons the free media repository, Maccabi Haifa Supporters during the Qualifier vs. FC Salzburg," 2009, https://commons.wikimedia.org/wiki/File:Bengalische_Feuer,_Maccabi_Haifa_Fans_im_EM-Stadion_Wals-Siezenheim.jpg.
[33] RESCUER App, "Interactive prototype," June 2017, http://xudtxg.axshare.com/home.html.
[34] M. Schuette, "Bestimmung der bedingungsbezogenen Messgenauigkeit der Anstrengungsskala. Determination of the condition-related reliability of the Anstrengungsskala (Effort Scale)," *Zeitschrift fuer Arbeitswissenschaft*, vol. 56, no. 1-2, pp. 37–45, 2002.
[35] W. Albert and T. Tullis, *Measuring the User Experience*, Morgan Kaufmann, Burlington, MA, USA, 2010.
[36] J. Prümper, "Der Benutzungsfragebogen ISONORM 9241/10: Ergebnisse zur Reliabilität und Validität," in *Software-Ergonomie'97*, pp. 253–262, Vieweg+ Teubner Verlag, 1997.
[37] G. H. Bower and J. P. Forgas, "Affect, memory, and social cognition," in *Cognition and Emotion*, E. Eich, J. F. Kihlstrom, G. H. Bower, J. P. Forgas, and P. M. Niedenthal, Eds., pp. 87–168, Oxford University Press, New York, NY, USA, 2000.
[38] F. Faul, E. Erdfelder, A. Buchner, and A.-G. Lang, "Statistical power analyses using G*Power 3.1: tests for correlation and regression analyses," *Behavior Research Methods*, vol. 41, no. 4, pp. 1149–1160, 2009.
[39] R. Verleur, "Affective video on the web: exploring presentation and task effects," Doctoral dissertation, University of Twente, Enschede, Netherlands, 2008.

15

Runtime Detection Framework for Android Malware

TaeGuen Kim,[1] BooJoong Kang,[2] and Eul Gyu Im[1]

[1]*Hanyang University, Haengdang-dong, Seongdong-gu, Seoul, Republic of Korea*
[2]*Center for Secure Information Technologies, ECIT, Belfast, UK*

Correspondence should be addressed to Eul Gyu Im; imeg@hanyang.ac.kr

Academic Editor: Elio Masciari

As the number of Android malware has been increased rapidly over the years, various malware detection methods have been proposed so far. Existing methods can be classified into two categories: static analysis-based methods and dynamic analysis-based methods. Both approaches have some limitations: static analysis-based methods are relatively easy to be avoided through transformation techniques such as junk instruction insertions, code reordering, and so on. However, dynamic analysis-based methods also have some limitations that analysis overheads are relatively high and kernel modification might be required to extract dynamic features. In this paper, we propose a dynamic analysis framework for Android malware detection that overcomes the aforementioned shortcomings. The framework uses a suffix tree that contains API (Application Programming Interface) subtraces and their probabilistic confidence values that are generated using HMMs (Hidden Markov Model) to reduce the malware detection overhead, and we designed the framework with the client-server architecture since the suffix tree is infeasible to be deployed in mobile devices. In addition, an application rewriting technique is used to trace API invocations without any modifications in the Android kernel. In our experiments, we measured the detection accuracy and the computational overheads to evaluate its effectiveness and efficiency of the proposed framework.

1. Introduction

With the growing popularity of mobile devices, many malware authors have been targeting the mobile devices [1]. According to the mobile malware report in G DATA [2], security experts discovered 750,000 new Android malware during the first quarter of 2017. Android malware is expected to be developed continuously and to be spread to attack Android devices. As a result, a lot of studies on Android malware detection have been conducted in order to defend against Android malware.

Static analysis can be used to analyze malware with low computational overheads without executing malware itself. However, static analysis-based detection methods have a problem that malware authors can apply various transformation techniques to avoid the detection. Rastogi et al. [3] developed a system that applies obfuscation techniques to generate various Android malware variants. They tested ten commercial antivirus scanners with the newly generated malware variants, and they showed that no scanners could detect any variants.

In contrast, dynamic analysis is useful to detect the transformed malware variants, because most dynamic features are preserved while the static features are changed by the transformation. In addition, many malware variants share common behaviors that can be extracted through dynamic analysis.

Despite this advantage, existing malware detection methods based on dynamic analysis have some shortcomings. First, most of the previously proposed methods require kernel modifications to analyze malware [4–10]. To monitor application behaviors, the existing methods usually insert monitoring code or instrumentation code to the Android kernel. The kernel modification has a drawback that the Android kernel needs to be rebuilt and the new kernel needs to be deployed to mobile devices. In addition, there is a concern that the kernel modification makes the process tedious for a mobile device user to update the new kernel and new security mechanisms. Secondly, many malware detection methods do not consider the limited resources of the devices. These existing methods are designed to be deployed and performed on users' devices, and these methods would degrade the performance of the user device because of their expensive detection processes. In addition, most methods use the algorithms that are not suitable to be applied in real time because they focus on dynamic analysis for malware detection when the sufficient time for the analysis is given. Therefore, it is necessary to analyze applications using the efficient algorithms while reducing the analysis overhead in the user devices.

In this paper, we propose an Android malware detection framework which utilizes the application rewriting technique to monitor application behaviors instead of injecting the monitoring code into the kernel. In addition, our framework is designed as the client-server model in order to alleviate the overhead of malware detection processes in user devices. The user devices are only responsible for monitoring application behaviors and any other detection processes performed on the external server. To reduce the computational overheads of detection, we also designed the malware detection model based on the suffix tree and developed a probabilistic model using the suffix tree so that the model can distinguish between malware and benign applications. In addition to the efficiency of the model, the accuracy of the detection mechanism is an essential factor for the malware detection. Our proposed detection method uses a scoring method to measure and to capture the maliciousness of invoked APIs. We developed this runtime detection method and selected the suitable parameters that produce the best performance in terms of the malware detection accuracy.

2. Related Work

This section explains the previous studies that aim to secure Android devices, such as malware detection or malicious behavior analysis. For malware detection, various approaches have been proposed so far, and these approaches could be classified into two categories: static analysis-based approaches and dynamic analysis-based approaches.

2.1. Static Analysis-Based Approaches. There are various static analysis-based methods [11–19] proposed so far. Droidlegacy [11] is a tool to extract malware family signatures by decomposing malware into loosely coupled modules and extracting the API invocation frequencies of each module. Droidmat [12] and Drebin [13] use various features, including API invocation traces, intents, hardware/software components, and permissions, to distinguish between malware and benign applications. Yerima et al. [14] proposed a system that uses API frequency and permission frequency of an application as features in the Bayesian classifier to detect malware. DroidSIFT [15] classifies malware and benign applications by analyzing weighted API dependency graphs of applications. In Android, APIs are invoked in two ways: direct invocations and indirect invocations. Android provides a JAVA reflection mechanism to invoke APIs indirectly, and invoked APIs can be determined at the runtime. These aforementioned approaches may miss the APIs that are invoked at the runtime. In addition, these approaches would analyze the garbage APIs that are not even invoked at runtime because it is impossible to predict the application behaviors through static analysis. This may cause the false alarms as well as the analysis overheads.

There are also some studies to protect benign applications from malicious applications. Chex [16], Droidchecker [17], AAPL [18], and Amandroid [19] use various methods to check if applications have component-hijacking vulnerabilities. These methods perform the data-flow analysis to find out whether an application can be exploited maliciously. These researches focus on how to prevent unauthorized accesses through component-hijacking attacks on vulnerable applications. The purpose of their researches is different from ours.

2.2. Dynamic Analysis-Based Approaches. Many dynamic analysis-based methods [4–10, 20] have been also proposed so far. Isohara et al. [4] proposed a system that analyzes kernel-level behaviors for malware detection. System call invocations are monitored and checked whether the system call traces are matched with preidentified malware signatures, that is, system call traces. A signature is expressed as a regular expression that is related to specific malicious activities. Droidscope [5] provides the semantic views of malware through dynamic analysis. Their system monitors invoked system calls as well as changes of processes, threads, and memories to generate kernel-level semantic views. In addition, Dalvik instruction traces are extracted and used in taint analysis to generate Dalvik-level semantic views. Andrubis [6] is a dynamic malware analysis system that can analyze application behaviors. This system monitors system call invocations, and API invocations, and also performs the taint analysis. Taintdroid [7] conducts the data-flow analysis of applications and detects the information leakage of sensitive data. These approaches modified the Android kernel to monitor the behaviors of applications such as instruction sequences, system call invocations, API invocations, and changes of system resource usages.

Shabtai et al. [8] and Schmidt et al. [9] proposed anomaly detection systems in mobile devices. Their systems monitor and analyze mobile devices using various features to identify abnormalities of device usages. The features used in their systems include CPU usages, the number of sent packets, battery levels, and so on. There are some malware that conducts stealthy attacks such as user information leakages, and this kind of malware can avoid these detection systems because the stealthy attacks hardly affect the system metrics used in their systems.

AASandbox [10] is a malware detection system that uses both static analysis and dynamic analysis, and system call frequencies of both benign applications and malware are collected and compared for dynamic analysis. There is no classification nor detection algorithm used in their paper, and its results just show the possibility to detect malware using this proposed system.

Crowdroid [20] is a framework that uses system call traces to distinguish between benign applications and malware using the k-means algorithm. The framework uses the client-server model as our framework does. The k-means algorithm has high overheads to examine application behaviors at

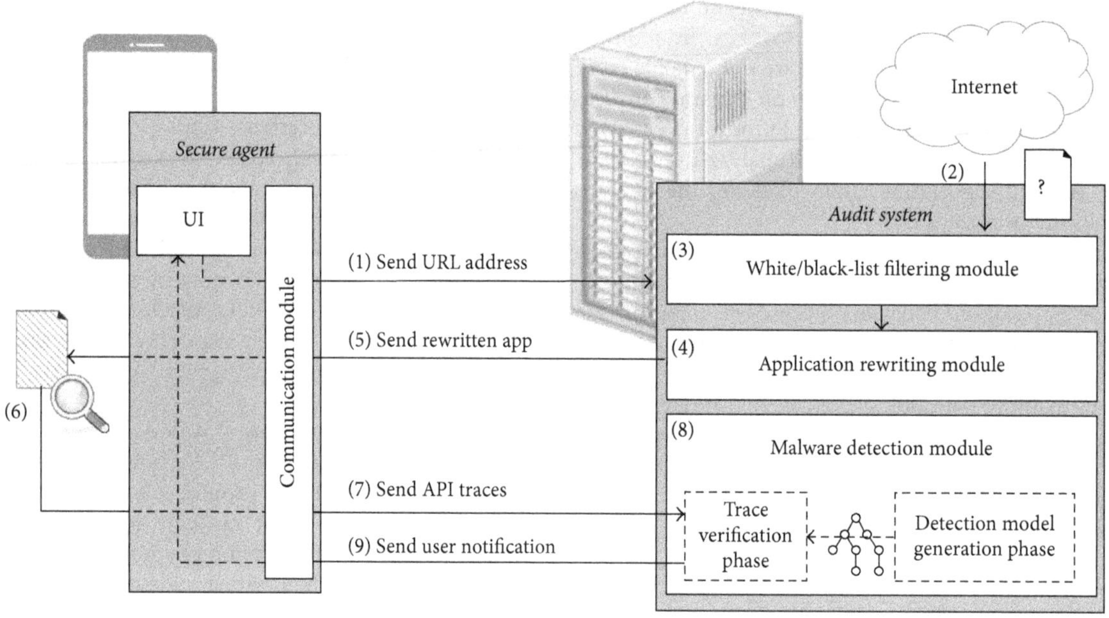

FIGURE 1: The overall architecture of malware detection framework.

runtime, and the framework cannot be used in real-time detection because the framework needs a certain amount of time to collect the system call frequencies. Our proposed framework uses a detection model to examine application behaviors finely at runtime while perserving high efficiency.

There exist researches [21–25] that use application rewriting techniques to enforce security policies in Android. I-ARM-Droid [21] inserts a reference monitor into an Android application, and the method to embed the reference monitor code in an application is introduced in [21]. Hao et al. [22] proposed an application rewriting system for API-level access control, and they studied on the effectiveness of the application rewriting technique for access control. Schreckling et al. [23] proposed a system that uses the application rewriting technique to prevent Android applications from abusing unnecessary permissions. The system analyzes what permissions should be declared for the specific application components and provides users a tool to activate or to deactivate permissions of applications. Aurasium [24] repackages applications to observe behaviors for the purpose of detecting information leakage of sensitive data. A rewritten application is isolated in its own sandbox, and its API invocations are monitored by their proposed system. Boxify [25] supports application sandboxing to prevent malicious activities such as privilege escalation. The system hooks Bionic libc to intercept system calls and to check whether each system call is valid or not. These researches based on the application rewriting technique only focus on how to monitor applications for the reference monitoring, and the collected information from the monitoring system is not delivered securely, which can be used as an additional attack vector.

3. Proposed Malware Detection Framework

Our framework consists of two parts: the server-side component (called as an audit system) and the client-side component (called as a secure agent). The overall architecture of the framework is shown in Figure 1. When a user wants to download an application, the user informs the secure agent of URL (uniform resource locator) information. Then, the secure agent delivers the URL information to the audit system. The audit system downloads an application from the URL, and checks whether the application is a known application or not, using whitelist and blacklist. If an application is known as a benign application, the application is transmitted to the secure agent and is allowed to be installed on the user device. If an application is known as malware, the audit system raises an alert to the secure agent. Otherwise, the application is rewritten in the audit system for further analysis.

The purpose of the application rewriting mechanism is to insert self-monitoring code into an application. After the audit system rewrites an application, the rewritten application is transmitted to the device, and the secure agent installs the rewritten application. The self-monitoring code extracts and delivers the API invocation traces of the application to the secure agent, and the traces are transmitted to the audit system. Each transmitted trace is examined with the detection model in the audit system. If any parts of the traces are matched with traces of malware, the audit system notifies results to the secure agent, and the secure agent generates alarms on the device.

3.1. Whitelist/Blacklist Filtering Module. After the audit system downloads an application, the audit system checks whether the application is included in the whitelist or the blacklist of known applications or not. The purpose of the filtering module is to avoid unnecessary inspections of known applications or known malware. We designed the filtering module using a hash algorithm for fast comparison. We calculate a hash value of each known benign or malicious application and store it in the whitelist database or the blacklist database. The file size of each application is also stored together to mitigate the hash collision problem. By comparing the hash value and the size of an application with those in the whitelist database or the blacklist database, the audit system can identify the known applications.

3.2. Application Rewriting Module. The application rewriting process consists of three major steps: (1) the application package is unzipped and the dex files are decompiled, (2) self-monitoring code to monitor invocations of dangerous APIs is inserted into the decompiled dex files, and (3) the rewritten files are recompiled and repacked to a new application package. The dex file is decompiled with the modified version of the baksmali tool [26] to produce smali files, and Apktool [27] is used to repack the files into a new application package. The dangerous APIs and the self-monitoring code are explained in the next subsections.

3.2.1. Dangerous APIs. In our framework, an application behavior is defined as a part of the API invocation trace of an application.

Monitoring all the API invocations can cause high overheads on the applications, and the size of the rewritten application will be increased significantly due to a large number of injected monitoring codes. Therefore, our framework decided to monitor only certain APIs that are mostly utilized to perform malicious activities, and, as a starting point, we selected APIs that were presented in [28] as dangerous APIs. Additionally, we investigated APIs that are explained in the Android Developers reference [29] and selected dangerous APIs which might be used in malicious activities.

Our selected dangerous APIs are classified into seven categories: APIs that access users' sensitive data, APIs that perform network activities or file activities, APIs that modify or access the device information, APIs that access or send SMS messages or Emails, APIs that access or execute services or installed applications, APIs that are used in API reflection or dynamic code loading, and APIs that affect the permission information.

3.2.2. Self-Monitoring Code. We adapted the hooking technique of the application rewriting framework introduced in [21] to monitor API invocations. The application rewriting module generates a class that contains the self-monitoring methods, and the class is imported into the original application.

The example of a smali file that contains the definition of the class is depicted in Figure 2.

The self-monitoring methods replace dangerous APIs in the original code. The self-monitoring method is invoked instead of the specified dangerous APIs, and it conducts three principal functions: delivery of the names of the invoked dangerous APIs to the secure agent, identifying the types of dangerous API, and the invocation of the original dangerous APIs.

(1) *Invocation trace delivery to the secure agent*: Each self-monitoring method contains code to deliver the names of the invoked dangerous APIs to the secure agent. Invocation

```
.class public Lsf/containterCls;
.super Ljava/lang/Object;
...
.method static sfMonitor_1(LContext;LString;)LObject;
 //get parameters of the original API
   ...
   sget-object v0, Lsf/containterCls;->buffer
   //get buffer obj
   const/16 v1, 0x01
   invoke-static {v1}, LInteger;->valueOf(I)LInteger;
   move-result-object v1
   invoke-virtual {v0, v1}, LArrayList;->add(LObject;)Z
   //add the dangerous api number to the buffer
   invoke-static {}, Lsf/containterCls;->IsBufferFull()Z
   move-result v0
   if-eqz v0, :cond_0
   //check whether the buffer is full
   invoke-static {}, Lsf/containterCls;->sendIntent()V
   // the trace in the buffer is sent to the secure agent using intent msg
   ...
   :cond_0
   invoke-virtual {p0, p1}, LContext;->getSystemService(
                                              LString;)LObject;
   //original API call
   move-result v0
   return-object v0
   //pass the return value
.end method
...
```

FIGURE 2: The example of the self-monitoring code.

traces are delivered using intent message communications. Sending an intent message whenever a dangerous API is invoked can degrade the performance of the mobile device, so the imported class contains a trace buffer to collect the traces to reduce the communication overhead. When the buffer becomes full, the buffer is wrapped in an intent message, and the message is sent to the secure agent. If a rewritten application communicates with the secure agent using intent messages insecurely, then the communication can be modified by malware. To prevent the attack on vulnerable intent-based communications, explicit intent messages with random tags are used to deliver the API invocation traces securely. The explicit intent message can specify its destination, and only the specified application can receive the messages while the other applications cannot eavesdrop the messages. By checking random tags in the messages, the secure agent can identify applications and also avoid the denial of service attacks.

(2) *Identifying the types of APIs*: The application rewriting module scans the dangerous API invocations and replaces invocation code of the APIs with the self-monitoring methods. Naturally, the dangerous APIs that originally supposed to be executed can be identified without any additional process. However, in the self-monitoring method, the name of API to be invoked is delivered as described previously. In the case of the JAVA reflection API, the additional process is needed to identify the name of API because the name of API is dynamically determined at runtime. To monitor the reflected API, the self-monitoring method for java.lang.reflect.Method.invoke() method gets the first parameter that contains the reflected API name. A lot of Android malware use the Java reflection API to hide their behaviors, so this process must be performed to improve the accuracy of API monitoring.

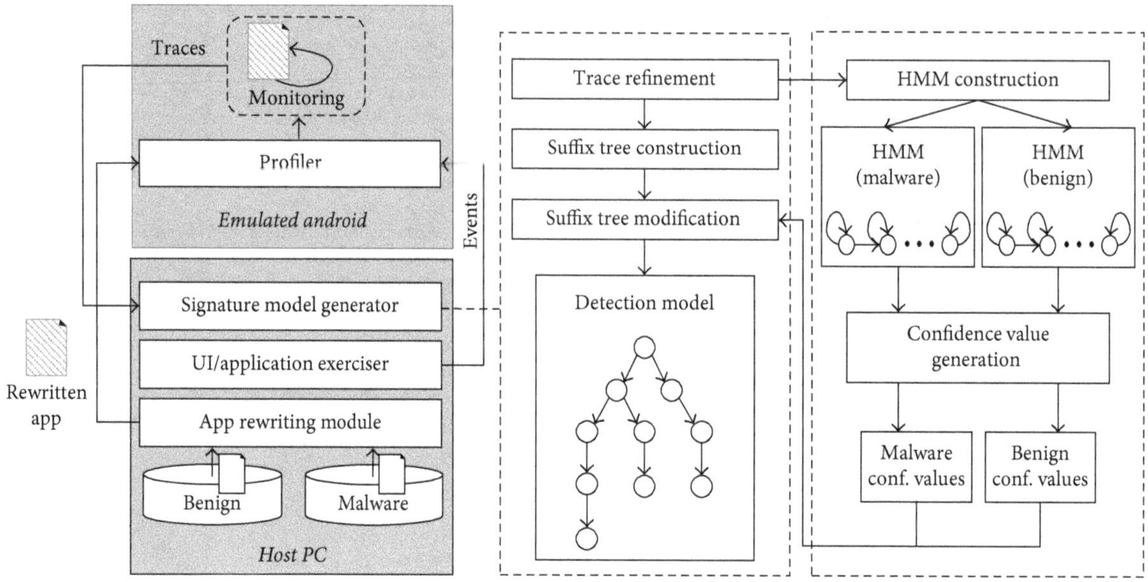

FIGURE 3: Malware detection model generation process in the detection model generation phase.

(3) *Invocation of the original APIs:* After the name of the dangerous APIs is collected, the original dangerous APIs are invoked and their return-values are also recorded by the self-monitoring code.

3.3. Malware Detection Module. The audit system examines the API invocation traces of an application using the detection model, and the detection model is built from known malware and known benign samples. We designed the detection model based on the suffix tree which represents the dangerous API invocation traces' degree of maliciousness and benignness. The suffix tree [30] is a data structure that is used to index all the substrings which are contained in a set of strings. The suffix tree can provide linear-time solutions to search substrings, and the suffix tree can only indicate which malware or benign samples are included in the input trace. So, in our detection method, we altered the suffix tree to have not only trace labels, but also each trace's confidence value that can be used in the malware detection. The malware detection module is mainly composed of two phases: the detection model generation phase and the trace examination phase. The details about these phases are explained in the next subsections.

3.3.1. Detection Model Generation Phase. To generate the detection model, we extracted API invocation traces from known malware and benign applications. The left part of Figure 3 shows how the API invocation traces of applications are extracted. Each application is rewritten by the application rewriting module, and it is executed by the UI/Application exerciser. While an application is executed, its API invocation traces are stored. The explanation about the application execution for the API invocation trace collection is given in Section 4.2.

After the API invocation traces of all the application samples are extracted, the detection model is generated as described on the right side of Figure 3.

All the API invocation traces are refined before the suffix tree is constructed. There are many consecutive elements in each trace, and those repeating elements are merged. For example, $[API_1, API_1, API_3]$ is refined to $[API_1, API_3]$.

The trace refinement process reduces the size of the API invocation trace of applications, and the decreasing rate of trace size of 1,000 malware and 1,000 benign samples was about 70.34% and 68.72% on average in our experiment.

In addition, the refinement process alleviates the effects of consecutive APIs on detection accuracy. Traces containing many consecutive elements were frequently extracted from both malware and benign applications. So, the repeating elements in a trace could cause false alarms in malware detection. After the trace refinement, the suffix tree is constructed with the refined traces using Ukkonen's algorithm [31].

In the left side of Figure 4, the example of the suffix tree is described. Each edge in the suffix tree is labeled with a subtrace of traces, and it is used as a transition condition. Each node has a list of trace labels whose subtraces are same with a certain prefix of the concatenation of all the edge labels in the path from the root to the node. If a trace is inputted to search the same trace in the suffix tree, nodes are visited by following the edges that are same with prefixes of the input trace. When a certain leaf node is reached, the node traversal ends. This means that the traces that reach a leaf node include a subtrace of samples indicated in the node. For example, if $[API_2, API_3, API_4, API_5]$ is the input trace, then the transition state will move along the shaded nodes in the Figure 4, and its destination node will be the leaf node that has mt.1 trace which means that the input trace is a subtrace of mt.1 $(= [API_1, API_2, API_3, API_4, API_5])$.

It is noted that even if a trace is smaller than an edge's trace and is a prefix of an edge's trace, a transition also occurs. The node followed by the edge is reached, and the node traversal ends. It happens because of the path compression in the Ukkonen's algorithm that merges the consecutive nodes that have only a single descendant to

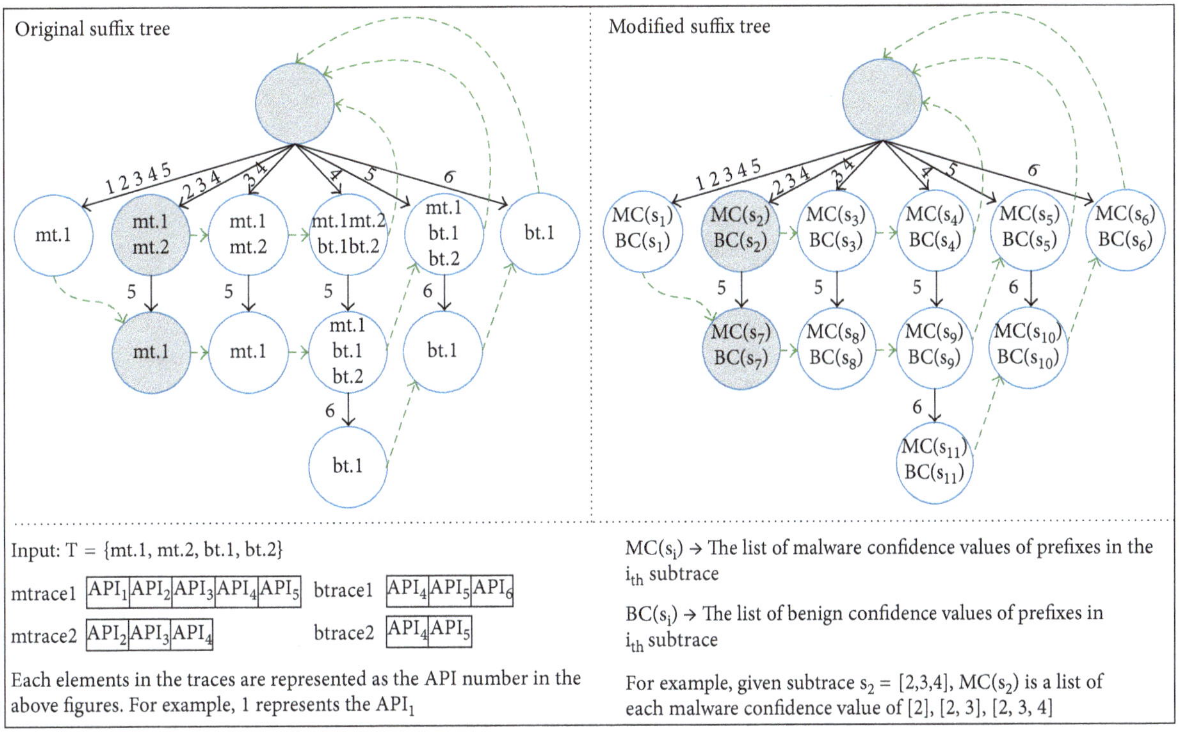

FIGURE 4: Example of suffix tree generation/modification: dotted line edges denote failure transition, and solid line edges denote state transitions. Ukkonen's algorithm is utilized to construct the suffix tree.

one. So, it is considered that the trace is matched to a certain implicit node.

After the initial suffix tree is constructed, the suffix tree is modified for the further malicious trace detection. While traversing the nodes in the suffix tree, the list of trace labels is replaced with the confidence values of prefixes in each edge's trace for both malware and benign applications. The example of the modified suffix tree is described in the right side of Figure 4.

We defined two confidence values to measure the input trace's maliciousness, and they are called as a malware confidence value and a benign-ware confidence value each. The document frequency ratio and the log-likelihood value are used to calculate the confidence values. The document frequency ratio is the proportion of malware or benign-ware invocation traces that include the subtrace to the total malware or benign-ware traces, and the malware or benign-ware log-likelihood value of a subtrace is determined using malware or benign-ware's HMM (Hidden Markov Model).

The HMM [32] is an algorithm which is widely used in various fields such as speech recognition, bioinformatics, and so on. For given observed variables, the HMM algorithm can predict the state transition as well as can estimate the probability of matches between the given data and the predefined model.

The HMMs for both malware and benign applications should be prepared to evaluate the log likelihood of each subtrace. For the HMM generation, the parameters such as the number of states, emission probabilities, transition probabilities, and initial probabilities are specified first.

Generally, there is no rule to specify the number of states, but we tried to find the number of states whose transition

TABLE 1: Notations for the confidence value calculation.

s	An observed sequence of dangerous API invocations, $(x_1, x_2, \ldots, x_T)$
dfr	Document frequency, the number of documents the term occurs in
A	Transition probability distribution, $\{a_{1,2}, \ldots, a_{N-1,N}\}$
B	Emission probability distribution, $\{b_1(x_1), \ldots, b_N(x_P)\}$
π	Initial probability distribution, $\{\pi_1, \ldots, \pi_N\}$
λ	Hidden Markov model, (A, B, π)
N	The number of states
K	The number of traces
P	The number of dangerous API invocations

probabilities are evenly distributed after the model generation. As a result, the number of states is specified as five. The emission probabilities and initial probabilities are randomly selected. The observed variables are the dangerous API invocations in the trace.

Once the model is initialized, the parameters are reestimated using the Baum-Welch algorithm that is widely used as EM (expectation maximization) algorithm. The Baum-Welch algorithm [33] searches a locally optimal model that fits with the observed data. The parameters are updated until the maximum likelihood values of the updated models are converged. After the parameters are estimated completely, then the likelihoods of each subtrace are computed using the HMMs.

The malware and benign-ware confidence values are computed as described in (1-5), and the notations in the equations are explained in Table 1. As described in (5), the

```
procedure Trace Examination (trace, gmc_sum, gbc_sum, lmc_queue, lbc_queue)
  trlen←len (trace)
  swlen←len (sw)    ▷ get the len. of the sliding window
  i←0
  while i < trlen − swlen do
    sw←trace[i : i + swlen − 1]
    mc, bc←get_conf_from_suffixtree (sw)
    lmc_queue.enqueue (mc)
    lmc_queue.dequeue ()
    lbc_queue.enqueue (bc)
    lbc_queue.dequeue ()
    if lmc_queue.isfull () and lbc_queue.isfull () then
      lms←sum (lmc_queue)/lmc_queue.size
      lbs←sum (lbc_queue)/lbc_queue.size
    end if
    gcount←gcount + 1
    gmc_sum←gmc_sum + mc
    gbc_sum←gbc_sum + bc
    gms←gmc_sum/gcount
    gbs←gbc_sum/gcount
    if lms − lbs > lth or gms − gbs > gth then
      send_alarm (clientscok)
    end if
    i←i + 1
  end while
end procedure
```

ALGORITHM 1: Trace examination procedure.

confidence values of a trace, $\mathrm{mc}(s)$ and $\mathrm{bc}(s)$, are the results of the multiplication of the normalized log-likelihood values and the document frequence ratios. The min-max normalization is used to scale the log-likelihood values, $\mathrm{llh_m}(s)$ and $\mathrm{llh_b}(s)$. The document frequency, $\mathrm{dfr_m}(s)$ or $\mathrm{dfr_b}(s)$, is the proportion of the malware or benign-ware traces to the whole traces that contains s as a subtrace. The log likelihood, $\mathrm{llh_m}(s)$ or $\mathrm{llh_b}(s)$, is the probability of the observed sequence for the malware model or the benign-ware model. The log likelihood is computed in a recursive way using the forward algorithm. $\alpha_T^{(\log)}$ in (2) is a forward vairable, and the initial forward variable and the recursive forword vairable in the forword algorithm are defined in (3) and (4).

$$\mathrm{dfr_m}(s) = \frac{\mathrm{df_m}(s)}{\mathrm{df_T}(s)}, \qquad \mathrm{dfr_b}(s) = \frac{\mathrm{df_b}(s)}{\mathrm{df_T}(s)}, \tag{1}$$

$$\begin{aligned}\mathrm{llh_m}(s) &= \log P(s|\lambda_\mathrm{m}) \\ &= \left[\log \mathrm{sum}_{i=2}^{N-1}\left(\alpha_T^{(\log)}(i) + \log a_{i,N}^{\mathrm{m}}\right)\right], \\ \mathrm{llh_b}(s) &= \log P(s|\lambda_\mathrm{b}) \\ &= \left[\log \mathrm{sum}_{i=2}^{N-1}\left(\alpha_T^{(\log)}(i) + \log a_{i,N}^{\mathrm{b}}\right)\right],\end{aligned} \tag{2}$$

$$\alpha_1^{(\log)}(i) = \log a_{1,i} + \log b_i(x_1), 2 \le i \le N-1, \tag{3}$$

$$\begin{aligned}\alpha_{t+1}^{(\log)}(j) = &\left[\log \mathrm{sum}_{i=2}^{N-1}\left(\alpha_T^{(\log)}(i) + \log a_{i,j}^{m}\right)\right] \\ &+ \log b_j(x_{t+1}), 1 \le t \le T, 2 \le i, j \le N-1,\end{aligned} \tag{4}$$

$$\begin{aligned}\mathrm{mc}(s) &= \frac{\mathrm{dfr_m}(s) * \left(\mathrm{llh_m}(s) - \min_{0\le i\le k}(\mathrm{llh_m}(s_k))\right)}{\max_{0\le i\le k}(\mathrm{llh_m}(s_k)) - \min_{0\le i\le k}(\mathrm{llh_m}(s_k))}, \\ \mathrm{bc}(s) &= \frac{\mathrm{dfr_b}(s) * \left(\mathrm{llh_b}(s) - \min_{0\le i\le k}(\mathrm{llh_b}(s_k))\right)}{\max_{0\le i\le k}(\mathrm{llh_b}(s_k)) - \min_{0\le i\le k}(\mathrm{llh_b}(s_k))}.\end{aligned} \tag{5}$$

3.3.2. Trace Examination Phase. Once the detection model is generated, the audit system can examine the trace transmitted from the secure agent using the detection model. The confidence values in the detection model are used to compute the scores for the trace examination. The trace examination process is described in Algorithm 1.

First, the trace from the client is scanned with a sliding window, and each subtrace within the sliding window is searched in the suffix tree. The confidence values of each subtrace are retrieved from the suffix tree, and the malware and benign-ware confidence values are accumulated to calculate local scores and global scores.

TABLE 2: Malware samples used in the experiment.

ADRD	Anserverbot	Asroot
BaseBridge	BeanBot	Bgserv
CoinPirate	CruseWin	DogWars
DroidCoupon	DroidDeluxe	DroidDream
DroidDreamL	DroidKungFu1	DroidKungFu2
DroidKungFu3	DroidKungFu4	DroidKungFuS
DroidKungFuU	Endofday	FakeNetfilx
FakePlayer	GGTracker	GPSSMSSpy
GamblerSMS	Geinimi	GingerMaster
GoldDream	Gone60	HippoSMS
Jifake	KMin	LoveTrap
NickyBot	NickySpy	Pjapps
Plankton	RogueLem	RogueSPP
SMSReplic	SndApps	Spitmo
Tapsnake	Walkinwat	YZHC
Zitmo	Zsone	jSMSHider
zHash		

The local malware score or the local benign score represents the maliciousness or benignness of the traces in the prespecified local scoring period, and it is calculated by averaging the confidence values in the local scoring period. The global malware score or benign-ware score represents the maliciousness or benignness of the traces in the whole period, and it is also calculated by averaging the confidence values from the beginning to the present. After the local or global scores are calculated, it is checked whether the difference between the local scores and the global scores exceeds a certain threshold. If either of the comparison results exceeds the threshold, then the trace matching procedure concludes that the application is malware, and the alarm is sent to the client.

There are three parameters that affect the malware detection in the trace examination procedure: the size of sliding window, the local scoring period, and the score threshold. We have some experiments in order to find parameters that make the malware detection accuracy high, and the experimental results are explained in Section 4.3.1.

4. Experiments

4.1. Dataset. We evaluated our framework in terms of detection accuracy, runtime overhead, and the size overhead of rewritten applications. 1,260 malware samples from the Mal-Genome project [34] were used in the experiments, and the malware families are listed in Table 2. Meanwhile, 1,033 benign applications were downloaded from Google Play Store [35] between February and March 2016, and these applications are used in the experiments as benign samples. We implemented a tool that downloads applications from the official Google market, and to find out whether an application is benign or not, the online malware scanning service of VirusTotal [36] is used. The collection tool uses the VirusTotal APIs to request the tests and to get the scanning results. If all the virus-scanning tools consider an application as benign, then the application is stored as a benign sample.

4.2. Automatic Application Executions. The experiment to select suitable parameters that affect the accuracy was conducted first, and the experiment to measure performance metrics in best case was conducted to show the final results. In the experiments, 1,033 benign applications and 1,260 malware were used, and 80% of applications (826 benign applications and 1,008 malware) were used as the training dataset in the detection model generation, and the others were used as the test dataset.

Before conducting the experiments, we measured code coverages of applications to find out how many system events are required to observe application behaviors. To measure the code coverage, we rewrote 107 applications and added additional code in the applications. The code is inserted in each basic block, and the code logs unique values that are assigned to basic blocks. Whenever a basic block is executed, its unique value is logged. The rewritten applications were executed by MonkeyRunner [37], and by increasing the number of the events from 1,000 to 15,000, the number of the executed basic blocks was measured. The code coverage might be affected by the number of repeated executions. Therefore, we also executed the applications up to three times while measuring the number of unique basic blocks executed.

Figure 5 shows the results of the code coverage measurements. As shown in Figure 5, the code coverage rate is not increased linearly with the increasing number of the system events or the increasing number of executions. Even though there are possibilities of increasing number of events when applications are executed multiple times, the training time for malware detections is limited in practice. To find out practical parameters, we investigated the minimum number of events that produce the maximum code coverage rates for each application. We also computed the code coverage difference by the number of executions. As a result, the average of the minimum number of the events was 8,604, and when the applications were executed twice, the code coverage was increased significantly; 1.4% of the code coverage rate is increased. Therefore, in the training phase for malware detections, we executed the applications twice up to 9,000 system events.

4.3. Malware Detection Accuracy

4.3.1. Parameter Selection. To select the suitable parameters for detection, we measured the accuracies of our framework with different parameter values. The size of sliding window, the local scoring period, and the score threshold are parameters that are related to the detection accuracy. We specified the range of each parameter, and the malware detection accuracies were measured with the combinations of possible parameter values in the specified range. The size of the sliding window was in the range of 3 to 20, the local scoring period was in the range of 1 to 50, and the score threshold was in the range of −1 to 1. In addition, the malware detection accuracies were measured in various cases where the local score is only used or both local and global scores are used. Hereafter, the case that the local score is only used in the detection is referred to the local-only case, and the other case is referred to the local-global case.

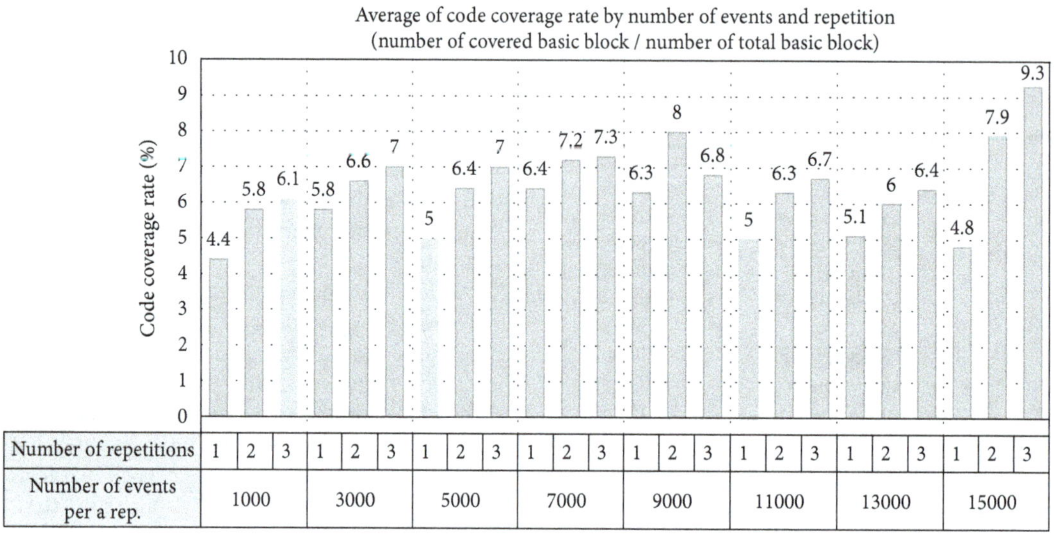

Figure 5: Basic block code coverage rate of the 107 applications.

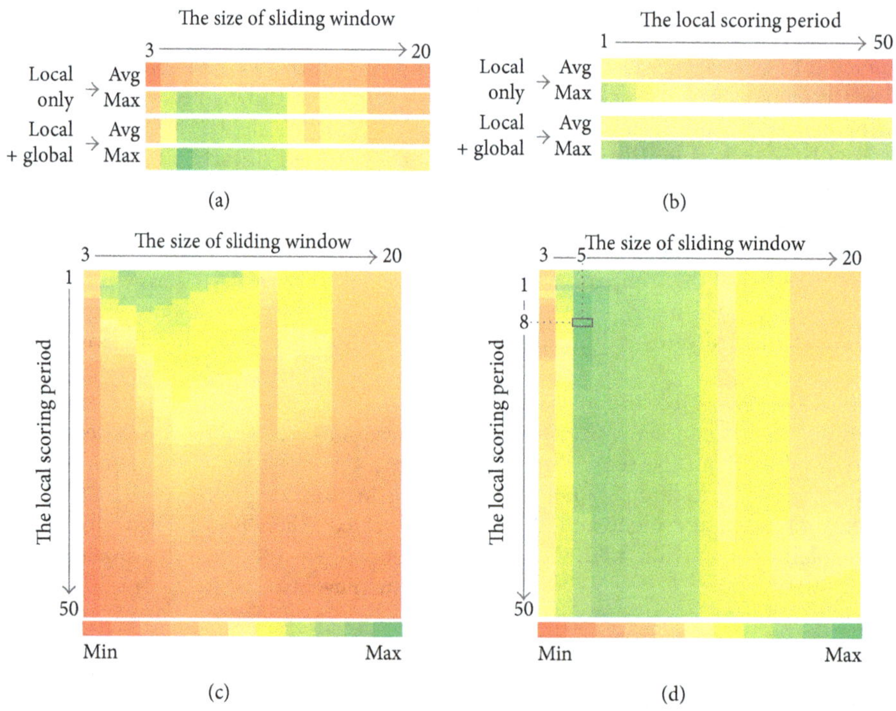

Figure 6: Heap maps of the malware detection accuracy by the size of sliding window and the local scoring period (the min acc, 0.705, is expressed as red, and the max acc, 0.96, is expressed as green). (a) ACCs by the size of sliding window. (b) ACCs by the local scoring period. (c) ACC heat map of the local only case. (d) ACC heat map of the local-global case.

As shown in Figures 6(c) and 6(d), the two large heat maps show how the malware detection accuracy is changed with different parameter values of the size of the sliding window and the local scoring period. Each heat map is the result either when both of the local and global scores are used or when only the local score is used. To express the color of each cell in the heat maps, the maximum value among the accuracies of the different threshold is selected, while the size of the sliding window and the local scoring period is fixed. The selected maximum value is converted to the color of each cell. The highest value is represented by the green color, and the lowest value is with the red color.

The heat map from the accuracy measurement of the local-global case has more green-shaded cells compared to the other heat map. This means that malware detection accuracies are higher when both local and global scores are used together. When the size of the sliding window was 5 and the local scoring period was 8, the highest accuracy value, that is, 0.96, was obtained among the accuracy results in all the cases that we observed.

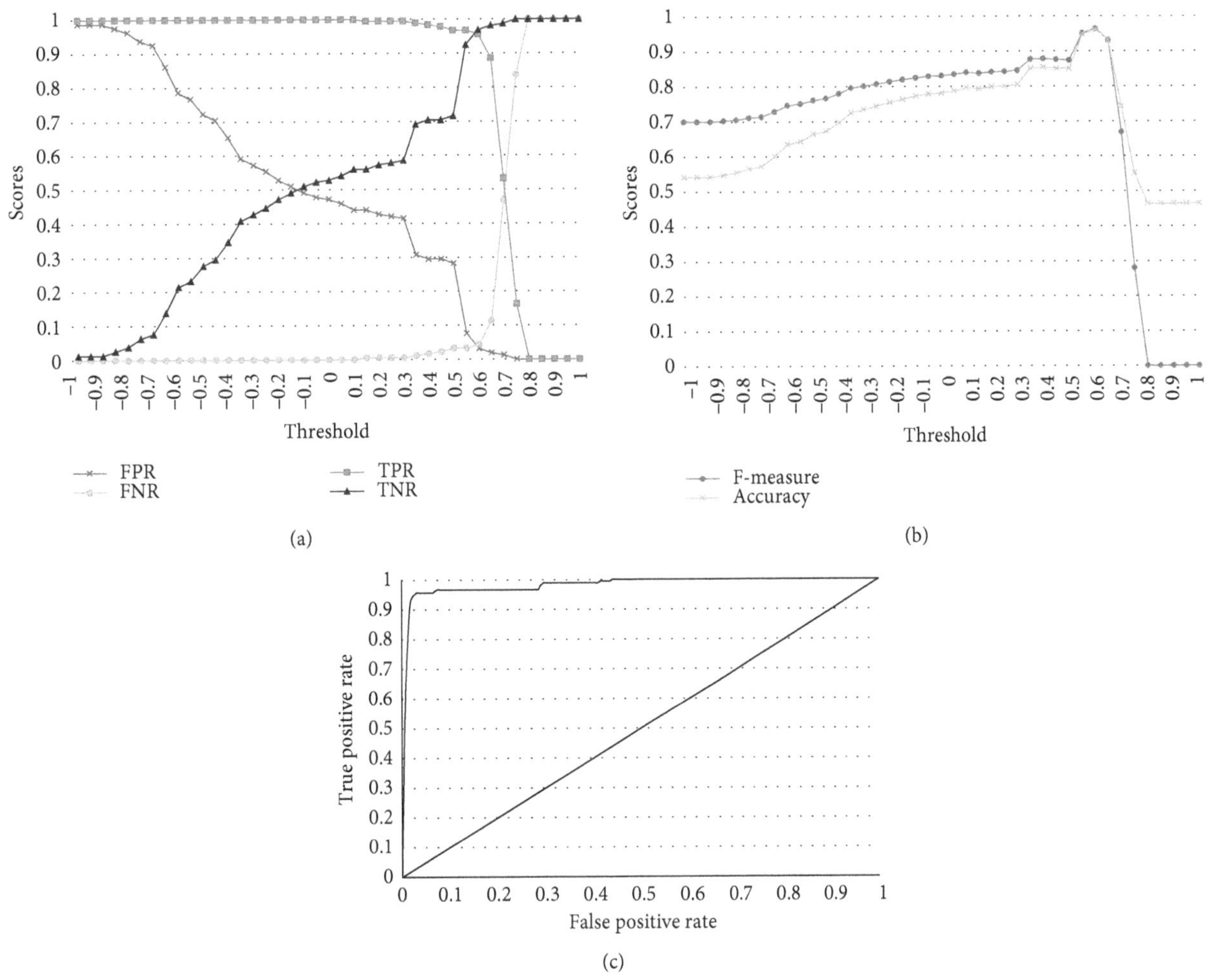

FIGURE 7: The evaluation results in terms of the malware detection performance. (a) FP/FN/TP/TN rate by the threshold. (b) F-measure/Accuracy by the threshold. (c) ROC (Receiver Operating Characteristic) Curve.

To figure out the impact of the size of the sliding window and the local scoring period, the measured detection accuracies are rearranged according to each parameter. Figures 6(a) and 6(b) show the rearranged heat map results. In order to show the results in a comprehensive manner, we expressed each cell in each heat map using the average or maximum value of the accuracies measured when a parameter is fixed to a certain value. The average and maximum values are calculated by scanning horizontally or vertically large heat maps represented in Figures 6(c) and 6(d).

There is no big difference between the local-only case and the local-global case in terms of the accuracy change by different size of the sliding window. The overall accuracy is increased by including the global score in the detection, but the accuracy change pattern still remains similar. The accuracy tends to be high at a certain size of the sliding window in both cases. The global score is less affected by the size of the sliding window, whereas the local score is highly associated with the size of the sliding window.

The sliding window whose size is in between 4 and 11 gives high accuracies in our malware detection. The detection process produced the best accuracy when the sliding window size was 5. In addition, the detection accuracy is lowered when the sliding window size is approaching 20, and this tendency is similar when the sliding window size is above 20.

In the local-only case, the accuracy increases as the local scoring period decreases. After the global score is applied, a range of the local scoring period showing a relatively high accuracy was discovered from the results. In detail, when a local scoring period in the range of 4 to 13 was used, the detection accuracy was measured as relatively high. The best local scoring period was 8.

4.3.2. Malware Detection Performance. We evaluated the malware detection performance of our proposed framework from various aspects when the size of the sliding window and the local scoring period were set to 5 and 8, respectively. Figure 7(a) shows the false positive rate, false negative rate, true positive rate, and true negative rate as the score threshold is varied. In addition, the F-measure and accuracy results are also described in Figure 7(b). When the score threshold was 0.6, our proposed framework produced the best result in the evaluation metrics. In the best case, the false

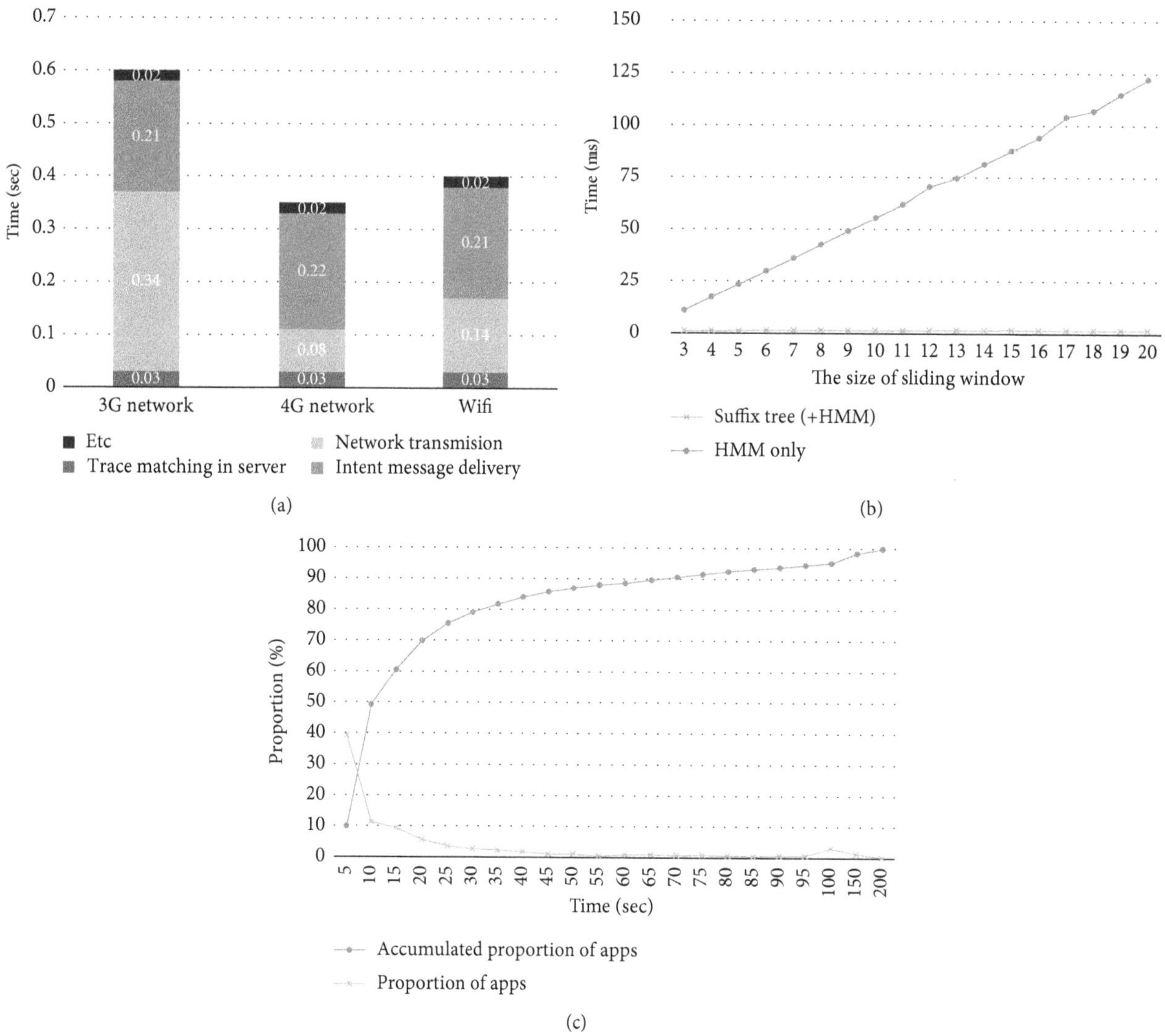

Figure 8: The evaluation results in terms of time overhead. (a) Trace processing time measurement. (b) Local and global scoring time measurement. (c) Application rewriting time measurement.

positive rate, false negative rate, true positive rate, and true negative rate were about 0.03, 0.04, 0.96, and 0.97, respectively, and the F-measure and accuracy were about 0.96.

As shown in Figure 7(a), when the score threshold was in between −1 and 0.1, the false negative rate and the true positive rate were 0 and 1, respectively. When the score threshold was less than 0.1, all the samples were determined as malware.

As the score threshold increased from 0.1, the accuracy was improved gradually. The false negative rate slightly increased from 0, and the true positive rate slightly decreased from 1. At the same time, the false positive rate decreased rapidly from 0.44 and the true negative rate rapidly increased from 0.56. When the score threshold was close to 0.6, the true positive rate and true negative rate became 0.96 and 0.97 each. At the same time, the false positive rate and false negative rate also became 0.03 and 0.04 each. When the score threshold was 0.7 or more, the true negative rate and false positive rate remained 1 and 0, respectively, due to the fact that most of the samples were classified as benign samples.

In addition to these evaluation metrics, the ROC (Receiver Operating Characteristic) curve is depicted in Figure 7(c). The ROC curve is a graphical plot that illustrates the robustness of malware detection in our proposed framework as the score threshold is varied. The ROC curve is created by plotting the true positive rate against the false positive rate with various score thresholds. As shown in Figure 7(c), the plotted points are above the diagonal, which represents good detection results. The curve line is close to the upper left corner of the ROC area, which indicates that the malware detection test of the framework has high discriminating capabilities. In detail, the AUC (area under the curve) was about 0.98. The AUC is the evaluation metric to summarize the performance of the classifier.

4.4. Run-Time Overheads. The end-to-end overhead, API monitoring overhead, and application rewriting overhead were measured to evaluate the efficiency of our framework. The end-to-end overhead shows how efficiently our client-server based framework works to process a trace generated from an application, and the API monitoring overhead shows the time added by the self-monitoring in the rewritten application. From these two evaluation factors, it is possible to estimate the time required for the detection process in our framework. Even though the application rewriting is performed in the preprocessing phase, we also measured the application rewriting time to show the usability of the framework. We used Samsung Galaxy S4 with an Exynos 5 Octa 5410 CPU and 2 GB RAM. The operating system of the device was Android 5.0.1. We also used a Windows 7 machine with an Intel Core i5-4570 CPU and a 32 GB RAM as the audit system.

(1) *End-to-end time*: We had two kinds of experiments to evaluate the end-to-end time overheads of our framework. In the first experiment, we measured the end-to-end time from when the secure agent receives the dangerous API trace to when the secure agent receives the examination result from the audit system. One hundred applications were used in the experiment, and the time overhead was measured in three network environments: 3G network, 4G network, and wifi network. Since the secure agent and the audit system communicate over the public network, the SSL (Secure Socket Layer) protocol is used in order to protect the communications from potential attacks. The detailed setting of SSL is TLS_RSA_WITH_AES_128_CBC_SHA. In addition, the size of trace which is transmitted from the client was set to 20, and each transmitted trace is scanned using the sliding window of length 5. The local scoring period was set to 8.

As shown in Figure 8(a), the average time to process a transmitted trace whose length is 20 did not exceed 0.6 seconds in all the network environments. The average time for the network transmissions in 3G network was longer than the others, and there were no significant differences between 4G network and wifi network. Naturally, the difference between the end-to-end times of different network environments depends on the network transmission time. Except for the network transmission, the trace processing in the server took the longest time. The server, that is, the audit system, performs many jobs such as the reception of the dangerous API invocation traces, encryption and decryption for the communications, the trace examination, the transmission of the examination results, and so on. Among these jobs in the server, the trace examination procedure was evaluated through the additional experiment.

Our framework detects malware based on the local and global scores, and each score is calculated based on the confidence values. Therefore, the execution time of examination procedure depends on the computation of the confidence values of the given trace. In order to measure the effect of the suffix tree, we repeatedly measured time to calculate the local score and the global score in the second experiment. We assumed that the audit server has the trace of length 296 which is delivered from the client, and the trace from the client is examined with the different size of sliding windows. The trace in the audit server was the real trace of the AnserverBot malware. In addition, the local scoring period was set to 8, and the number of symbols in the HMMs was specified as five. The scoring time was measured 20 times for each size of the sliding window, and the average value of scoring time according to the size of sliding window is shown as graphs in Figure 8(b).

TABLE 3: Scoring time using suffix tree and HMM.

The size of sliding window	Suffix tree-based scoring time, $t1$ (ms)	HMM-based scoring time, $t2$ (ms)	$t1/t2$ (ratio)
3	1.1	10.9	9.6
4	1.1	17.3	15.5
5	1.2	23.4	19
6	1.3	29.8	23.2
7	1.3	36.1	27.1
8	1.5	42.6	29
9	1.3	49.1	37.3
10	1.4	55.5	38.7
11	1.3	61.8	48.3
12	1.4	70.4	49.5
13	1.6	74.6	47.3
14	1.5	81.3	52.9
15	1.7	87.6	52.6
16	1.6	94.1	58.2
17	1.6	104.1	66.1
18	1.6	107	66.3
19	1.7	114.9	67.4
20	1.7	122.2	71.5

When the suffix tree that contains the condence values is used, all the measured times are reduced significantly. The scoring time when the HMM is used alone was about 10–72 times of the scoring time when the suffix tree is used. Table 3 shows the measured average time of both cases. The suffix tree-based scoring time ranged from 1.1 to 1.7 ms, whereas the HMM-based scoring time ranged from 10.9 to 122.2 ms. Assumed that the amount of the traces from many clients can be very large, the trace examination should be performed as fast as possible. Therefore, the suffix tree containing the precomputed condence values is suitable to be utilized in the trace examination procedure.

(2) *Application rewriting time*: We measured the time for the application rewriting including the application package decompiling and the self-monitoring code insertion. 1,617 applications were used in these experiments. The measured time ranges from 1.52 seconds to 197.41 seconds, and its average was 23.81 seconds. As shown in Figure 8(c), about 80% of the applications were rewritten within 30 seconds.

(3) *Self-monitoring time*: We also measured the overheads of the rewritten applications which are generated with the added self-monitoring methods. To evaluate the performance degradations caused by API invocation monitoring, we implemented a program that repeatedly invokes FileWriter.append() to write a string to a file. The buffer size which is same with the transmission trace unit in the secure agent was set to 20. Table 4 shows that the time for API invocations are increased by the self-monitoring method. As a result, the average of the increased time per one invocation was about 0.013 ms.

4.5. Size Increase of Rewritten Applications. The application rewriting module inserts the self-monitoring code into the dex file of the application. 1,645 applications were used in our experiments. We measured the size increases of the dex

Table 4: API execution time overhead by self-monitoring.

		Number of repetitions				
		1000	2000	3000	4000	5000
Original app	Total time (ms)	28	63	150	145	197
	Average time per an invocation (ms)	0.028	0.032	0.05	0.036	0.039
Rewritten app	Total time (ms)	59	75	152	205	255
	Average time per an invocation (ms)	0.059	0.038	0.051	0.051	0.051

Table 5: Size increase by app rewriting.

Percentage increase (=size increase/original size)	Number of applications
0% ≤ P.I. ≤ 0.5%	1285
0.5% ≤ P.I. ≤ 1.0%	237
1.0% ≤ P.I. ≤ 1.5%	68
1.5% ≤ P.I. ≤ 2%	13
2.0% ≤ P.I. ≤ 2.5%	7
2.5% ≤ P.I. ≤ 3.0%	1
3.0% ≤ P.I.	34

files in the rewritten applications and calculated the percentage of the increased size of each dex file. The average of the size increase was 25.17 kB. The minimum size increase and the maximum size increase were 12.56 kB and 47.13 kB, respectively. The percentage increases of the applications are shown in Table 5, and the percentage increases of most applications were less than 1, which is negligible.

4.6. Comparison with Other Methods. We had an experiment to compare our framework with other detection methods. In the experiment, we evaluated n-gram frequency-based frameworks using general classification methods which are widely used in dynamic analysis-based detection. The n-gram frequency of the API trace is measured, and it is inputted to the classification algorithm. The bi-gram whose size of n-gram is 2 was used in the frameworks.

Table 6 shows the performance result of our framework and the n-gram frequency-based detection methods. As shown in the table, we used three different kinds of classification algorithm in the n-gram based frameworks, and the sliding window size and the local scoring period of our framework were set to 5 and 8 each. 1,033 benign applications and 1,260 malware were used in this experiment, and 80% of total applications (826 benign applications and 1,008 malware) were used as the training dataset, and the others were used as the test dataset.

The results show that our framework produces higher detection accuracy compared to the other methods. In addition, the n-gram frequency methods using classification algorithms are not designed to be used in the runtime detection, and it needs the sufficient amount of API traces for more accurate detection. To achieve the accuracies presented in Table 6, 475 API invocations on average were collected and used in the audit system. It means that traces of length 20 should be transmitted through the network at least 20 times. In contrast, our framework can detect the malware in near real time with the relatively small amount of traces while showing higher detection accuracy. On average, our framework was able to detect the malware using only 4% of traces that were used in n-gram based methods.

Table 6: Comparison with n-gram frequency-based detection methods.

Method	TPR	FPR	TNR	FNR	Accuracy
n-gram (RandomForest)	0.95	0.05	0.95	0.05	0.95
n-gram (DecisionTree)	0.93	0.07	0.93	0.07	0.93
n-gram (NaiveBayes)	0.76	0.25	0.75	0.24	0.76
Ours	0.96	0.03	0.97	0.04	0.96

5. Discussion

Rewritten applications are resigned with a new signature. Therefore, if the rewritten application tries to update itself, then it would be failed because of the different signatures. If the updated version of an application is available to download from the online market, then it is possible to apply our mechanism to the updated application for further analysis. In another case, to mitigate the update problem, a simple heuristic solution can be applied to our framework, as follows: the secure agent prohibits the autoupdate of applications, and if the new dangerous API traces from the applications are not generated in a certain period of time or a user wants to update the application directly, then the original applications are provided by the audit system. There is no proper period of time that guarantees that an application is not malicious, and so the period should be specified by considering two factors: usability and security.

6. Conclusion

In this paper, we proposed a novel Android malware detection framework that uses the application rewriting technique for application behavior monitoring without any kernel modifications. Furthermore, we designed the framework as a client-server model and deployed the analysis processes for malware detection on the server-side. The client component only monitors applications, and the server component examines the traces using the suffix tree algorithm. In the detection process, we designed the suffix tree-based model to have the confidence values that are precomputed using the document frequency ratios and the likelihood values from the HMM. In addition, we also proposed the scoring method to make the final decision for the malware detection. We made efforts for our framework not only to have the high detection accuracy, but also to be efficient enough to examine the application behaviors at runtime.

In the evaluation, we conducted the experiments to show the performance of our proposed framework in terms of detection accuracy, runtime overhead, and size overhead of rewritten application. The framework could achieve about 96% malware detection accuracy in the best case, the average time for processing each trace from the secure agent and the audit system was less than 0.6 seconds in three different network environments. In addition, the newly proposed detection method only took 1-2 ms to process about three hundred subtraces. This means that our score based detection barely affects the end-to-end time overhead.

7. Future Work

The event generations to trigger behaviors of applications still remain as the future work. We figured out the appropriate way to improve the code coverage with MonkeyRunner. However, if there is an elaborate way to generate the events considering execution paths in an application, the performance of dynamic analysis can be improved significantly. Some existing researches on automated testing [38, 39, 40] have been proposed, but their approaches use heuristic ways to generate events. We plan to apply the symbolic execution methods to our framework to trigger application behaviors.

Acknowledgments

This work was supported by the Institute for Information & Communications Technology Promotion (IITP) grant funded by the Korea government (MSIT) (no. 2017-0-00388, Development of Defense Technologies against Ransomware) and the National Research Foundation of Korea (NRF) grant funded by the Korea government (MSIP) (no. NRF-2016R1A2B4015254). This work was supported by the Human Resources program in Energy Technology (No. 20174010201170) of the Korea Institute of Energy Technology Evaluation and Planning (KETEP) grant funded by the Ministry of Trade, Industry and Energy of the Korea government.

References

[1] N. Islam and R. Want, "Smartphones: past, present, and future," *IEEE Pervasive Computing*, vol. 4, no. 13, pp. 89–92, 2014.

[2] G DATA Security Blog, *8,400 New Android Malware Samples Everyday*, G DATA Software, Bochum, Germany, 2017.

[3] V. Rastogi, Y. Chen, and X. Jiang, "Droidchameleon: evaluating android anti-malware against transformation attacks," in *Proceedings of the 8th ACM SIGSAC Symposium on Information, Computer and Communications Security*, pp. 329–334, Hangzhou, China, May 2013.

[4] T. Isohara, K. Takemori, and A. Kubota, "Kernel-based behavior analysis for android malware detection," in *Proceedings of the Seventh International Conference on Computational Intelligence and Security (CIS)*, pp. 1011–1015, Hainan, China, December 2011.

[5] L. K. Yan and H. Yin, "Droidscope: seamlessly reconstructing the OS and dalvik semantic views for dynamic android malware analysis," in *Proceedings of the 21st USENIX Security Symposium (USENIX Security 12)*, pp. 569–584, Bellevue, WA, USA, August 2012.

[6] M. Lindorfer, M. Neugschwandtner, L. Weichselbaum, Y. Fratantonio, V. Van Der Vee, and C. Platzer, "Andrubis 1,000,000 apps later: a view on current android malware behaviors," in *Proceedings of the 2014 Third International Workshop on Building Analysis Datasets and Gathering Experience Returns for Security (BADGERS)*, pp. 3–17, Wroclaw, Poland, September 2014.

[7] W. Enck, P. Gilbert, S. Han et al., "Taintdroid: an information-flow tracking system for realtime privacy monitoring on smartphones," *ACM Transactions on Computer Systems*, vol. 32, no. 2, pp. 1–29, 2014.

[8] A. Shabtai, U. Kanonov, Y. Elovici, C. Glezer, and Y. Weiss, ""Andromaly": a behavioral malware detection framework for android devices," *Journal of Intelligent Information Systems*, vol. 38, no. 1, pp. 161–190, 2012.

[9] A. D. Schmidt, F. Peters, F. Lamour, C. Scheel, S. A. Camtepe, and S. Albayrak, "Monitoring smartphones for anomaly detection," *Mobile Networks and Applications*, vol. 14, no. 1, pp. 92–106, 2009.

[10] T. Blasing, L. Batyuk, A. D. Schmidt, S. A. Camtepe, and S. Albayrak, "An android application sandbox system for suspicious software detection," in *Proceedings of the 5th International Conference on Malicious and Unwanted Software (MALWARE)*, pp. 55–62, Nancy, France, October 2010.

[11] L. Deshotels, V. Notani, and A. Lakhotia, "Droidlegacy: automated familial classication of Android malware," in *Proceedings of the ACM SIGPLAN on Program Protection and Reverse Engineering Workshop 2014*, , San Diego, CA, USA, January 2014.

[12] D. J. Wu, C. H. Mao, T. E. Wei, H. M. Lee, and K. P. Wu, "Droidmat: Android malware detection through manifest and API calls tracing," in *Proceedings of the Seventh Asia Joint Conference on Information Security (Asia JCIS)*, pp. 62–69, August 2012.

[13] D. Arp, M. Spreitzenbarth, M. Hubner, H. Gascon, and K. Rieck, "Drebin: effective and explainable detection of android malware in your pocket," in *Proceedings of the Network and Distributed System Security Symposium (NDSS)*, San Diego, CA, USA, February 2014.

[14] S. Y. Yerima, S. Sezer, and G. McWilliams, "Analysis of Bayesian classication-based approaches for android malware detection," *IET Information Security*, vol. 8, no. 1, pp. 25–36, 2014.

[15] M. Zhang, Y. Duan, H. Yin, and Z. Zhao, "Semantics-aware android malware classification using weighted contextual API dependency graphs," in *Proceedings of the 2014 ACM SIGSAC Conference on Computer and Communications Security*, pp. 1105–1116, Scottsdale, AZ, USA, November 2014.

[16] L. Lu, Z. Li, Z. Wu, W. Lee, and G. Jiang, "Chex: statically vetting android apps for component hijacking vulnerabilities," in *Proceedings of the 2012 ACM Conference on Computer and Communications Security*, pp. 229–240, Raleigh, NC, USA, October 2012.

[17] P. P. Chan, L. C. Hui, and S. M. Yiu, "Droidchecker: analyzing android applications for capability leak," in *Proceedings of the 5th ACM conference on Security and Privacy in Wireless and Mobile Networks*, pp. 125–136, Tucson, AZ, USA, December 2012.

[18] K. Lu, Z. Li, V. P. Kemerlis et al., "Checking more and alerting less: detecting privacy leakages via enhanced data-ow analysis and peer voting," in *Proceedings of the 2015 Network and Distributed System Security (NDSS)*, San Diego, CA, USA, February 2015.

[19] F. Wei, S. Roy, X. Ou et al., "Amandroid: a precise and general inter-component data flow analysis framework for security

vetting of android apps," in *Proceedings of the 2014 ACM SIGSAC Conference on Computer and Communications Security*, pp. 1329–1341, Scottsdale, AZ, USA, November 2014.

[20] I. Burguera, U. Zurutuza, and S. Nadjm-Tehrani, "Crowdroid: behavior-based malware detection system for android," in *Proceedings of the 1st ACM Workshop on Security and Privacy in Smartphones and Mobile Devices*, pp. 15–26, Chicago, IL, USA, October 2011.

[21] B. Davis, B. Sanders, A. Khodaverdian, and H. Chen, "I-arm-droid: a rewriting framework for in-app reference monitors for android applications," *Mobile Security Technologies*, vol. 2012, no. 2, pp. 1–7, 2012.

[22] H. Hao, V. Singh, and W. Du, "On the effectiveness of api-level access control using bytecode rewriting in android," in *Proceedings of the 8th ACM SIGSAC Symposium on Information, Computer and Communications Security*, pp. 25–36, Hangzhou, China, May 2013.

[23] D. Schreckling, S. Huber, F. Hohne, and J. Posegga, "Uranos: user-guided rewriting for plugin-enabled android application security," in *Proceedings of the International Workshop on Information Security Theory and Practices*, pp. 50–65, Heraklion, Crete, Greece, June 2013.

[24] R. Xu, H. Sadi, and R. Anderson, "Aurasium: practical policy enforcement for android applications," in *Proceedings of the 21st USENIX Security Symposium (USENIX Security 12)*, pp. 539–552, Bellevue, WA, USA, August 2012.

[25] M. Backes, S. Bugiel, C. Hammer, O. Schranz, and P. von Styp-Rekowsky, "Boxify: full-edged app sandboxing for stock android," in *Proceedings of the 24th USENIX Security Symposium (USENIX Security 15)*, pp. 691–706, Washington, DC, USA, August 2015.

[26] Baksmali: https://baksmali.com.

[27] Apktool: http://code.google.com/p/Android-apktool/.

[28] Y. Aafer, W. Du, and H. Yin, "Droidapiminer: mining api-level features for robust malware detection in android," in *Proceedings of the International Conference on Security and Privacy in Communication Systems*, pp. 86–103, Sydney, Australia, September 2013.

[29] Package Index: https://developer.android.com/reference.

[30] M. Farach, "Optimal suffix tree construction with large alphabets," in *Proceedings of the 38th Annual Symposium on Foundations of Computer Science*, pp. 137–143, Miami Beech, FL, USA, October 1997.

[31] E. Ukkonen, "On-line construction of suffix trees," *Algorithmica*, vol. 14, no. 3, pp. 249–260, 1995.

[32] L. R. Rabiner and B. H. Juang, "An introduction to hidden Markov models," *IEEE ASSP Magazine*, vol. 3, no. 1, pp. 4–16, 1986.

[33] L. R. Welch, "Hidden Markov models and the Baum–Welch algorithm," *IEEE Information Theory Society Newsletter*, vol. 53, no. 4, pp. 10–13, 2003.

[34] Y. Zhou and X. Jiang, "Dissecting android malware: characterization and evolution," in *Proceedings of the 2012 IEEE Symposium on Security and Privacy*, pp. 95–109, San Francisco, CA, USA, May 2012.

[35] Google Play Store: https://play.google.com/store/apps.

[36] VirusTotal: https://www.virustotal.com/.

[37] MonkeyRunner: https://developer.android.com/studio/test/monkeyrunner.

[38] A. Machiry, R. Tahiliani, and M. Naik, "Dynodroid: an input generation system for android apps," in *Proceedings of the 9th Joint Meeting on Foundations of Software Engineering*, pp. 224–234, Saint Petersburg, Russia, August 2013.

[39] T. Azim and I. Neamtiu, "Targeted and depth-rst exploration for systematic testing of android apps," *ACM SIGPLAN Notices*, vol. 48, no. 10, pp. 641–660, 2013.

[40] S. Hao, B. Liu, S. Nath, W. G. Halfond, and R. Govindan, "Puma: programmable ui-automation for large-scale dynamic analysis of mobile apps," in *Proceedings of the 12th Annual International Conference on Mobile Systems, Applications, and Services*, pp. 204–217, Bretton Woods, NH, USA, June 2014.

Permissions

All chapters in this book were first published in MIS, by Hindawi Publishing Corporation; hereby published with permission under the Creative Commons Attribution License or equivalent. Every chapter published in this book has been scrutinized by our experts. Their significance has been extensively debated. The topics covered herein carry significant findings which will fuel the growth of the discipline. They may even be implemented as practical applications or may be referred to as a beginning point for another development.

The contributors of this book come from diverse backgrounds, making this book a truly international effort. This book will bring forth new frontiers with its revolutionizing research information and detailed analysis of the nascent developments around the world.

We would like to thank all the contributing authors for lending their expertise to make the book truly unique. They have played a crucial role in the development of this book. Without their invaluable contributions this book wouldn't have been possible. They have made vital efforts to compile up to date information on the varied aspects of this subject to make this book a valuable addition to the collection of many professionals and students.

This book was conceptualized with the vision of imparting up-to-date information and advanced data in this field. To ensure the same, a matchless editorial board was set up. Every individual on the board went through rigorous rounds of assessment to prove their worth. After which they invested a large part of their time researching and compiling the most relevant data for our readers.

The editorial board has been involved in producing this book since its inception. They have spent rigorous hours researching and exploring the diverse topics which have resulted in the successful publishing of this book. They have passed on their knowledge of decades through this book. To expedite this challenging task, the publisher supported the team at every step. A small team of assistant editors was also appointed to further simplify the editing procedure and attain best results for the readers.

Apart from the editorial board, the designing team has also invested a significant amount of their time in understanding the subject and creating the most relevant covers. They scrutinized every image to scout for the most suitable representation of the subject and create an appropriate cover for the book.

The publishing team has been an ardent support to the editorial, designing and production team. Their endless efforts to recruit the best for this project, has resulted in the accomplishment of this book. They are a veteran in the field of academics and their pool of knowledge is as vast as their experience in printing. Their expertise and guidance has proved useful at every step. Their uncompromising quality standards have made this book an exceptional effort. Their encouragement from time to time has been an inspiration for everyone.

The publisher and the editorial board hope that this book will prove to be a valuable piece of knowledge for researchers, students, practitioners and scholars across the globe.

List of Contributors

Yu Liu, Kai Guo, Xiangdong Huang and Yichi Zhang
School of Electronic Information Engineering, Tianjin University, Tianjin 30072, China

Zhou Zhou
Teachers College of Columbia University, New York, NY 10027, USA

Jaehyeong Wee and Wooguil Pak
Department of Computer Engineering, Keimyung University, Daegu 704-701, Republic of Korea

Ruxia Sun, Jinwen Xi and Chunyong Yin
School of Computer and Software, Jiangsu Engineering Center of Network Monitoring, Nanjing University of Information Science & Technology, Nanjing, China

Jin Wang
School of Computer & Communication Engineering, Changsha University of Science & Technology, Changsha, China

Gwang-jun Kim
Department of Computer Engineering, Chonnam National University, Gwangju, Republic of Korea

Tieming Chen, Qingyu Mao, Yimin Yang and Mingqi Lv
College of Computer Science and Technology, Zhejiang University of Technology, Hangzhou 310023, China

Jianming Zhu
College of Information Engineering, Central University of Finance and Economics, Beijing 100081, China

Yongchul Kim and Jungho Kang
Korea Military Academy, Seoul, Republic of Korea

Young-Hyun Oh
Department of Computer Science, North Carolina State University, Raleigh, NC, USA

Sandeep Gupta, Attaullah Buriro and Bruno Crispo
Department of Information Engineering and Computer Science, University of Trento, Trento, Italy

Luca Casati and Andrea Visconti
Department of Computer Science, Università degli Studi di Milano, Via Comelico 39/41, 20135 Milano, Italy

Bingfei Ren, Chuanchang Liu, Bo Cheng and Junliang Chen
State Key Laboratory of Networking and Switching Technology, Beijing University of Posts and Telecommunications, Beijing, China

Jie Guo
Beijing Liyun Technology Development Company, Beijing, China

Bang Liu, Nianbo Liu, Xili Dai and Ming Liu
University of Electronic Science and Technology of China, Chengdu, China

Guihai Chen
Shanghai Jiao Tong University, Shanghai, China

Yu Zhang and Yin Li
School of Computer and Information Technology, Xinyang Normal University, Xinyang 464000, China

Yifan Wang
Wayne State University, 42 W. Warren Ave., Detroit, MI 48202, USA

Sergi Delgado-Segura, Cristina Pérez-Solà, Jordi Herrera-Joancomartí, Guillermo Navarro-Arribas and Joan Borrell
Department d'Enginyeria de la Informació i les Comunicacions, Universitat Autònoma de Barcelona, 08193 Bellaterra, Catalonia, Spain

Ashish Rauniyar, Desta Haileselassie Hagos and Manish Shrestha
Network and Distributed System (ND) Research Group, Department of Informatics, University of Oslo, Oslo, Norway

Ashish Rauniyar and Desta Haileselassie Hagos
Autonomous Systems and Networks (ASN) Research Group, Department of Computer Science, OsloMet-Oslo Metropolitan University, Oslo, Norway

Manish Shrestha
eSmart Systems, Halden, Norway

Yousheng Zhou, Siling Liu and Min Xiao
College of Computer Science and Technology, Chongqing University of Posts and Telecommunications, Chongqing 400065, China

Yousheng Zhou and Shaojiang Deng
College of Computer Science, Chongqing University, Chongqing 400044, China

Yousheng Zhou and Min Xiao
School of Cyber Security and Information Law, Chongqing University of Posts and Telecommunications, Chongqing 400065, China

Xiaojun Wang
School of Electronic Engineering, Dublin City University, Dublin, Ireland

Margit Antal, László Zsolt Szabó and Tünde Tordai
Faculty of Technical and Human Sciences, Sapientia University, Cluj-Napoca, Romania

Hakima Khelifi and Senlin Luo
School of Information and Electronics, Beijing Institute of Technology, Beijing, China

Boubakr Nour
School of Computer Science, Beijing Institute of Technology, Beijing, China

Sayed Chhattan Shah
Department of Information Communication Engineering, Hankuk University of Foreign Studies, Seoul, Republic of Korea

Claudia Nass, Jessica Jung, Eduard C. Groen, Karina Villela and Konstantin Holl
Fraunhofer Institute for Experimental Software Engineering (IESE), Fraunhofer-Platz 1, 67663 Kaiserslautern, Germany

TaeGuen Kim and Eul Gyu Im
Hanyang University, Haengdang-dong, Seongdong-gu, Seoul, Republic of Korea

BooJoong Kang
Center for Secure Information Technologies, ECIT, Belfast, UK

Index

www.ingramcontent.com/pod-product-compliance
Lightning Source LLC
LaVergne TN
LVHW060608210926
848485LV00005B/1390